"Everyone who enjoys beer, esp... researched and captivatingly writ... formations that made possible the ...optopia will change how readers think and—most importantly—how they taste their favorite hoppy beers."

—Mark Fiege, Montana State University

"Peter Kopp has produced a masterful work in *Hoptopia*. He creatively brings together agriculture, agronomy, science, environment, labor, and market economics to tell this story of hop production in Oregon's Willamette Valley. What's beer without hops? What's the history of that crop without all of the important connections explained so well here? *Hoptopia* is a must-have book for all interested in the history of the Pacific Northwest and for all who love beer."

—Sterling Evans, University of Oklahoma

"*Hoptopia* finally encapsulates the noble role of the lowly hop cone in the world of quality beer. Peter Kopp deftly weaves the story of how American hops—and particularly Oregon hops—went from a laughingstock of the beer world to an ingredient highly sought after by brewers worldwide. Cheers to *Hoptopia!*"

—Karl Ockert, Director of Brewery Operations, Deschutes Brewery

"Imagine a Venn diagram with hops, a crucial ingredient in making beer, in the center, attached to circles containing farming, agronomy, climate, ecology, business, labor, gender, race, class, festivals, globalization, and utopias. As the title of Peter Kopp's entertaining and informative history of hop farming suggests, the story of hops is regional history placed in contexts of world history. Like the beers that hops make palatable, this book nourishes and stimulates. Imbibe!"

—Bernard Mergen, author of *At Pyramid Lake*

"Peter Kopp has taken seriously the advice of many environmental historians to begin with the natural world and ask questions about human engagement. In this fascinating history of a plant and its place in the Pacific Northwest, we get everything from transnational economic competition to indigenous labor to modern bio-scientific research—and all of it also packaged to give us a new viewpoint on the world's most popular alcoholic beverage."

—William L. Lang, Portland State University

"Cheers to this fascinating agricultural history of the aromatic hops that infused America's craft beer revolution. Kopp relates the rich biological, scientific, social, labor, and industrial history of the development of Oregon's Willamette Valley as a major hop producer. Along the way, he reveals the complex connections between global markets and the local landscapes and people who transformed the way many of us imbibe beer."

— Marsha Weisiger, Julie and Rocky Dixon Chair of U.S. Western History, University of Oregon

"Everyone who enjoys beer, especially craft beer, should read this book."

— Mark Fiege, Wallace Stegner Chair in Western American Studies, Montana State University

Hoptopia

CALIFORNIA STUDIES IN FOOD AND CULTURE

Darra Goldstein, Editor

Hoptopia

A WORLD OF AGRICULTURE AND BEER
IN OREGON'S WILLAMETTE VALLEY

Peter A. Kopp

UNIVERSITY OF CALIFORNIA PRESS

University of California Press, one of the most distinguished university presses in the United States, enriches lives around the world by advancing scholarship in the humanities, social sciences, and natural sciences. Its activities are supported by the UC Press Foundation and by philanthropic contributions from individuals and institutions. For more information, visit www.ucpress.edu.

University of California Press
Oakland, California

A version of chapter 1 first appeared as "The Global Hop: An Agricultural Overview of the Brewer's Gold," in *The Geography of Beer: Regions, Environment, and Societies,* edited by Mark Patterson and Nancy Hoalst-Pullen, pp. 77–88. © Springer, 2014.
Parts of chapter 3 and chapter 4 first appeared in slightly different forms as "'Hop Fever' in the Willamette Valley: The Local and Global Roots of a Regional Specialty Crop," in *Oregon Historical Quarterly,* vol. 112, no. 4, pp. 406–33. © Oregon Historical Society, 2011.

Library of Congress Cataloging-in-Publication Data

Names: Kopp, Peter Adam, author.
Title: Hoptopia : a world of agriculture and beer in Oregon's Willamette Valley / Peter A. Kopp.
Other titles: California studies in food and culture ; 61.
Description: Oakland, California : University of California Press, [2016] | Series: California studies in food and culture ; 61 | Includes bibliographical references and index.
Identifiers: LCCN 2016001734 (print) | LCCN 2016007187 (ebook) | ISBN 9780520277472 (cloth : alk. paper) | ISBN 9780520277489 (pbk. : alk. paper) | ISBN 9780520965058 (ebook)
Subjects: LCSH: Hops industry—Oregon—Willamette River Valley—History.
Classification: LCC HD9019.H72 U65 2016 (print) | LCC HD9019.H72 (ebook) | DDC 338.1/7382—dc23
LC record available at http://lccn.loc.gov/2016001734

Manufactured in the United States of America

25 24 23 22 21 20 19 18 17 16
10 9 8 7 6 5 4 3 2 1

In keeping with a commitment to support environmentally responsible and sustainable printing practices, UC Press has printed this book on Natures Natural, a fiber that contains 30 percent postconsumer waste and meets the minimum requirements of ANSI/NISO Z39.48–1992 (R 1997) (*Permanence of Paper*).

For my dad

CONTENTS

ACKNOWLEDGMENTS

Collaboration rests at the heart of historical scholarship. This being the case, I would like to thank many people and institutions for their support in making this project possible. First and foremost, I owe a debt of gratitude to Bill Rowley, my doctoral adviser at the University of Nevada, who oversaw this project as a dissertation and subsequently helped transform it into the book in your hands. Bill asked compelling questions, assisted in editing at all stages, and helped me to find research funding. I cannot thank him enough for his generosity and good humor.

For their guidance during graduate school (both at Nevada and at Portland State University), I would like to thank Alicia Barber, Katy Barber, Mike Branch, Scott Casper, Linda Curcio-Nagy, Martha Hildreth, David Johnson, Bill Lang, Elizabeth Raymond, Patricia Schechter, Hugh Shapiro, Scott Slovic, Tom Smith, Paul Starrs, and Barbara Walker. My appreciation also stems to my interdisciplinary graduate school cohort, including Jim Bishop, Paul Boone, Kyhl Lyndgaard, Andrew McGregor, Nick Plunkey, and Ryan Powell. Additionally, I was very lucky to enjoy the friendship of Lawrence Hatter, Meredith Oda, Ned Schoolman, and Erica Westhoff during my graduate studies. At New Mexico State University, where I joined the history department in the fall of 2012, my colleagues have been incredibly supportive. I would like to thank Jamie Bronstein, Nathan Brooks, Bill Eamon, Iñigo García-Bryce, Ken Hammond, Liz Horodowich, Jon Hunner, Margaret Malamud, Elvira Masson, Andrea Orzoff, Dwight Pitcaithley, and Isa Seong Leong Quintana. A variety of research grants from the University of Nevada and New Mexico State University allowed me to complete my research, and I have been grateful for those opportunities.

Throughout my research, many scholars, librarians, and archivists helped immensely. Larry Landis and Tiah Edmunson-Morton of Oregon State University, Pat Ragains of the University of Nevada, Geoff Wexler of the Oregon Historical Society, Greg Shine and Doug Wilson of the Fort Vancouver National Historical Site, Dennis Larson of the Ezra Meeker Historical Association, Mary Gallagher of the Benton County Historical Society, Patrick Harris of the Old Aurora Colony, Anne Barrett of Imperial College London, and Peggy Smith of the Independence Heritage Museum all went out of their ways to provide access to resources. Along with these individuals, I would like to thank the librarians and archivists who assisted me at the Lane County Historical Society, Marion County Historical Society, Multnomah County Library, New Mexico State University, Polk County Historical Society, Portland State University, University of Kent, University of Oregon, and University of Washington. I also could not have completed this book without guidance from members of the scientific community. I am in tremendous debt to Al Haunold and Peter Darby for spending days with me talking about hops. John Henning, Kim Hummer, and Jim Oliphant also offered valuable assistance. Thanks to all of you.

In the hop-growing and craft-beer crowd I would first like to thank Michelle Palacios and Nancy Frketich of the Oregon Hop Commission, who assisted with research and introduced me to the Willamette Valley's hop farmers. For helping me to understand their farms, operations, and family histories, I would like to offer my appreciation to the Annen, Coleman, Goschie, and Weathers families, all of whose hop stories run deep. The same goes for the craft-brewing community who offered me their time and, yes, good beer. Christian Ettinger, Jamie Floyd, the late Jack Joyce, Art Larrance, and Karl Ockert were all particularly generous. Tim Hills and John Foyston also offered a wealth of knowledge on regional brewing history. Thank you.

Beyond research, many people assisted with writing and editing. Matt Becker and Marianne Keddington-Lang helped guide the transition from dissertation to book. Eliza Canty-Jones, Peter Darby, Bill Eamon, Sterling Evans, Mark Fiege, Lawrence Hatter, Tim Hills, Nancy Hoalst-Pullen, Joseph Kopp, Bill Lang, Mark Patterson, and Isa Seong Leong Quintana all looked at various chapters. Finally, Gayle Goschie, Al Haunold, Stan Hieronymus, Sarah Roberts, Bill Rowley, Michael Tomlan, and Jerry Wallace went out of their way to read every word of my initial manuscript. I cannot thank you all enough. At the University of California Press, I would like to thank my editor Kate Marshall, along with Zuha Kahn, Carl Walesa, Victoria Baker, and two peer reviewers who offered important feedback. My excellent graduate students Ben Craske and Derek Travis also helped tremendously with research and fact-checking. Thank you.

These acknowledgments would be incomplete if I did not offer a deep appreciation for my family. My wife, Sarah, has offered unending patience and support, as has our son, James, in his own right. I will never be able to thank you two enough. My mother, Sue, as well as Lucy, Ryan, and Joe and the rest of our family, have been encouraging and enthusiastic all along, and I am grateful. Finally, I am indebted to my father, James Joseph Kopp (1952–2010), who introduced me not only to the world of craft beer and hops, but also to the world of history and writing many years prior. Though he passed away before I completed this book, his influence carries on, not the least in his suggestion of the title *Hoptopia* early on in the project. For these reasons and more, I dedicate this book to him.

Introduction

DEFINING HOPTOPIA

AMONG THE FIELDS AND ORCHARDS of Oregon's Willamette Valley grows a unique agricultural crop called hops. In this rural farming region, the climbing plant stands out for its vigorous vertical growth up high trellises and for the bright green cones that peek out from the plant's leaves, top to bottom. (See figure 1.) During the summer months, when the hop reaches maturity, the cones exude aromas reminiscent of fruits, grasses, spices, and herbs. Upon harvest, the hop's yellow resin, called *lupulin,* dusts the air. Given the plant's distinctive physical characteristics, inquisitive passersby often stop and exit their vehicles for a closer inspection. But botanical curiosity is probably only part of the reason. For those in the know, the real intrigue lies in the fact that those hop cones partake in a nearly singular purpose: almost all of the harvested crop is used to make beer.[1]

Hops serve beer makers unlike any other ingredient. When added to the kettle early in the brewing process (whether as whole cones or, more commonly in the recent past, in a pelletized or extract form), they provide a bitterness to counteract the syrupy sweetness of malted grain boiled in water, called *wort.* If added toward the end of the boil or during cooling, certain hops (often simply referred to as *aroma hops*) impart pleasant aromas—spanning grapefruit, pine, currants, mint, and more—that will one day delight a beer drinker's palate. But those are not the only reasons that brewers covet the plant. Antibacterial agents within hop cones act as natural preservatives to sustain the shelf life of beer, and the chemical makeup of the cones also helps clarify the beverage and stabilize its foamy head. No other ingredient can claim such versatile worth to beer makers, and for that reason the plant has been dubbed the brewer's gold.[2]

FIGURE 1. Hops on the bine. Courtesy Rogue Ales and Spirits.

In the past thirty years, the craft beer revolution that swept across the United States brought attention to hops more than at any other moment in the past. As brewers from California to New England eschewed bland American lagers in favor of more complex recipes, they almost universally featured more of the ingredient in their malted concoctions. The hopped-up vats created more flavorful and aromatic beers, unlike anything seen in U.S. markets since before Prohibition. In the process of winning over the senses of American beer drinkers, the hop also became an effective symbol of the craft beer industry, and, in effect, a marketing tool. Breweries showcased hop plants and cones on their beer labels and branded new brews with names such as Hop Czar, Hop Henge, Hop Jack, and Hopportunity Knocks. By the early twenty-first century, it was arguable that, whether beer drinker or not, Americans confronted hops on a daily basis via television commercials, billboards, and grocery-store displays. The hop truly achieved star status.

Yet the plant's significance to brewers and beer drinkers has a rich history that transcends the craft beer craze. Prior to being the featured ingredient in today's pilsners and pale ales, the hop took part in a global journey that has been in motion for millions of years. The story began with the evolution of the species and its quest to spread forth across the world, and took a turn much later when European beer-making traditions expanded the plant's territory as an agricultural commodity essential in brewing. Like a climbing hop bine (a shoot that uses sticky hairs to pull itself upward around a host, as opposed to a vine that ascends via tendrils, suckers, or runners), the plant's journey entangled the stories not only of brewers and beer drinkers, but also of farmers, businesspeople, scientists, wage laborers, and governmental agencies that engaged with one another and the agricultural landscape.

This book tells that story as it unfolded in the Willamette Valley, Oregon—a region that, unbeknownst to even many craft beer lovers today, is one of only a handful in the world where farmers grow hops commercially. Twenty-first-century beer enthusiasts know well that Portland (which sits at the northern edge of the valley) has emerged as the Craft Beer Capital of the World, because it houses the most breweries per capita on the planet. Some residents simply called their home Beervana. But many do not know that the Pacific Northwest is the only region in the country that currently produces hops on a large commercial scale.[3] Competing yearly with Germany as the world's leading producer, Washington, Oregon, and Idaho collectively cultivate one-third of all hops on earth.[4] That means that approximately one in three beers consumed across the world contain essential ingredients from the Pacific Northwest. Since the mid-1940s, Washington's Yakima Valley has been the most productive region of the three states. But Oregon has been just as significant both in the past and present for its contributions to hop agriculture. This proved especially true during the 1890s, when Willamette Valley promoters proclaimed the region as a "garden spot" for hop cultivation, and it intensified in the following decades, when the area identified itself as the "Hop Center of the World"—long before Portland's similar claim about craft beer.[5] Despite world wars, Prohibition, and the constant threat of botanical pests and diseases, success has lasted into the present and integrated into the more recent craft beer revolution.

Specialty crops like hops offer a valuable way to understand the identity of the places in which they grow. In contrast to staple crops, such as wheat and cotton, that span immense tracts of land across the world, specialty crops flourish because of a more specific set of conditions. Environmental factors offer a starting point, since plants prosper only in certain soils and climates. Upon that foundation, novel farming and harvest practices unfold to transform the landscapes and the lives of the people that inhabit them.

But specialty crops cannot thrive amid these actions alone. Farmers need infrastructure and must work with a wide range of experts and organizations to find success. These connections offer broad insight into regional environments, economies, society, politics, and culture. Specialty crops also unite local places to the wider world, because growers depend on the global marketplace and the importation and exchange of plant material and agricultural knowledge. All of this injects a unique tapestry of local, national, and global meanings into those places where specialty crops grow.[6]

In weaving this history together, *Hoptopia* argues that the current revolution in craft beer is the product of a complex global history that converged in the hop fields of Oregon's Willamette Valley. The chapters ahead explain that what spawned from an ideal environment and the ability of regional farmers to grow the crop transformed rapidly into something far greater because Oregon farmers depended on the importation of rootstock, knowledge, technology, and goods not only from Europe and the eastern United States but also from Asia, Latin America, and Australasia. They also relied upon a seasonal labor supply of people from all of these areas as a supplement to local Euro-American and indigenous communities to harvest their crops. In turn, Oregon hop farmers reciprocated in exchanges of plants and ideas with growers and scientists around the world, and, of course, sent their cured hops into the global marketplace. These global exchanges occurred not only during Oregon's golden era of hop growing in the late nineteenth and early twentieth centuries, but through to the present in the midst of the craft beer revival. All along, the history of Willamette Valley hop growing wrapped itself in the diverse lives of those associated with the industry, not to mention a thirsty public. The history that unfolded upon the region's environment integrated the rural and the urban, capital and labor, and economic, biological, and technological changes over time. All offer a layered meaning about the nature of this place, and an explanation of why the beers of the craft beer revolution taste and smell the way they do.

The title of this book, *Hoptopia,* is a nod to Portland's title of Beervana and the Willamette Valley's claim as an agricultural Eden from the mid–nineteenth century onward.[7] But the story is fundamentally about how seemingly niche agricultural regions do not exist and have never existed independently of the flow of people, ideas, goods, and biology from other parts of the world. To define *Hoptopia* is to define the Willamette Valley's hop and beer industries as the culmination of all of this local and global history. With the hop itself as a central character, this book aims to connect twenty-first-century consumers to agricultural lands and histories that have been forgotten in an era of industrial food production. In other words, the story hopes to connect consumers to the agricultural origins of their beer in an era when people more broadly want to know where their foods come from and why. All of this is to say that the contents of your pint glasses have a much richer history than you might have imagined.

ONE

Wolf of the Willow

MILLIONS OF YEARS BEFORE AMERICAN BREWERS opened the taps of the craft beer revolution, a climbing plant of the *Cannabaceae* family evolved in Asia. The hop, characterized by herbaceous bines, vigorous growth, and cylindrical green cones, made its home in river bottomlands and forest margins. It grew best in temperate climates with ample spring rains to inspire rapid growth, dry summers to help stave off pests and disease, and enough of a winter freeze to allow for a period of dormancy. The plant also preferred deep, fertile soils that allowed for an extensive root system to support its ascent up trees or shrubs. If all of these conditions could be met, the perennial hop might subsist for two or three decades, with its bines growing annually each spring and then dying back to the permanent root system in the autumn.[1]

Over the course of millennia, the dioecious plant (one with two distinct sexes) set out roots in the soil and pollen in the wind. The gene pool widened as the hop successfully spread to temperate regions not just in Asia, but all over the Northern Hemisphere at approximately latitudes thirty to fifty-five degrees. Eventually, three distinct species populated the planet. Two of those, *Humulus scandens* (until recently called *Humulus japonicas)* and *Humulus yunnanensis,* remained isolated in Asia. But the most ubiquitous of the three, *Humulus lupulus L.,* or the common hop, established itself across Eurasia and found a path to North America. Still, the plant's adapting and evolving was far from over. Across all of the regions that the plant colonized—whether present-day England, Germany, Russia, or New Mexico—the hop took on variances based on climates and soil types. Local varieties evolved with different rates of growth, resistances to diseases, and, most noticeably, unique shapes, sizes, and colors of cones.[2]

In many ways, the hop's early story unfolded like countless others during the angiosperm revolution, a period that began over a hundred million years ago when flowering plants repopulated a world of conifers, moss, and ferns. It was a time when nature embellished the earth with the botanical biodiversity with which we are more familiar today. To survive, all plants confronted an evolutionary path rife with geologic and climactic change, not to mention competition among other species. Taking a moment to consider *Humulus lupulus L.* as part of that grand narrative is worthwhile. Envision how the plant interacted with its environments and how it tested new ones as it traveled the world. How did the common hop entangle itself in unfamiliar places in the face of shifting continents and floods, droughts, and competition from other flora and fauna? Why did it succeed? When did it fail, and why? Scientists and archeologists have mapped out some answers, including the expansion and contraction of populations as the earth warmed and cooled.[3] Fossil records and DNA help in this task, and the plants provide clues about their past via the places they inhabit and in the physical manifestations of their evolution (whether color, size, shape, or fragrance). But most of our understanding of the hop's evolution and movement across land and sea, the trial and error, remains to the imagination.

While the details of the distant past may seem a far-off place to begin this book, knowledge of deep time provides an essential backdrop for understanding what unfolded much later in the far west of North America. Plants have much older stories than do humans. They simply lack an easy way of telling them. The fibers that make up the fabrics we wear and materials from which we build our homes, as well as the foods and drinks we consume, all have long, intricate histories. At the very least, even if we lack clarity of the day-to-day details throughout the millions of years that those plants carried forth, it is worth acknowledging that they all have a long and complex past. Perhaps all plants should be revered simply for the fact that they have persisted and continue to do so as human populations have expanded in the past millennia. Of course, the common hop deserves particular distinction among all of those botanical stories: for it was the plant that evolved and modified to become the chosen ingredient to flavor and preserve beer.

A deep-time overview of the hop is useful for several other reasons. It is important to know that the first brewers to add hops to their vats used varieties that had adapted to their regional environments. Those hops that evolved with different physical characteristics also had different flavors and aromas. The hop's distant past helps us understand how and why agriculturalists developed specific cultivation methods. In simple terms, farmers have long sought to replicate and improve on the growing conditions of wild plants.

Such activities required individuals to study when hop shoots emerged from the soil and latched themselves onto trees and shrubs; it required cultivators to test which soils the hop would grow best in and how to provide the nutrients that the plants needed to achieve an abundance of cones each harvest. Farmers also studied the length of growing seasons and harvest times in accordance with their local environment. This botanical and evolutionary knowledge has been gathered and passed down through generations, and it remains vital to successful hop raising in the twenty-first century. But this is getting ahead of the story.

For most of its existence, *Humulus lupulus L.* carried forth in a world absent of extensive human interaction. That changed in the last fifty thousand years when *Homo sapiens* reached a stasis of behavioral modernity, or took on the physical and mental traits with which we are familiar today. Although evidence is sparse, it is likely that even long before agricultural revolutions around the world and the rise of sedentary civilizations, people discovered various uses for wild hops. Gatherers of the plant used bines for twine and the tender shoots for food. Perhaps, most prominently, they used the cones for medicines, believing in the plant's power to heal a variety of ailments ranging from insomnia to digestive issues.[4] Undoubtedly, someone somewhere tried to eat hop cones right off of the plant. As similarly curious people find today, a taste of raw hops offers a bitter and unpleasant experience. Amid this process of human botanical discovery, the marriage of hops and beer was still far off because beer making arose much later in the course of human civilization. And even then, hops were a relatively late addition to the brewer's trade.

THE MARRIAGE OF HOPS AND BEER

The discovery of all forms of alcohol, including beer, coincided with various agricultural revolutions that unfolded across the world around eight thousand to fifteen thousand years ago, when nomadic hunting and gathering societies transitioned to sedentary farming civilizations. Most archeologists attribute the transition to a warming climate that allowed for the raising of plants and stock animals after the cold Pleistocene epoch transitioned into the warmer Holocene.[5] Others embellish that story by suggesting that alcohol proved the motivating factor in this process, because people around the world discovered the intoxicating delights of fermented grains, fruits, and honey and wanted to reproduce them with regularity. Whatever the truth of these origins, it is vital to know that brewing and distilling knowledge matured over time in tangent with agricultural expansion.[6]

The world's first beer makers discovered their craft in Mesopotamia quite accidently when they found that baked grain left to the elements might create an inebriating substance. The Sumerians get credit for the culinary innovation, since they were the first to replicate the process on their own terms. Eventually, the ancient brewers added water to the sweet grainy substance and flavored the malty beverage with dates or honey. The beer-making process took time to develop and perfect. But it is clear the Sumerians achieved incredible success. They even preserved a recipe for beer in the Hymn for Ninkasi, named after their goddess of the beverage. As one historian noted, the hymn, captured in writing around six thousand years ago, provided an expansive overview of the brewing process, from the gathering and treatment of grain to the types of vessels used in transferring and storing the delightful liquid.[7] Over time, the knowledge of brewing passed through generations of empires, from the Babylonians to the Egyptians, to the Greeks, and to the Romans. While the Greeks and Romans always preferred wine, they developed a brewing culture and spread it to the rest of Europe, which became the beer-making center of the world.[8] This is not to say that other regions of the globe missed out in establishing independent beer cultures. Around six thousand to eight thousand years ago beer brewing emerged in what is present-day Iran and Latin America. However, those regions would not have the same global influence as Europe on the history of beer.[9]

Throughout the course of thousands of years, the brewing process has remained simple and relatively similar. Brewers boil malted grain (most commonly, barley, wheat, or rye in the modern era) with water and, sometimes, lesser ingredients of one sort or another to add flavor. After cooling off that sweet concoction, called the *wort,* the brewer pitches yeast—a live culture that sets about digesting the sugars of the malted grain to create alcohol. One of the biggest differences between the first brewers and those in the more recent past is a matter of ingredients and flavor, particularly in regards to hops. The plant was not part of the original beer recipes, and its use would not be widespread in brewing until the late Middle Ages.[10]

If beer making proliferated throughout the ancient world, what did brewers use to flavor and preserve their beer if not hops? The answer is extensive. According to one scholar, they used nearly two hundred different flowers, spices, and herbs. Some of the most common ingredients added to the wort included dandelion and heather, but the list also included peat moss, cumin, willow, and juniper. Like good cooks anywhere, early brewers experimented with available ingredients and adjusted their recipes over time. Fundamental in this quest for the best beer were locally available plants. The earliest brewing pioneers foraged the countryside around their homes to find ingredients. The results contributed a multitude of beer flavors. In part because of the absence of hops, however, these early beers (often called *gruit beers*) tasted much different from the beers we have today.[11]

According to the best available records, the Roman naturalist Pliny the Elder first documented the common hop just after the time of Jesus Christ. In *Naturalis Historia,* he noted that the ancient Europeans called the plant *lupus salictarius,* most commonly translated as the "wolf of the willow"— perhaps because its climbing bines suffocated willow trees with their rapid growth throughout the spring and summer, or perhaps because of the gnarling twisting of the vines. Even during Pliny's lifetime, when beer was common in parts of Europe, there is no documentation of hops being used in the brewing process. Instead, gatherers of wild hops continued to find uses in the bines for twine, in the shoots for food, and in the cones for medicines, as had likely been done by various civilizations for thousands of years across the Northern Hemisphere.[12]

There are debates on the exact origins, but it is generally accepted that Western Europeans first added hops to their beer in the eighth and ninth centuries. At that time, the hop provided a remarkable addition to the medieval brewer's trade. The gastronomical alchemists came to rely on the bitter alpha acids of hops (known today primarily as *humulone, cohumulone,* and *adhumulone*) that helped balance the sweetness of malted grains and on the essential oils that infused pleasant aromas. The soft resins of *Humulus lupulus L.,* found in the cone's inner yellow lupulin glands, also exhibited strong antibacterial activity and thereby acted as a preservative for beer. (See figure 2.) Of course, it is only in the past century that scientists have been able to explain the chemical makeup of the plant. And even now, there is still much to learn. A thousand years ago, brewers gained the knowledge by testing hops in their beers.[13]

The first generations of beer makers who used hops did not cultivate the plant. Instead, they gathered it from the wild, just as had their ancestors. Brewers in Bavaria, likely the first to use hops, found the ingredient rather easily. Wild hops grew (and still grow) abundantly in German river valleys and on the edge of forests. Brewers likely added the whole cones to their vat

FIGURE 2. Cross-section of an experimental hop variety (with lupulin glands exposed). Photo by Stephen Ausmus. Courtesy U.S. Department of Agriculture.

upon collection in the late summer and early fall. Over time, beer makers also began to dry the cones and store them for uses throughout the year. These adaptations informed future practices and provided standards.[14]

One prominent English hop expert suggested that the A.D. 736 records of a "Wendish prisoner in the Hallertau district of Germany" offer the "earliest written evidence of hop cultivation."[15] Little is known of this account, and it is uncertain why the individual began his work. Nevertheless, sources indicate that shortly after that date, Bavarian monks began planting hops. Perhaps the plant added some charm to their gardens in the summertime, with running bines climbing high and the hop cones hanging throughout. More likely, these early cultivators harvested hop cones for medicinal purposes and brewing. Hop growing spread as the plant became revered for these reasons.[16]

By the end of the ninth century, hop growing for use in beer making expanded from Bavaria to Bohemia (in the present-day Czech Republic),

Slovenia, France, and other temperate regions of continental Europe. In the early spring, growers planted rootstock in small, evenly spaced hills.[17] After shoots emerged, growers trained the bines to climb clockwise on timber posts, since the plant's botanical makeup determined that the shoots would fall off if trained otherwise. Come summer the plants matured, and by early fall hop cones adorned the plant. Families then handpicked the cones after the posts and their attached bines had been laid to the ground. Success in the process, as in any other agricultural activity, depended on trial and error. Hop growers searched for and discovered better ways to encourage growth and productivity, whether those were improvements in training bines or methods of fertilization. Intercontinental travelers helped the agriculturalists by spreading both knowledge and plant material in efforts to improve cultivation. That cascading process continued over generations and would significantly improve hop farming.[18]

The most important activity in early hop agriculture pertained to the selection of hops for planting. Early horticulturalists found that the hop did not breed true from its seed, but rather from its rhizomes, or underground stems that send shoots above the soil. This is not an uncommon agricultural phenomenon; many fruit trees behave similarly. Local geography, or *terroir*—the term used in viticulture to describe the environmental features in which specific grapes grow and which impart unique tastes—proved essential. Although the common hop could be found across Europe, individual regions had unique plants that had adapted to distinctive climates, elevations, and soils. Such variations are called *landraces*. Beer makers and agriculturalists selected the hardiest and most productive of these, as well as those that offered the best qualities in flavoring and preserving beer. Along with the use of local grains and yeasts, the hop selection contributed to regionally specific beers. The first German hops under cultivation included the Hallertauer and Spalter, and the first in Bohemia was the Saazer—all named from the region in which they grew (i.e., Hallertau, Spalt, and Saaz, respectively). These hops have been long considered the world's finest, particularly because of their pleasant, aromatic attributes. For that reason they have been deemed "noble hops." All of the noble aroma varieties are as revered in the early twentieth-first century as they were in the era of the Crusades.[19]

Around one thousand years ago, the hop began an accelerated period of territorial expansion, unlike anything in its deep past. People began to transport the plant by rhizome cuttings. Following the successful rise of hop agriculture across Bavaria, Bohemia, and surrounding regions of western and central Europe, this practice spread to other temperate parts of the Continent. In the thirteenth century, the Hanseatic League played a crucial role in transporting hop agriculture, after the German trading organization adopted hops as the standard preservative in beer. The decision affected not only German beer makers but also those who traded with the German states.[20] While brewers in some regions relied on the importation of what they saw as the ideal German product, many began to cultivate local hops for their own supply. Hop agriculture spread to Scandinavia and Russia, and, by the sixteenth century, English brewers also embraced the hop as an essential ingredient in their brews. By 1700, English growers dedicated approximately twenty thousand acres of land to hop cultivation, largely in Kent, Sussex, Surrey, and Hampshire.[21] At that time the preferred hop variety was the Farnham Pale, later appropriated in Kent and renamed the Canterbury Whitebine. At the century's end, the Golding variety—selected from a field of Canterbury Whitebines—became the standard hop used in English beer.[22] Similar stories explain the nomenclature of hop varieties grown across the world.

In England and across Continental Europe, the expansion of hop growing coincided with the expansion of beer culture. As populations recovered from the trying years of the Black Death in the fourteenth century, the number of brewers who relied upon the hop increased. A general approach to brewing also changed. In the late Middle Ages, a major gendered transformation unfolded: the cottage industry—largely run by rural women, called *alewives*—transitioned into larger-scale urban operations run by men.[23] This significance cannot be overstated, because men dominated the history of professional brewing in the Western world from that point until the very recent past. Amid these developments, beer makers became more professional, joining brewing guilds and adhering to specific codes that included the requirement of using quality ingredients. Beer also established itself as an important part of northern European culture—namely, because the fermented beverage offered a safe alternative to polluted water supplies. Production grew as populations increased, and hop growing also expanded. Farmers began to dedicate more land to the crop, far more than could be used by a household or small community, which had been the previous practice. As a result of increased volume, large hop-trading networks emerged that funneled hops to brewers across Europe. Nuremburg and London arose as two of the largest centers of the hop trade, where formal inspectors judged hops for quality and began to offer local seals of approval. At the same time, unique beer styles cemented themselves into European cultures, including German lagers and English ales.[24]

The professionalization of brewing and the commercialization of the hop trade created greater competition among growers to cultivate quality products and inspired more-intensified exchanges of agricultural knowledge. Universities and agricultural societies in the central hop-growing regions assisted with research that helped develop more productive cultivation methods. Nothing aided farmers more than the proliferation of print culture from the sixteenth century onward, when hop growers began publishing guides. One of the most famous of these publications was Reynolde Scot's *A Perfite Platform of a Hoppe Garden* (1574), an English treatise that provided detailed advice on the preparation, cultivation, and harvesting of hops. The written discourse outlined the nuances of preparing hills for planting, selecting timber poles, training bines clockwise for upward growth, and combating various pests and diseases. Scot also described the *oast house,* the English term for the building where growers dried their cones. He outlined the construction of the two-storied structure where growers laid hops across the top floor to dry from the heat of a kiln underneath. Finally, Scot emphasized the best ways to dry uniformly for transport.[25] His guide and similar publications played an essential role in improving hop growing for generations. The result of these works could be seen in the physical environment as Europeans planted ever more hop gardens and constructed hop dryers by the thousands. Farmers and businesspeople published similar guides in other regions and languages.

The benefits of new agricultural knowledge and increased productivity in hop growing across Europe came with its share of problems; chief among them was the strain on labor resources for the harvest. Once able to rely on family members and neighbors, hop growers came to depend on hiring seasonal help toward the end of the summer and early autumn. As hop growing became more commercialized, most European growers solved the problem by hiring a temporary pool of lower-class laborers. The workers camped for the duration of the harvest and engaged in the monotonous daily task of pulling cones from the bine. Because the work was unskilled, entire families participated, with men, women, and children of all ages earning wages according to the weight of their hauls. The harvest labor situation remained an integral aspect of hop agriculture across the world, up to the present day in some areas, always connecting the hands of agricultural workers to the brewer's gold.[26]

So what does the history of hop growing and beer making in Eurasia have to do with a relatively small agricultural valley in western Oregon?

Hops and beer have substantial origin stories across time and place. For millions of years, *Humulus lupulus L.* evolved and traveled around the world via wind and bees and the forces of nature. Then, very recently in the plant's own past, humans found a use for hops in beer making. This inspired the development of specialized agriculture. Production of knowledge and technologies centered on hops and brewing spread, and the hop plant spread physically, too, into gardens and fields across temperate Europe. In other words, hops and humans have been participants in a grand experiment of agriculture and beer that had traveled from Europe across the world in the previous thousand years, and in nature and the earth millions of years prior. The characters and events introduced are part of the Pacific Coast hop industry's global heritage, playing essential roles in the flavors and aromas that emanate in the craft beer revolution. The next section brings this story closer to home.

THE EUROPEAN BEER AND HOP DIASPORA

The hop's territorial migration accelerated once again between the sixteenth and nineteenth centuries, when Europeans introduced their brewing culture across the globe. German, British, Dutch, French, and Scandinavian immigrants hauled brewing kettles and beer recipes with them to settlements in Africa, Asia, Australia, and the Americas. Hopped beer offered colonists a significant source of calories and a reminder of home. More important, the beverage continued to be safer to drink than water from contaminated sources in colonial settlements. It should then not come as a surprise that European colonists in many areas of the world constructed breweries as some of the first buildings within forts or town sites. Along with the planting of grains, fruit orchards, and other European crops—not to mention the importation of cattle, sheep, and other nonnative animals—the process became part of the Europeanization of the globe.[27] But how did colonial brewers obtain the spice of their brew in regions distant from commercial hop production? This was a question that the Dutch would face in South Africa by the 1650s and German colonists in China during the late nineteenth century. It was the English during this period, however, who left the biggest imprint. Today's popular India pale ale (IPA) traces its roots to over two hundred years ago, when it earned its name because of the large quantities of hops that English brewers used to preserve beer for long oceanic voyages to colonial India.[28]

The greatest expansion of British hop growing during that period occurred in North America. Records of the Massachusetts Bay Company indicate that along with hopped beer, hop plants arrived with Puritan immigrants as early as the 1620s.[29] From then on, brewing beer caught on across the colonies. Access to water and baker's yeast never proved cause for concern. Although the initial malts arrived from England because of inadequate supplies of barley in North America, that situation did not last long.[30] By the 1640s, English and Dutch settlers planted barley and other familiar grains. They also imported and planted more European hops in small plots. Some ambitious beer makers sought out American subspecies of wild hops. But those plants never won over imported domestic varieties.[31]

While it might seem strange that colonial brewers did not embrace local wild hops, they had good reason. The recipes upon which they learned their craft called for specific hops with specific profiles from their homelands. From the beginning, European beer makers complained that American hops gave off flavors and aromas that were too strong and that, in turn, altered their end product. Brewers faced the choice of importing familiar hop varieties from Europe or trying to grow European hop varieties themselves. Eventually the latter option became preferable given the expenses of importation and because colonists discovered favorable growing conditions. Not only did the temperate regions of eastern North America have climates similar to those in Europe, but they also benefited from virgin soils that had not been under intensive long-term cultivation. To their luck, colonists discovered that growing conditions in North America could produce more hops per acre than many of the hop-growing regions in Europe.[32]

Throughout the colonial era, hop raising mostly occurred at the household level, with families growing small plots for their use and possibly that of their neighbors. The most universally grown hop was probably the Farnham Pale, commonly grown in England. Over time, hop farmers turned toward a hybrid hop, called Cluster, born accidently of an English hop and a wild North American variety. The hop appeared perfect for colonial growers because it produced flavors familiar to English stock but had the benefit of genes better suited to the North American agricultural landscape. Different varieties of the Cluster that matured earlier or later in the season also developed, and the hop variety became the preeminent feature of the English ales, stouts, and porters that Americans brewed and consumed well into the nineteenth century. While strong in flavor and alcohol content, suitable beer could now be made with ease in most American colonies.[33]

By the early 1800s, New England and New York farmers had established the first commercial hop operations in the United States.[34] Initially, growers selected native hops. As one New England farmer noted, they could be found in abundance "growing spontaneously on the banks and intervals of our large rivers."[35] Eventually the commercial farmers recognized the need to supply brewers with hops that they preferred. Along with the Cluster hops, these agricultural pioneers grew mostly English varieties, which made sense given climactic similarities and the fact that American growers found a significant portion of their buyers in the British marketplace. Yet, competition was fierce. Brewers in England and elsewhere often decried the American product as inferior. This occurred in part because of the amount of stems and leaves that producers accidently packed in with each shipment of hop cones and that serve no purpose in brewing, and in part because North American farmers had not yet mastered cultivation techniques. Both of these realities probably justified the reputation of early American hops as a substandard product. Throughout the nineteenth and early twentieth centuries, European buyers more commonly complained about taste and fragrance, claiming that American hops were "coarse," "pungent," or otherwise unpleasant and not suitable for quality beer. An 1857 report from the British House of Commons, for example, noted that American hops were "grown too rank; they are very powerful in their flavour, and very bitter; they are naturally grown of a certain quality, which cannot be changed by reason of the soil and climate being the only cause of it."[36] Official and unofficial reports of all types shared similar language well into the twentieth century.[37] Still, when prices were right or European supplies ran low, American hops served the global marketplace as an acceptable backup.

Equally prohibitive for the development of commercial hop agriculture in the United States was that fact that hop farmers did not yet have the advantage of a vibrant domestic beer market. Although New York and Philadelphia had emerged as centers of brewing during the colonial era and early republic, consumers of alcohol favored readily available corn whiskey, hard apple cider (courtesy of the likes of Johnny Appleseed), and rum made in New York and New England from imported Caribbean sugar. During this period, production of hops for the domestic beer market remained modest.[38]

But the hop industry would rapidly expand during the nineteenth century, a period when the United States transitioned into a beer-drinking nation. Two important factors contributed to this growth. First, the 1840s and 1850s witnessed a substantial influx of German and Irish immigrants, who brought distinct beer cultures and beer-making techniques. The German brewers, importantly, brought with them a different kind of beer: lager, which won American favor for its lightness and drinkability compared with heavier ales in the English style.[39] Second, an expansive temperance movement arose by midcentury that sought to reduce the volume of alcohol consumed by Americans. As one historian notes, "During the first third of the nineteenth century the typical American drank more distilled liquor than at any other time in our history."[40] Leaders of the temperance movement, who tended to be women and Christian, equated alcohol with detriment to family and society; they fought against the stigma that the United States was a nation of drunkards. The arrival of the German lager during this time period was, perhaps, serendipitous. The temperance movement embraced lager as an alternative to more-potent kinds of booze. Immigration and temperance, it can be said, turned American alcohol consumers into lager-beer drinkers.[41]

It was under these circumstances that the Midwest, with its large German populations, became a center of American brewing. German-American brewers, including Adolphus Busch, Frederick Miller, and Frederick Pabst, successfully set up shop in Milwaukee (Miller and Pabst) and Saint Louis (Busch) just before the Civil War. Not surprisingly, states in the Midwest began commercial hop production thereafter in the hopes of catering to these emerging beer empires.[42] Many of the large German-American brewers initially shunned American hops in favor of continuing imports from Europe. But gradually they conceded that the purchase of local hops proved both cost effective and qualitatively competitive on the world market.[43] This would be an uphill battle not only because of favored European hop varieties, but also due to the perceived low quality of hops produced by American farmers, who were new to the business.

By the outbreak of the Civil War in 1861, the rise of the German-American lager altered the beer scene in the United States for brewers and consumers, and opened new opportunities for the country's farmers to grow hops. Across the country, the sparsely populated Pacific Northwest also felt the repercussions. There, south of the Columbia River in a valley drained by the Willamette River, stood the future "Hop Center of the World."

TWO

Valley of the Willamette

LONG BEFORE THE HOP MADE its appearance on this planet, a small valley in western North America charted a history of its own. It was around thirty-five million to forty-five million years ago that volcanic forces and tectonic shifts crashed a seafloor into the emerging Cascade Range. Shallow seas initially inundated this new part of the continent. But after another ten million years, a renewed period of geologic activity lifted the Coast Range on the western portion of that landmass to help drain its new terrestrial neighbor. A one-hundred-by-fifty-mile region with a bedrock of basalt remained with elevations approximately forty to four hundred feet above sea level. Eventually, the valley settled, with the Cascades looming to the east, the Coast Range to the west, the Calapooya Range to the south, and the Columbia River to the north. At the center of it all lay a northward-flowing river fed from drainage of the surrounding mountains that would one day be called the Willamette.[1]

It was not until much later—at the end of the last ice age, around thirteen to fifteen thousand years ago—that the Willamette Valley took on its current physical character. After the cold Pleistocene epoch, inland glaciers melted, and magnificent torrents of water, ice, and debris pushed westward from what is today Idaho and Montana toward the Pacific Ocean. The violent floods, more powerful than anything witnessed by recent civilizations, carved part of the Columbia River Gorge, which separates what is today Oregon and Washington.[2] More important to this story, the floods carried debris that frequently dammed the Columbia River near its mouth, back-flooding waters into the Willamette Valley. That activity changed the soil and topography. When the floods subsided, a deep layer of silt rested upon the bedrock. Combined with a constant flow of alluvium, or sediment, from

MAP 1. The Willamette Valley in the context of the North American West.

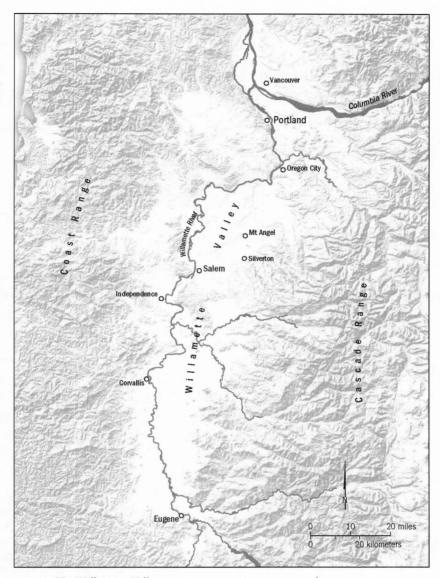

MAP 2. The Willamette Valley.

the surrounding volcanic mountains, the valley settled with a nutrient-rich sandy-loam soil—the stuff of a farmer's dream. (See maps 1 and 2.)

The valley's moderate climate also took shape after the last ice age. Though some observers have characterized it as Mediterranean, it is best described as windward marine, or a climate dependent on an ocean's prevailing winds and currents to regulate heat, cold, and precipitation. In this way, the Willamette Valley climate resembles parts of Great Britain and northern France, whose nearby ocean currents bring warm tropical air to their northern latitudes. Residents in the twenty-first century know well that rain is the defining attribute of the Pacific Northwest's climate west of the Cascades, with the Willamette Valley receiving approximately forty inches of annual precipitation depending on the specific area. But there are also somewhat distinct seasons, including rainy autumns, mild winters, wet springs, and pleasant, dry summers.[3]

That temperate climate in the Willamette Valley suited a wide range of life. During the Pleistocene, the valley provided fertile ground for stands of Douglas fir, western hemlock, western red cedar, and white oak. The vegetation and climate also provided homes for countless prehistoric mammals, including mammoths, bison, giant sloths, and dire wolves. As the valley dried after the last ice age, an oak savanna dominated, with other common vegetation including fir, Oregon grape, salmonberry, elderberry, and many kinds of grasses. This was all good nourishment for the variety of animals that also made the valley home such as beaver, deer, elk, wolves, and rodents, birds, and insects of many types. Salmon, trout, sturgeon, and several other species of fish laid claim to the Willamette River and its tributaries.[4]

The environmental conditions of the Willamette Valley also proved ideal for the region's first human inhabitants, who had crossed Beringia, the famous land bridge that connected North America to Asia, around the same time as the melting glaciers transformed the landscape. These first peoples benefited from the wealth of nature, finding abundant food and resources in the valley's forests and waterways. Their animistic religions and notion of reciprocity differed from those of the peoples who would arrive in the more recent past, and this guided their approach to harvesting nature's resources. The valley's indigenous populations respected the plants and animals on which they lived and reaped only what they would need for the foreseeable future, so as to maintain a balance in nature.[5]

As anthropologists and historians have emphasized in the past thirty years, however, the first American Indian tribes of the Willamette Valley contributed to transformations of the landscape. Even if it is unclear whether those initial groups of nomadic hunters introduced new animals or plants (intentionally or not) from Central Asia to North America, some believe that by hunting they contributed to the destruction of the famed Pleistocene megafauna, though that hypothesis is increasingly challenged.[6] But that was not the only way they modified the landscape. Within a few thousand years of their arrival, the valley's first peoples regularly burned the dense forests to encourage habitat for their most important game animals, vegetables, berries, and herbs. As native peoples across the continent did, many turned toward agriculture within a few thousand years of their arrival. Camas, tarweed, and wapato (a tuber), for example, complemented the salmon-based diet of the Columbia River tribes. These burning and agricultural practices cannot be ignored in understanding the formation of the Willamette Valley over the course of millennia.[7]

Like the deep-time story of hop evolution, understanding the origin story of the Willamette Valley is important for several reasons. The existence of nutrient-rich sandy-loam soil, a moderate climate with ample spring rains and summer heat, and a landscape modified by burning all went hand in hand to create an ideal location for commercial farming to arrive in the nineteenth century. While many crops grew well in the region, the environmental conditions proved particularly perfect for hops—a species that needs deep, fertile soil and a specific temperate climate. But herein lies an important detail. Throughout the millions of years that the Willamette Valley existed, the wolf of the willow remained absent from the region's ecosystems.

To the best of our knowledge *Humulus lupulus L.,* the common hop, never established itself in the Willamette Valley, or anywhere else in the Pacific Northwest, during its prehuman global travels. This was an anomaly across temperate North America. Scientists have determined that around a million years ago, the progeny of Asian hops spread to temperate regions east of the Rocky Mountains, as well as in the Southwest and Great Basin. Across the continent, four subspecies—*neomexicanus, pubescens, lupuloides,* and *lupulus*—evolved from plants that had migrated eastward from Asia long before humans.[8] Genetic testing of plant materials found during scientific hop-seeking excursions and in packrat middens suggests this evolutionary path. But, as hop expert Jim Oliphant has explained, there is no evidence that wild hops found a home in present-day Oregon, Washington, or California prior to human introduction. Having studied intently the migrations of plants, particularly *Humulus,* across the millennia, he has determined that climatic conditions in the deep past prohibited the hop from taking root. Namely, glaciation prevented the plant from finding habitats in those regions or destroyed populations in colder periods. At the present time, Oliphant's research offers the clearest perspective on the subject.[9]

THE PACIFIC NORTHWEST'S FIRST HOP GROWERS

So when did *Humulus* appear in the region that would one day claim to be the "Hop Center of the World"? That, too, is up to speculation. But historical context offers us a better idea of how to answer the question. Given what historians and archaeologists know about the importance of beer in European culture and the propensity of transoceanic travelers to pack plenty of the beverage in their ships' hulls, one might surmise that the first hops to arrive in the Pacific Northwest did so in bottles of English ale during the late eighteenth century. It was at this historical moment that English, French, American, and Russian trappers sought out the prized pelts of otters and beavers for the global marketplace and began charting and mapping the Columbia River and its tributaries. Though no historical marker designates the site of the first ale consumed in the Willamette Valley or larger Columbia Basin, it is clear that by the time the Euro-American trappers and traders established settlements in the early nineteenth century, beer also took hold. Aside from what they could acquire from trading with Indians, employees of the American John Jacob Astor's Pacific Fur Company and the Canadian North West Company relied on the importation of foods and drinks from Europe and the East Coast of the United States. Among the earliest surviving artifacts of this period include bottles of English pale ale. In all likelihood some travelers carried malt and hops with them, but those details as of now are unclear. In any case, the main source of hops found across the Pacific Northwest through the first two decades of the nineteenth century most likely would not have been plants or even dried cones, but in a transmogrified state in the liquid that contained them via the chemical processes of brewing.

After 1824, the story changed. Under the regional governorship of George Simpson, the Hudson's Bay Company established a headquarters at Fort Vancouver, near the confluence of the Columbia and Willamette rivers (what is today Vancouver, Washington). The primary objective of this first permanent Euro-American settlement in the region continued to be the extraction of animal pelts and furs for the global marketplace. But Simpson selected the location specifically for the potential of agricultural enterprise and therefore freedom from expensive imports of victuals and the volatility of Indian trade. The Hudson's Bay Company had championed this idea for its outposts across North America as early as the 1670s, and the other fur traders of the Pacific Northwest dabbled in farming prior to the development of Fort Vancouver, planting barley, potatoes, cabbage, turnips, and other vegetables to supplement the proteins they could acquire via hunting, fishing, and trade. But those agricultural experiments in Oregon Country achieved limited success and did not prove lasting.[10]

Aside from the direct order of their superiors, chief factor John McLoughlin, who oversaw day-to-day operations, and the residents at Fort Vancouver realized early on that transforming the landscape and championing familiar European crops would be vital for sustaining their new community. In this way, they followed in the familiar footsteps of the Pilgrims in Massachusetts and settler societies across the world, setting in motion the process of Europeanizing their new home. The company men at Fort Vancouver cleared forests, built living quarters, mills, and other structures, and fenced land and cultivated it by introducing Eurasian plants and animals. By remaking the landscape in the vision of their own culture, they hoped to improve their quality of life and attract others to join them.[11]

The agricultural success at Fort Vancouver was staggering. By the early 1830s, cattle and pigs thrived, producing more than enough dairy and meat products for the residents of the fort. Wheat became the most abundant of the field crops, with barley, peas, potatoes, and a host of other grains, starches, and vegetables also growing magnificently. The fort's orchard and garden produced apples, grapes, cherries, peaches, plums, melons, a variety of berries, and ornamental flowers to the delight of residents near and far. Over time, as one geographer noted, "the fort assumed more the appearance of a farming community than that of a fur trade establishment."[12] The success resulted in large part from the advantageous climate and environment that resembled agriculturally productive regions in Europe. It also resulted from McLoughlin's leadership, the work of hundreds of Euro-American and American Indian farm laborers over the years, and the expertise of those who passed through the fort.

During the mid-1820s, the famed Scottish naturalist David Douglas (namesake of the Douglas fir) spent a significant amount of time at Fort

Vancouver amid his botanical investigations of the Pacific Northwest. Not only did he offer his expertise to the fort's agriculturalists, but he also arranged for his sponsor, the Horticultural Society of London, to send a variety of seeds for planting at his temporary home on the Columbia. McLoughlin and other friends of the Hudson's Bay Company also obtained a wide array of seeds and plants from around the Pacific Basin. While not all transplants fared well (citrus, for example), McLoughlin and the fort's residents could not have been more pleased with the agricultural diversity and abundance.[13]

The vitality of the Fort Vancouver farm and garden allowed Simpson and McLoughlin to supply other fur-trading outposts and the bands of missionaries and other Euro-Americans who trickled in during the 1830s and 1840s. It also allowed them to fatten their bottom line. Around the same time, Simpson and McLoughlin arranged for the periodic sale of agricultural goods to Sitka, Alaska, where Russian traders required provisions. Although this relationship may seem tangential to this story of hops and beer, it actually proves quite central. For it is in the records of shipping from Fort Vancouver to Sitka that we learn of the first commercial hops, and probably the first cultivated hops in all of the far west of North America. Along with the wheat, potatoes, cabbage, peas, and corn transported to Alaska, the wolf of the willow contributed to Fort Vancouver's agricultural portfolio.[14]

In all likelihood, hops arrived at Fort Vancouver along with the importation of botanical specimens from around the world during the 1820s, and with certainty by the 1830s.[15] But the Sitka exchange offers the first evidence of the plant sold as a commodity. Shortly after the record of hops in the Alaska shipping manifesto, there are more frequent mentions of the plant from residents and visitors, including an early-1840s note from Lt. Charles Wilkes that a Mr. D., perhaps David Douglas, "introduced Hops by bringing a few living plants with him from Canada some 3000 miles & the plants now are to be seen in abundance" in the fort's gardens.[16] A subsequent stream of letters requested increased volumes of hop exports, revealing the continual success of hop production in the fort's garden.[17]

Given the presence of hops and ample supplies of barley and wheat at Fort Vancouver, one might assume that McLoughlin sponsored the establishment of a brewery to help quench the thirst of the many soldiers, trappers, and traders who passed through. This was not the case. From the onset, the Hudson's Bay Company maintained specific limits on alcohol consumption, in large part to prevent local American Indians, who had not yet been exposed to alcohol, from imbibing. Additionally, in 1844, even before the United States laid sole claim to the land, American residents of Oregon Country enacted a temperance law. At Fort Vancouver, the high-ranking officials could have access to alcohol, but both British and American authorities did not want to see the privilege abused. Maintaining those rules offered a different story. Despite regulations, common residents had access to beer, even if the site lacked a formal brewery.[18]

As was the case with earlier fur-trading outposts, the initial brews arrived at Fort Vancouver in the form of imported English ales, porters, and stouts. In particular, archeological digs have found multiple bottles of Read's India Pale Ale, the hoppy style of beer brewed especially for long oceanic travels— in this case around Cape Horn, at the southern tip of the Americas.[19] It was not the only beer that the fort's residents imbibed. By 1826, records show that McLoughlin himself dedicated part of the barley harvest for malting and use in small-scale beer making. Little is known about these experiments. But it becomes clear that by the end of the decade, brewing occurred regularly at the fort. One visiting missionary wrote a letter that not only commented on the impressive agricultural success at Fort Vancouver, but also mentioned that "they raise their own breadstuff, cultivate barley, malt it and make beer which they will soon be able to export in small quantity."[20] Similar types of evidence throughout the 1830s and 1840s suggest that beer making on a small scale existed for consumption and export. Still, because of the restriction on alcohol use, no official brewery emerged at or near the fort until John Muench (alternatively spelled "Maney," "Menig," and "Minich") opened one in 1857, toward the end of the Hudson's Bay Company's authority in the region.[21]

Overall, there are many questions surrounding the status of *Humulus lupulus L.* at Fort Vancouver. Lack of evidence prevents us from knowing what types of hops grew, or how the residents even learned the specific details of growing the crop. Lack of evidence also prevents us from knowing how hop cultivation might have spread throughout the Pacific Northwest following its origins inland on the Columbia River. Given the fundamental desire for beer in settler societies, it is possible that retirees of the Hudson's Bay Company, who began to live in the Willamette Valley by the end of the 1820s, brought hop rootstock with them for their individual brewing purposes.

What we do know is that by 1850 the U.S. Agricultural Census reported only eight total pounds of hops harvested in the entirety of the Pacific

Northwest.[22] It might be fair to question whether that number underrepresented the actual harvest of hops from individuals or families who acquired rootstock from Fort Vancouver or elsewhere for household beer making. After all, a number of new Euro-American settlements had sprouted up by that time, with many people likely wanting beer. It is unfortunate that there are not more records that provide a clearer picture. Nevertheless, social, economic, and technological conditions would soon arise to transform the entire Pacific Northwest into a producer and exporter of hops and specialty crops of all types.

Prior to getting to that point, however, it is first important to understand how residents of the Willamette Valley engaged in efforts similar to those McLoughlin and others at Fort Vancouver were making to create an agricultural infrastructure upon which hops and other specialty crops could enter commercial production. That is, we cannot understand the rise of hop growing without understanding the broader Europeanization of the Willamette Valley and the growth of a new agricultural economy and society that connected to the rest of the nation and world.

THE EUROPEANIZATION OF THE VALLEY

By the 1840s, English, French-Canadian, and American employees of the Hudson's Bay Company and other travelers expanded their presence outward from Fort Vancouver on the Columbia River to the surrounding country, including southward to the Willamette Valley. French Prairie, about thirty miles south of present-day Portland, became the first of these settlements on the Willamette River and an important hub of early Euro-American farming. At the same time, substantial groups of emigrants had set out from eastern regions of the United States and Canada with religious motivations and enthusiastic prospects for farming and other enterprises. On the famous Oregon Trail, pioneers of all types looked westward to construct new lives and transform the land west of the Cascades.[23]

While the stories are many, the efforts of Methodist missionary Jason Lee offer some insight into these transformations. Along with a small community, Lee settled in the Willamette Valley during the early 1830s, eventually founding of the city of Salem in 1842. Lee's first objective was to convert the local Indians to Christianity, but he also set out to transform the landscape and to promote further missionizing efforts. The process entailed the development of a farming community (with the assistance of John McLoughlin and others at Fort Vancouver) as well as the creation of a school, now Willamette University. Lee wrote glowing letters of the valley to audiences back east that would inspire further emigration, and McLoughlin himself supported the settlements south of the Columbia River so as to not interfere with the Hudson's Bay Company's authority to the north.[24] One thing Lee was not keen on was the consumption of alcohol. As early as 1834 he created a temperance society, and he labored mightily to pass an Oregon prohibition act in the next decade, the first of its kind in the nation.[25] For these reasons, Oregon's Euro-American settlements of the 1830s and 1840s were decidedly dry.

Two major political events further affected early settlement of the Willamette Valley. First, in June 1846 the United States and Great Britain signed the Oregon Treaty, establishing political boundaries of the Pacific Northwest between the United States and British Canada on the forty-ninth parallel. By 1853, the lands south of the Columbia River that encompassed fertile valleys, together with basin and range lands north of the border with California and Utah Territory (eventually Nevada and Utah), became part of Oregon Territory, with the lands north of the Columbia becoming Washington Territory. Second, in 1850 Congress signed the Donation Land Act (preceded in 1843 by a similar land act under the provisional government) to encourage settlement in the Willamette Valley and thus solidify the American hold on the land. The act generously offered individuals 320 acres and married households 640 acres of land upon the condition of "improvement" within four years. The parcels were much larger than the 1862 Homestead Act and were unique in that they recognized married women and half-blood Indians as property owners. The measure achieved success for the federal government. Settler farming communities spanned the entire valley with their large estates by the end of the decade and helped Oregon become the nation's thirty-third state in 1859. As a result, family and commercial farming supplemented the early missionary efforts.[26]

Many of those who settled the Willamette Valley in the mid–nineteenth century believed that they had found an agricultural utopia. Not only did the climate, soil, and landscape offer ideal farming conditions, but the major river networks provided for efficient transportation, and the valley's surrounding foothills and mountains provided ample fishing, hunting, and logging opportunities.[27] Some settlers referred to the valley as a "horn of plenty," not only because of its agricultural potential, but because the physical boundaries of the valley resemble a horn—beginning in a narrow portion in the southern mountains and growing wider by the time it reaches the Columbia River. Whereas the abundant and incessant rain of the long winter and spring seasons prevented some from declaring this place a "second Eden," impressive agricultural progress occurred by midcentury.[28]

Although Willamette Valley farmers recognized early the myriad advantages in their chosen valley, life remained challenging. They lived far from the population centers in the Ohio and Mississippi valleys, and their communities were sparsely populated. Furthermore, integrating into the American economy was not an easy task given their isolated location. These problems resembled those of earlier Euro-American colonists in North America and their connections across the Atlantic. But the new Oregonians drew upon a familiar story on the opposite coast. According to one scholar, "Settlers sought not so much the creation of a new society as the re-creation of an older, familiar one which allowed them to advance their own personal fortunes."[29] Along with forming familiar community and religious ties, farmers set about "improving" the agricultural landscape by clearing forests and draining wetlands. They fenced off land and diverted water. They eradicated species they viewed as noxious, including predators: wolves and bears that posed threats to livestock as well as ground squirrels and rabbits that ate crops. The end result was the imposition of an order upon the environment for the purpose of economic and community development.[30]

At the same time that farmers "improved" the lands and waters of the Willamette Valley, they successfully introduced familiar and marketable flora and fauna. As was the case at Fort Vancouver, grain crops and livestock flourished. Pigs became particularly abundant as they roamed freely and thrived in an environment plentiful in food and newly lacking in predators. But, as did the Massachusetts Bay Company immigrants in New England, the initial settlers emigrating to the Willamette Valley were limited to the seeds and plants they could carry on their westward journeys. The choices for agricultural crops expanded after Henderson Lewelling and William Meek established the region's first horticultural and nursery supply company in 1848 after bringing wagons full of plants and seeds from the east. By the early 1850s, settlers delighted in creating a landscape similar to that of their homelands with nonnative shade trees, vegetables, and specialty crops that mostly included fruits and nuts. This transformation of the landscape cannot be understated. Not all specialty crops, however, proliferated by this time.[31] Early nurseries did not appear to offer hops. Settlers may have acquired the plants from Fort Vancouver; or, more likely, they imported European root-stock from the East Coast or Midwest if they were to have hops at all.

By the end of the 1850s, Euro-American agricultural settlements spanned the entire Willamette Valley, and the region began to earn praise for its economic developments in farming. Grain and livestock became central to the livelihoods of the new farmers.[32] California's gold rush initially provided outlets for these commodities, but by the 1860s Willamette Valley growers established a transoceanic grain trade, particularly the white winter wheat that grows well in temperate climates and can be harvested in early summer. The farmers used the ample tributaries of the Willamette and Columbia rivers to ship grain from their homes to locations all over the world. Liverpool became a central trade destination, but markets extended to the East Coast, East Asia, and Australia as global populations rose.[33] Historian William G. Robbins suggests that the wheat trade offered an early "metaphor for the ever-expanding and complex interrelationships that developed between Pacific Northwest landscapes and national and international places."[34] This early farming and marketing infrastructure proved indispensable for later farmers selling a more diversified array of commodities, including hops.[35]

Although rural agriculturalists cultivated goods for the international marketplace, urban areas were also essential for the growth of Willamette Valley farming. It was in a group of small cities that Oregonians processed, marketed, and shipped the bounties of the farms. By the 1840s, Oregon City became a center of gristmill and sawmill activity, as well as a center of marketing for shipments that came northward on the Willamette River from all parts of the valley. While far from the size of San Francisco after the California Gold Rush, the Willamette Valley's cities grew throughout the rest of the nineteenth century. By the 1870s Salem, Portland, and Oregon City had the largest populations. Robbins again provides important context for what this meant not only in grain shipping, noting that "Portland developed a strategic infrastructure that fronted two worlds, one that faced outward to the Pacific and oceanic markets and sources of capital, and a second that looked inland to a vast hinterland linked to Portland by the magnificent Columbia River and its tributary, the Willamette."[36] In other words, the business and infrastructure that developed in Portland and other cities became the arteries to distribute the products of commercial agriculture in the Willamette Valley. Country and city intertwined early and often in the region's agricultural expansion.[37]

Along with the benefits of agricultural success, residents of the Willamette Valley celebrated their quality of life. Historians have suggested that the

Willamette Valley in particular differed from other settlements in the Far West. In contrast to largely individualistic and male-centered mining settlements of California and Nevada, the Willamette Valley tended to be a destination for families. This sprang, in part, from the prevalence of missionizing activity, as well as from the desire of farming families to travel westward as a whole. Tightly knit communities and towns arose as a result, and violence was far less prevalent than in other western settlements. Further adding to the idealized quality of life in the Willamette Valley were medical factors. Doctors believed that the environment was exceptional not just for farming, but also for health. They saw salubriousness in the valley in its mild winters and dry summers, and in its lack of mosquitoes and other pests that plagued the Central Valley of California.[38]

As a whole, the Willamette Valley drew praise in the mid–nineteenth century. The booster Hall Jackson Kelley wrote of the environmental advantages of Oregon soon after his ventures to the area in the 1830s, before major settlement. Residents and onlookers did, too. One temporary resident from the East Coast remarked in 1848, "[T]he universal expression of the . . . population, is, that Oregon excels the world for health, variety of beautiful scenery, certainty of good crops, excellence of water and water privileges . . . and for strength and depth of the soil." He would go on to note, "I know of no people so generally, or so highly pleased with their locations, or homes, as those of the Willamette valley."[39] Similar sentiments appeared across local and national media outlets throughout the nineteenth century.[40] The Willamette Valley even became a destination for utopian communities, popular in New England and the Midwest as sites of social and cultural resistance to the industrializing and modernizing world. Seeking the benefits of community and agricultural promise in 1856, William Keil of Bethel, Missouri, established the Aurora Colony in Marion County. The settlement was one of the longest-lasting utopian experiments of the nineteenth century (until 1883) and continues today as a town populated by some descendants of the original settlers. Though shorter in duration, many other intentional communities sprang up in the Willamette Valley for its perceived Edenesque qualities.[41]

Eventually, the resettlement of Oregon would also see the introduction of a permanent brewing culture that had been largely absent at Fort Vancouver and forbidden from the missionary settlements. In 1849, the territorial government repealed the earlier temperance law.[42] Then, during the 1850s, an expanding economy led to growing urban centers, and more diverse groups of people arrived in Oregon. As a result of these events, a lasting beer and saloon culture began to emerge. Along with it came an associated rise in hop growing. Upon this backdrop of the Willamette Valley's early agricultural success and favorable reputation among residents and travelers, Portland's most famous brewer, an ingenuous promoter, and a whole cast of characters would come together to help usher in the rise of a new global center of hops and beer.

THREE

Hop Fever

BY THE TIME OREGON ACHIEVED STATEHOOD in 1859, a few local breweries served the Pacific Northwest's major towns and cities. The quality of beer varied, and beer-making establishments often came and went, not unlike many settlements in the boom-and-bust Far West. In Portland, a town of around three thousand residents by 1860, a handful of brewers had set up shop. According to historian Tim Hills, British immigrant Charles Barrett was probably the city's first professional brewer, operating his Portland Brewery from 1852 to 1855. There, he sold traditional British ales and porters, as opposed to the German lagers that would dominate the regional and national marketplace by the end of the decade. A year after Barrett closed his brewery to reinvent himself as a bookseller, German immigrant Henry Saxer opened Liberty Brewery on the west bank of the Willamette River near what is present-day Old Town. The lager-producing brewery stepped in to help quench the thirst of the working-class town of loggers, millers, fishermen, and merchants. And it fared quite well, as evidenced by the expansion of its facilities over the decade of its existence.[1]

The most significant brewer operating in Portland prior to the Civil War, however, was none other than Henry Weinhard, a name familiar to even Oregon's non-beer-loving residents for almost as long as there has been beer culture. After arriving in the region from Germany in the mid-1850s, Weinhard found work at the Vancouver Brewery, established by John Muench and the first to have opened alongside Fort Vancouver. At the same time, Weinhard built a business relationship with George M. Bottler, who established the City Brewery on the other side of the Columbia River in Portland. In a short time, Weinhard built a brewing empire. By 1858, he acquired Muench's establishment, and four years later, in a partnership with

Bottler, he purchased Saxer's Liberty Brewery plant in Portland and used it as a new Vancouver Brewery outlet. Amid this flurry of activity, Bottler had also relocated the City Brewery from its original riverfront location into a much larger state-of-the-art facility he had built at what today is Northwest 12th Avenue and Burnside Street.[2]

When Bottler fell ill and eventually died in 1865, Weinhard took full control of the business, and it became the most productive and famous of Oregon's breweries for more than a century. Weinhard's flagship beer during his lifetime, Columbia Lager, would not only become the beverage of choice for many residents of the American West but, by the turn of the century, also reach markets in East Asia, in Russia, and around the Pacific Basin. While the brewery underwent substantial transformations throughout its existence, one thing never changed. Portland's most famous brewery contributed a heavy malty and hoppy aroma that enveloped all who passed by.[3]

As Hills has explained of the early brewing crowd, Barrett, Saxer, Muench, Bottler, and Weinhard were far from the only professional beer makers in the Pacific Northwest. Into the late nineteenth century, it appears, a cadre of German-born brewers navigated from town to town, either establishing their own breweries or working for a financial backer as head brewer. These individuals might have started in Portland but moved around to Oregon City, Eugene, Tacoma, or Spokane as various opportunities arose.[4] There were undoubtedly differences in the flavors of the beers these brewers produced. But most were German lagers, the beer style that had won broad popularity in mid-nineteenth-century North America. Along the Willamette Valley, some of the other lasting operations included the Oregon City Brewery, the Salem Brewery (later the Capital City Brewery), and the Albany Brewing and Bottling Company. In Portland, the Gambrinus Brewery joined Weinhard's City Brewery as a major regional producer during the late nineteenth century. Elsewhere in the region, John Kopp (no relation to the author) made a name for himself after founding the North Pacific Brewery in Astoria, eventually opening an establishment in Portland.[5] Prior to that, Kopp had partnered with Andrew Hemrich to build the Bay View Brewery in Seattle, famous for launching the Rainier Beer brand. That brewery and the Washington Brewery (later called the Seattle Brewery) controlled a large percentage of the Seattle market, with dozens of smaller brewers doing good business on both sides of the Cascades. At the same time, dozens of saloons and European-style beer gardens emerged to serve those suds to local clientele.[6]

The Pacific Northwest's brewing pioneers utilized the locale's natural resources that craft brewers still covet today. That is, they profited from an abundance of clean, fresh water and regionally grown grains. Similarly, they could rely on baker's yeasts or live yeast cultures brought from Europe. The ingredient that was not readily available by the time that Barrett, Saxer, and Weinhard opened their shops was hops. Despite the obvious success of hop growing and marketing at Fort Vancouver in the preceding decades, no one else had seized the opportunity to grow the crop extensively. The lack of a local hop supply proved a conundrum for regional brewers at times, or disastrous at others. Pacific Northwest brewers relied upon shipments from Europe or the East Coast into San Francisco and then brought northward to port cities. At best, the hops were always more expensive and would not be the freshest. At worst, hop shipments would not make it, thus interfering with the ability of Pacific Northwest brewers to craft their beers as they desired. The region's brewers needed a local hop supply.[7]

THE WOLF IN THE VALLEY

In the fall of 1865, the Pacific Northwest hop situation changed dramatically. It was then, to the north of the Willamette Valley, in Olympia, Washington, that brewer Isaac Wood became frustrated with his inability to acquire hops from European and East Coast centers of cultivation because of his distant location. Seeking a local hop supply, Wood (or "Uncle Isaac," as his community endearingly called him) asked a neighboring farming family, the Meekers, to plant a few hills of the crop, of which he had acquired some rootstock. It was a serendipitous decision.[8] Despite having no former experience in hop cultivation, Jacob Meeker and his sons Ezra and John achieved early success. As Ezra Meeker noted in his autobiography, the following September netted 180 pounds of hops that sold for over $150. "This sum," Ezra explained, "was more money than had been received by any of the settlers in the Puyallup valley, except perhaps two, from the products of their farms for that year."[9] From this unexpected accomplishment the Meeker family emerged as the earliest prominent hop growers in the Pacific Northwest. Given the ideal environmental conditions and an established agricultural infrastructure that offered possibilities to market and sell the product all over the world, Ezra Meeker touted the importance of the hop crop for the future of Pacific Northwest farming.[10]

In the same decade that Meeker began commercial hop raising in Washington, a couple of forward-looking Willamette Valley farmers shared a similar notion. Having secured European rootstock from Wisconsin farmers, William Wells and Adam Weisner of Buena Vista, Oregon (Polk County), planted their first hop acreages in 1867.[11] Likely because of the farmers' lack of knowledge about the crop, these efforts ended in failure. But other farmers in the region took note. Two years later, with rootstock obtained from Weisner, George Leasure of Lane County raised Oregon's first commercial hop harvest.[12] Inspired by this success and Meeker's outreach, several other farmers began growing hops on the banks of the McKenzie River near Eugene. The following decade witnessed the emergence of notable growers such as J. W. Kunoff and George E. May. Alexander Seavey also became a prominent figure in the area's hop culture, and his sons operated one of the nation's most successful hop companies well into the twentieth century.[13] In 1873, the *Oregonian,* Portland's newspaper of record, reported that yields in the Willamette Valley topped two thousand pounds of uncured hops an acre, or approximately double the output that other hop growers around the world produced on their land. As this news spread, acreage in the region multiplied. Hop farmers recognized that they not only were beneficiaries of an ideal climate, soils, and shipping opportunities, but also had the advantage of an absence of the European pests and diseases that reduced yields.[14]

During that promising beginning for the Pacific Northwest's hop industry, the 1870 Agricultural Census reported a combined 15,907 pounds harvested in the state of Oregon and the Washington Territory.[15] That crop supplied local brewers but had little impact on the national or international marketplace. Ten years later, the census emphasized that, "[o]f the 46,800 acres in this crop during the year 1879 New York reports 39,072 and Wisconsin 4,439. No other state besides California reports as many as a thousand acres."[16] These observations ignore the rapid growth of hop growing in the Pacific Northwest during the previous decade, not just in the Puyallup and Willamette valleys, but also in agricultural regions surrounding Puget Sound and in Douglas, Josephine, and Jackson counties in southern Oregon. Although many still struggled with the complexity of growing the crop, Pacific Northwest farmers had produced nearly a million pounds of uncured hops during that decade from 838 acres of fields. Once dried and ready for market, the amount served more than enough for regional brewers, allowing for a stream of the product to enter the national and global marketplace.[17]

Still, this was just the beginning. In 1882, a global hop shortage brought new and lasting attention to the Pacific Northwest. Unpredictable weather that year led to one of Europe's worst agricultural outputs of the late nineteenth century, and pests and disease attacked hop crops on the American East Coast and in the Midwest.[18] Along with California's expanding hop industry, Oregon and Washington arose as the most stable hop producers in the world. Prices skyrocketed as brewers from Latin America to Europe had few options but to purchase Pacific Coast hops. Ezra Meeker compared the agricultural windfall with California's gold rush, noting, "The high value of hops prevailing for the past four years, culminating in the unprecedented price of one dollar per pound for the crop in 1882, has naturally attracted a wide-spread interest." He then continued, "An article that can be produced in large quantities, and sold for nearly ten-fold its cost, engenders a speculative feeling akin to that of a veritable gold-mining furore [sic] of the palmy days of '49, when the discovery of gold in California was first made known to the multitude."[19] Although Meeker had a penchant for exaggeration, he did not overstate the revenues from the hop bonanza. What started as Isaac Wood's desire for a local hop supply in Olympia was beginning to transform into a lucrative agricultural industry with an international marketplace. In the closing years of the nineteenth century, anyone with land west of the Cascades could see an opportunity in the wolf of the willow. Meeker referred to the phenomenon as "hop fever."[20]

THE SPECIALTY-CROP REVOLUTION

Ezra Meeker, George Leasure, and the rest of the Pacific Northwest's hop growers entered the trade at an ideal historic moment. Prior to the Civil War, the region lacked a diversified agricultural economy and infrastructure that could bring greater wealth to its residents. Grain and livestock provided the crucial base, but most farmers failed to accrue wealth on these alone. After Henderson Lewelling and William Meek established their nursery in 1848 and others followed in the next decade, agriculturalists had the ability to grow higher-valued specialty crops, such as fruits, nuts, and ornamental plants. But they could not effectively market or ship these crops for lack of efficient transportation, storage, and refrigeration. While seafaring vessels remained a critical resource for commodity transport, the marketing for specialty crops in the Pacific Northwest changed in the 1870s and 1880s as transcontinental railroads provided new shipping opportunities. Railroads could transport goods from farms to cities and also across larger sections of the country at a faster pace. The new transportation revolutionized American commodity extraction by opening the vast natural resources of the United States to the world at incredibly accelerated rates.[21] By the 1880s, railcars refrigerated by ice proved vital for West Coast farmers, who could for the first time send successful shipments of fresh fruits, vegetables, and other specialty crops to the East Coast and other destinations across North America.[22]

In the Willamette Valley and the larger Pacific Northwest, the arrival of the Northern Pacific Railroad and a north–south connecting line all the way to the Southern Pacific Railroad in California had a direct effect on farming. Initially, farmers took advantage of the railroad to sell more grain and livestock. But then they took a page from their peers in California by diversifying their farmsteads. By the end of the 1870s, nearly all Willamette Valley farmers began testing their luck with fruits, nuts, and hops. Curiosity certainly inspired some, but it was the prospect of earning cash income that inspired most. The new crops redefined what it meant to live in the agricultural utopia. The reports of the Agricultural Census made clear that specialty crops became an increasingly important aspect of the state economy. Farmers integrated new knowledge of the crops and developed new relationships with urban processing facilities and marketing companies. Agricultural societies intensified their efforts to gather information for growing and marketing desirable new crops.[23]

Amid this specialty-crop revolution, perceptions of the Willamette Valley, along with its landscape, changed. The valley's booster literature of the late nineteenth century enthusiastically embraced the more diverse agriculture. Promotional literature from national media outlets, railroad companies, and international business interests referred to the cornucopia of crops that grew in abundance in the rich valley soils. In 1882, for example, a Northern Pacific Railroad publication reported of the Willamette Valley and the larger Pacific Northwest, "Whenever the plow is freely used, and the seed planted, the growth of grain and vegetables becomes luxuriant. . . . The practical benefit already is a larger variety of productions, and a grand harvest of cereals for home and foreign markets."[24] Locally, government agents, chambers of commerce, and regional media outlets such as the *Oregonian,* the *Willamette Farmer,* the *Overland Monthly,* the *Pacific Rural Press,* and the *West Shore* also promoted the new agricultural diversity. In 1886 the secretary of Oregon's Board of Immigration suggested, "We have every kind of fruit, vegetables in profusion, grow wheat in quantities and have large crops of cereals of all kinds. . . . Cattle and sheep raising are very profitable, our timber is of the best, we have hop raising, grape culture, salmon canning, mining . . . and in fact everything." He finished by noting, "Let me off with 'everything,' for the list is too long."[25] Although hyperbole abounded, the words genuinely captured what many residents believed: agriculture of all types thrived in the Willamette Valley and offered the potential for everyone involved to make money in a thriving society.

It was in this late-nineteenth-century atmosphere that a commercial hop industry arose from the Willamette Valley landscape. Success was immediate and expansive, and offered a different metaphorical picture than a farming region dominated by wheat with Portland as the entrepôt to the world. Playing its part in the promotional literature surrounding the Willamette Valley's Edenesque features, an 1887 *West Shore* article reported, "There is another profitable crop, to which more attention is being paid yearly. No less than four hundred acres of hops are now growing within the limits of Polk County. The deep, rich, alluvial bottom lands along the Willamette and its tributaries are splendidly adapted to hop culture." The article continued by underscoring the importance of hops as the most significant specialty crop of the valley:

> There are hundreds of acres of land adapted to hops, which are now not in cultivation. A low estimate is fifteen hundred pounds to the acre, though some fields have produced twice that amount. Taken for a series of years, the average price of hops is twenty cents per pound, though in the past few years it has risen as high as one dollar, and fallen as low as seven cents. The cost of raising hops is about eight cents per pound. A yield of fifteen hundred pounds per acre, at twenty cents per pound, gives a total of $300.00 per acre. This is a good profit at average price, and in the seasons of high prices, some growers have become comparatively rich on one crop of hops.[26]

While other regions in the West made similar claims, Oregon charted a path that would soon transform the global hop industry. In a matter of decades, growers would claim to reside in "the Hop Center of the World." Underscoring the rise of a commercial hop industry lay a wide set of connections that blossomed between the local landscape and consumers across the globe. And it is here that the multifaceted local and global identities of the Willamette Valley hop industry more clearly began to take shape.[27]

Expansion of the Willamette Valley hop industry first left its footprint on local landscapes. Farmers initially preferred to plant hopyards (more commonly called "hop gardens" in Europe) on bottomlands, claiming incredible yields from nutrient-rich soils. They created evenly spaced rows of the plants that climbed ten or more feet high. For the first generation of growers, timber poles offered support for the bines to climb. Each spring they would set these poles upright in the ground. Like their predecessors across the centuries, they trained the bines clockwise to the host poles. Knowledgeable growers would not train the first growth of the year, but rather cut it back and wait for more mature and hardy bines to emerge. After that point, the hop plant spent its summer climbing and producing cones. Come late August or early September, depending on the hop variety, workers pulled out the timber posts with bines attached so the hops could be easily picked. Harvest was both expensive and labor intensive. In the first couple of years, when the young plants produced fewer cones, some farmers also planted corn or potatoes in between the hop rows, both to take advantage of the worked soil and to keep back unwanted weeds. Compared with orderly hopyards in the next century, these early efforts would have qualified as somewhat messy agriculture.[28]

Aside from these markers of change in the fields, the other noticeable imprint on the changing farm landscape of the Willamette Valley was the arrival of large barnlike structures that housed kilns for drying hops. Those buildings, often referred to as *hop dryers* or *hop houses*—or *oast houses* in England—had long sturdy ramps on the exterior so wagonloads of fresh hop cones could be unloaded onto an upper-story drying area. (See figure 3.) Inside, large wood-burning kilns reduced the moisture content of the cones by up to 90 percent. This process extended the shelf life of the crop considerably and aided with shipping. Some growers also burned sulfur in their hop dryers in an effort to promote a greenish-gold color in the hops, though brewers eventually discouraged that practice. Once the drying was complete, workers compressed the hops into bales wrapped with burlap for shipping. Needless to say, the entire operation represented a significant financial investment. To save on costs, many growers shared hop-drying facilities, indicating that, although important, the crop was not the prime endeavor of most farms.[29]

Save for a handful of farmers in the Willamette Valley who cultivated the crop extensively, the hopyard was only one part of a larger, diversified

FIGURE 3. Benton County Hop Dryer, circa 1890s. Courtesy Benton County Historical Society.

farmstead. In 1895, a writer for the *Oregonian* described one of the new fields: "On a broad slope, near Fulton, about one mile and a half south from Portland is a model hopyard. It is 14 acres in extent, and each year of the six years since it was planted, has averaged 200 pounds of cured hops to the acre."[30] Although the size of Willamette Valley farmsteads had been shrinking from the 320-acre Donation Land Act claims as the population increased and competed for land, the Fulton farm's fourteen acres of hops likely would have been a small fraction of the entire operation.[31] A typical farmstead would have included grain and hay fields, pastureland for cattle, orchards, and a vegetable garden. A view of the landscape in 1895 would not have suggested that hops had become the central crop of the Willamette Valley. Rather, it was a widespread specialty crop that supplemented the rest of a farm's operations. Most farmers grew between five and twenty-five acres of hops. Their goal was modest: to generate cash income via the hop to buy consumer goods in the modernizing economy.[32]

Far less obvious than the poles, bines, and dryers that marked changes across the Willamette Valley landscape was how the cured product integrated into the local and international economy. Unlike Ezra Meeker, who had a close relationship with the brewer Isaac Wood and who signed a long-term contract with Portland's Henry Weinhard in 1869, most regional farmers did not fraternize or deal directly with the end users of their harvest.[33] Several hop companies arose in the Pacific Northwest, but a group of grower-dealers dominated the organized sale and transport of hops. Those individuals and companies—such as Colin Carmichael, of Yakima, Washington, and Krebs Brothers, of Salem—usually grew hundreds of acres of their own crop and contracted with smaller farmers for sale. The cured hops most often ended up in the ports of San Francisco and were destined for the East Coast, Europe, and markets across the Pacific.[34]

By the 1890s, the two most influential grower-dealers on the Pacific Coast were Ezra Meeker of Puyallup, Washington, and Emil Clemens Horst of Sacramento, California. Both men stationed representatives in the Willamette Valley during the summer growing and harvest seasons to scout out the choicest and most productive hopyards. The competition for hop contracts was cutthroat. Even during a period when brewing expanded in Portland, Salem, Tacoma, and Seattle, the majority of Willamette Valley hop farmers sold their crops to middlemen. Letters from Meeker family members stationed in the Willamette Valley frequently mentioned Horst's aggressive tactics, such as offering a penny more per pound. The spirit of this competition favored small growers because it brought buyers to their hopyards. But the arrangement also put pressure on them to produce high-quality products, which was sometimes difficult given their unfamiliarity with the crop.[35]

Once contracted for sale, a Willamette Valley farmer's hops entered the international marketplace just like any other commodity. A great deal ended up in the lagers and ales of Pacific Northwest breweries, which numbered nearly fifty by the 1890s—an all-time high prior to the craft beer revolution.[36] Other Willamette Valley hops ended up in beers of Anheuser-Busch, Pabst, Miller, or one of the many other rapidly expanding breweries of the Midwest. Millions of pounds also landed overseas, in the brewing kettles of beer makers in Mexico, China, or the largest American hop importer of the era: Great Britain. The U.S. Department of Agriculture (USDA) and state agricultural experiment stations gave some assistance to Pacific Coast hop growers in expanding their markets during the 1890s, but it was mostly the larger grower-dealers and other hop distributors who negotiated international trade.[37]

Further underscoring the globalized nature of Pacific Coast hop growing, Ezra Meeker visited Europe four times during the 1880s and 1890s, learning the details of the trade. He also stationed his son in England to maintain relationships and acquire new knowledge about the industry.

FIGURE 4. Ezra Meeker and Eliza Jane (Sumner) Meeker, front right, at the Washington Territory Collective Exhibit, American Exposition, New Orleans, 1885–1886. (The event used the same grounds as the world's fair of 1884.) Notice the hops hanging in the background. Courtesy Oregon Historical Society.

Meeker attended world fairs throughout the late nineteenth century, occasionally acting as a judge of hops and always promoting the bounty of the Pacific Northwest. (See figure 4.) In a more concrete sign of his global connections, the E. Meeker Company became one of the most prominent outfits to seek shipments of hops on the same oceangoing ships that carried Pacific Northwest wheat to Liverpool.[38] Emil Clemens Horst was a native of Germany who immigrated to New York to open a hop company before moving to California in the late 1890s. His connections to both the eastern United States and Europe also helped expand the market for the region's hop crops, particularly in the early decades of the twentieth century.[39]

The Pacific Northwest's global connections to the hop industry were not only directed outward. Aside from land, water, timber, manure fertilizer, and the raw materials needed in the construction of hop dryers, most hop growers' supplies arrived from beyond the region. Growers continually imported hop rootstock from Europe, and they became fascinated with finding new varieties that provided high yields and had properties valued by brewers. Most were Cluster hops, the open-pollinated hybrids of English and North American parents that produced high yields; farmers grew both Early Cluster (selected from an Oregon field in the late nineteenth century) and Late

Cluster (an older variety selected from the East Coast), which matured at different points in the season. Other specimens arrived from Bohemia, Bavaria, Canada, Russia, and even New Zealand and Tasmania.[40] Records show that the Canadian Red Vine and the English Fuggle varieties were common, but never as prolific as the Cluster varieties.[41] Aside from the rootstock, growers used a variety of materials for sprays and other applications used in pest and disease prevention. Tobacco from the American South, whale oil from around the Pacific Basin, and the bark of quassia trees from Brazil all found their way into Pacific Northwest hop fields. Growers found that when applied topically, the chemicals in these products deterred pests such as hop aphids. For drying and packaging farmers also relied on sulfur and burlap acquired from various merchants who were both inside and outside the region. All of those other commodity industries from outside of the region and the people instrumental in them wove into the tapestry of Pacific Coast hop farming.[42]

Not surprisingly, perhaps, it was Ezra Meeker who originally orchestrated relationships with Seattle and Portland merchants for acquisition and distribution of these materials. His E. Meeker Company lay at the center of these local and global commodity exchanges, and it therefore acted as a point of exchange between urban suppliers and rural farmers. Meeker's efforts in the hop industry allowed him to become one of the Pacific Northwest's first millionaires in the closing decades of the nineteenth century. Residents of the region anointed him the "hop king."[43]

But Meeker's influence reached beyond the marketplace. Because regional hop growers had little to no agricultural familiarity with the crop prior to securing rootstock and planting, they needed to build knowledge before entering the business. Meeker understood that and positioned himself as a matrix for knowledge acquisition and dissemination.[44] He first learned about hop culture by reading newspapers and trade journals published in Europe and on the American East Coast, and a range of books on the subject. English volumes available during his business career included E.J. Lance's *The Hop Farmer* (published in London in 1838) and H.M. Manwairing's *A Treatise on the Cultivation and Growth of Hops, in the Kent Style* (published in London in 1855). These works built upon centuries-old literature to describe agricultural methods and the state of markets. As American production rose during the mid–nineteenth century, domestic publications also began to aid Pacific Northwest growers. The most prominent books included Andrew Fuller's *Hop Culture* (published in New York in 1865) and D.B. Rudd's *The*

Cultivation of Hops and Their Preparation for Market (published in Wisconsin in 1868). New York's *Emmet Wells' Weekly Hop Circular* was the most important weekly journal. Because Meeker also spent considerable time corresponding with hop growers from near and far, his sources of knowledge reflected both local variances of cultivation and firsthand encounters with the most important representatives from around the world. That information allowed Meeker to succeed along with the regional growers who depended on his knowledge. He gained a reputation as an expert in the field, and he was willing and eager to share information with others. His eagerness reflected desires to both improve the overall quality and perception of Pacific Northwest hops and generate more business for his company.[45]

As the hop industry took firm root in the Pacific Northwest, Meeker recognized the need for an updated informational work. In 1883, the year after the global hop shortage that put the Pacific Northwest on the map, he published *Hop Culture in the United States: A Practical Treatise on Hop Growing in Washington Territory from the Cutting to Bale.* The work became the fundamental source of knowledge for the region's hop growers. While careful to declare the dangers of hop raising and the whims of the market, Meeker outlined clear plans for cultivating and preparing the crop for buyers.[46] He offered basic information, such as the need for planting rootstock in the spring, cutting back the initial shoots to encourage a second growth of more stable and vigorous stock, and training those climbing bines to timber posts. He also described many variances specific to the Pacific Northwest. In contrast to the setup preferred by many European growers because of their limited space, for example, Meeker advised: "Where but one pole to the hill is intended, the plants should be seven feet apart, set in squares. Some growers prefer to set two poles to the hill and in such case usually plant seven feet and a half or eight."[47] One German expert suggested that the difference could amount to nearly half as many plants per acre in the United States as compared to some Continental growers.[48]

Meeker included advice on other regional business matters, including available timber resources for poles and local sources of fertilization. He also made sure to note that, for first-year growers, "[i]t is customary to plant corn or potatoes between the hop-hills . . . sometimes one row and sometimes two between the rows. . . . The young vines are allowed to lie on the ground, as they produce no hops, but it will cost very little."[49] Finally, Meeker also shared his wisdom on acquiring sufficient labor for seasonal picking, the need for quality controls in the picking process, and the intricacies of kiln drying. His work is the best evidence of both the flow of new knowledge into the region and its creation there and dissemination elsewhere. Surviving records of the E. Meeker Company reveal a steady string of requests for *Hop Culture in the United States* from Pacific Coast hop growers as well as others from all over the world.[50]

As the hop industry matured, Meeker became just one of many regional farmers who were integrated into a local and global knowledge exchange. Other growers approached the business with the same verve as the "hop king" and engaged in widespread information-seeking activity. They scoured available print resources, acquiring and contributing to the broad creation and dissemination of hop knowledge. Like Meeker, many made journeys to the East Coast and Europe to better understand the industry's details. Additionally, local newspapers offered advice on where to find supplies and acquire loans. They also reported findings from around the region. In 1874, for example, the *Oregonian* reported that "T. W. Spencer, of McMinnville, in this State, has just returned from the hop regions of Puget Sound. He informs us that the people of that region are all in a fever of excitement about hops, and everyone who can secure a spot of ground is preparing to start in the business."[51] Another *Oregonian* article from two years later noted that "[t]he hop raisers are in the midst of their picking this week, and we are informed that there will be a fair crop, a good quality, and a good price per pound."[52] Perhaps one of the best sources for information across the Willamette Valley and the entire Pacific Coast was the California-based *Pacific Rural Press*. During its run from 1871 to 1922, articles abounded on everything from cultivation techniques to drying methods to assessing the international marketplace.[53] Although the journal focused on California more than the other Pacific Coast states, it presented a valuable resource, particularly as many of the Oregon and Pacific Northwest agricultural journals came and went.

During the late nineteenth century, the Willamette Valley hop industry's success was undeniably the product of goods and knowledge from around the globe. It required whale oil from the Pacific Ocean whaling industry, tobacco grown in the American South, hop treatises written by English and East Coast growers and merchants, and lumber harvested from Pacific Northwest forests, to name just a few. In turn, this local and global integration of resources allowed the Willamette Valley hop industry to shine as an emerging world leader. An 1898 publication of the Oregon State Board of Agriculture, for example, proclaimed, "The garden spot of the world for the cultivation of hops is the Willamette valley, and the city of Salem sits in the center of the greatest hop district in the world." It continued, "Every year the large hop dealers of London and the Eastern cities have their representatives occupying offices in Salem, and the money for an average of 75,000 bales passes through the banks of this city."[54]

Across the country in New York, Herbert Myrick, an expert in hops, agreed, noting the possible "monopoly of the world's hop market by the United States, and especially by our Pacific Coast States."[55] Across the Atlantic, European interest also intensified. In 1900, Emanuel Gross, a German expert, warned his readers of a transformation in the industry: "The introduction of hop-growing in the United States," he suggested with an eye to the Pacific Coast, "marks a turning-point in the history of the industry, the consequences being adverse to the interests of European growers."[56] Though Gross remained unsure of the lasting power of American production, his writings revealed a new global awareness of America's far corner. At the time of all of these comments, however, there were major concerns about the sustainability of the Willamette Valley hop industry.

HOPPING OBSTACLES

Despite all the advantages that the Willamette Valley offered, hop growers faced a range of obstacles. First and foremost, hop farming required significant financial investment up front, usually in the form of loans, to purchase rootstock, poles, and other necessities for cultivation. It also required a substantial amount of time and labor. Once invested, growers hoped for favorable weather and an absence of pests and disease.[57] Even if growers fared well in those regards, the whims of the global commodity market loomed large. In good years, growers could sell their hops for twenty to thirty cents a pound; in bad years those same hops might be worth only pennies.[58] If that was not bad enough, hop growing had an ugly litigious side. Contract disputes between growers and merchants, brewers, and laborers frequently ended in court. In short, the hop industry could make or break families, with far more facing bankruptcy than making it rich.[59]

American growers also faced the reality that brewers across the globe preferred European noble aroma hop varieties, including Hallertauer-Mittelfrüh, Tettnanger, and Spalter (Germany), Saazer (Bohemia/Czechoslovakia), and Fuggle (England). The Cluster and Fuggle hop varieties commonly grown in the Willamette Valley were acceptable, particularly in Great Britain, but generally were not a brewer's first choice. American growers brought further ills upon themselves because they continued to lack quality controls for "clean" picking of the crop without stems or leaves.[60] American hops also tended to include seeds. By planting a few male plants to pollinate the females (thereby fertilizing them and producing seeds in the cones), growers hoped to increase the volume of their harvest. But brewers did not value seeded hops, believing that they were a lower-quality product.[61] The disdain for male hops in Continental Europe even led to all-out bans in hop-producing regions, and a common refrain suggested that "a hop garden should resemble a nunnery, all males being excluded."[62] These stigmas continued to plague · American hop growers well into the twentieth century.

Despite all of these concerns, Pacific Northwest farmers continued to gamble on the crop in the fierce heat of "hop fever" and therefore contributed to the frequent cycles of overproduction and low prices. As early as the 1870s, farmers had formed hop-growing organizations to better educate farmers about the risks of overproduction. But even with Meeker as a leader, they made little headway.[63] By the onset of a major economic depression in 1893, the combined hop acreage of Oregon and Washington nearly topped that of New York, but in doing so created a glut in the market that affected growers large and small. Without any adjustments a year later, the *Oregonian* reported that the region's growers were "flooding . . . the market with surplus."[64] This proved ruinous for many. With prices staying low, some growers left hops on the bine, not wanting to pay for harvest labor costs. All growers were left wondering what to do, but the problem lingered. Two years later, Ezra Meeker penned a newspaper article asking that growers not only refrain from planting new acreages of hops, but also pull roots from the ground.[65] Only then did growers concede. In one of the first acts of an organized regional hop industry, farmers abandoned thousands of acres of the crop. Unfortunately, the timing was wrong. Within the same year, the global value of hops spiked, leaving hundreds of growers angered at the decision to abandon their crops.[66]

A final obstacle facing hop growers was that of labor: attracting enough workers for the harvest season recurred annually. Once met by family and neighbors, the need for seasonal workers sometimes reached crisis levels as the crop's popularity grew. As one historian argues, the seasonal-labor issue persisted throughout the nineteenth-century agricultural West because of the area's sparse population.[67] Yet the hop crop brought about uncommon challenges. One agricultural study, for example, reported that "in spite of the fact that the hop acreage is only a small fraction of that given to soft fruits, apples, and other intensive agricultural crops," the harvest required up to four times as many workers.[68] Hop plants produce huge quantities of cones that require immediate handpicking upon maturity. Although the three main hop varieties grown in the Pacific Northwest—Early Cluster, Late Cluster, and Fuggle—spread out the harvest season, there never seemed to be enough pickers on hand.[69]

In the Willamette Valley and across the West Coast, the potential for labor shortages lasted until mechanization of the harvest in the 1940s, challenging and stressing the region's farmers. But there is another side of that story. Ongoing labor demands were an obstacle for growers but a boon for temporary workers. The hundreds of thousands of individuals who participated in the annual hop harvest had experiences and memories that were quite different from those of the farmers, extending the meanings of this history beyond grower, brewer, and beer drinker. Those perspectives constitute an important part of the region's hop story, bringing regional and global forces together in still new ways.

FOUR

Hop-Picking Time

AS THE WILLAMETTE VALLEY GAINED RESPECT, if not fame, as a global leader of hop production into the twentieth century, the annual harvest became a major event for the region. The three-week season brought a festiveness and culture unique to the specific crop, and for that reason piqued the interest of Oregonians and outsiders alike. Whereas the world's brewers and local community boosters naturally kept their fingers on the pulse of agricultural innovation and productivity, general audiences delighted in the more titillating stories of those who labored during the harvest season. Initially, local newspapers carried stories of these diverse crowds that gathered in late August and early September, and then regional and national outlets such as *Sunset* magazine and the *New York Times* also began reporting on the curiosities of the season. What stories and dramas in the hop harvest drew such attention?

In 1907, University of Chicago–trained sociologist Annie Marion MacLean, for one, sought to answer that question by taking a hands-on approach.[1] Like her previous investigations—including "Two Weeks in Department Stores" and "The Sweat-Shop in Summer"[2]—her objective was to experience the day-to-day life of an "understudied industry" to help paint a picture of national labor, with a particular emphasis on women workers. Because MacLean "found [that] very little could be learned on the outside concerning the conditions of work" in the hopyards, she "decided to hire out as a picker and go with the crowd to learn something of the life first hand."[3] Even if MacLean might have had a muckraking agenda to expose labor abuses, her writings on this experience, published in the *American Journal of Sociology*, provide one of the most detailed snapshots of the Willamette Valley hop harvests during the early twentieth century.

After arriving in Portland in early September, MacLean began her undercover fieldwork with a perusal of recruitment advertisements from Willamette Valley hop growers. She recorded the details of pay, lodging, and transportation. Sponsored in part as a representative for the National Board of the Young Women's Christian Associations, she looked past one announcement reading: "We pay $1.10 per 100 pounds. . . . Perfect accommodations, good food at city prices, free whiskey, dances five nights in the week, evangelists on Sunday and a hell of a time."[4] But she soon found a prospective employer who, like other large operations, provided direct transportation from the city. Prepared for an adventure, MacLean joined nearly a thousand others aboard the *Hop Special* train bound for Independence from Portland.[5]

On the train MacLean recognized that her story would be not just about workingwomen, but about the diverse body of a western workforce. She observed "men and women and children, scores and scores of them belonging in family groups, and, in addition, several hundred young men and women off for a lark with a chance to make some money." In the four hours that the train headed southward through the valley, she also noted that "[m]any of the families were from the country, one woman having come a distance of two hundred miles with seven children ranging in age of from two to fifteen years. The other class, the unattached men and women, were mainly the city's floating working population." Given these differences, MacLean reported that the various social groups had different agendas. The rural families looked for supplemental income in an industry that allowed their children to participate. The urban "clerks and factory workers and servant girls" sought respite from the city and a chance to build new social ties (perhaps even romantic ones; MacLean observed the quickness with which many of the young workers settled in the railcars "pillowing their heads against each other"). At the very least, she noted, "[i]t was a picturesque gathering, with an air of expectancy about it. There was to be at least a change of occupation."[6] For some, the hop harvest offered an escape from the routine of urban life.[7]

After arriving in Independence, a six-mile wagon ride awaited the women and children who got off the *Hop Special,* while the men had to walk. All were tired after this last leg as they arrived to claim rudimentary accommodations of denim tents with straw for bedding. The promise of a dance reinvigorated the camp's collective spirit after the long day of travel, but the failure of the band's instruments to arrive caused disappointment. Nevertheless, MacLean and her companions socialized that Saturday night and the following Sunday before the long days of picking began. With these new acquaint-ances, including her helpful female tent mates, MacLean settled in for the late-summer labor and began her observations. She learned the process of yard assignments (185, Yard B, Company 4, in her case), the careful picking of hops free of stems and leaves for full payment, and the overall intricacies of eating and sleeping in the socially diverse camp. MacLean reported that the harvest had its own culture, even a distinct jargon. Calls of "wire down" rang through the hopyard when pickers moved to a new area. A shout of "box full" indicated to the yard's foreman the need for new picking containers. MacLean enjoyed the experience, as it differed markedly from some of her previous investigative sociological studies.

Among MacLean's most important remarks on the "Oregon tradition" of hop picking was its status as a "healthful occupation." The comment con-firms a positive image about the experience of working in Pacific Coast hop-yards portrayed in most published accounts, as well as remembrances by former hop pickers.[8] The seasonal work was almost celebratory, with ample sunlight and light physical labor that also brought social and financial ben-efits. While MacLean complained of the "air thick with pollen" that left her "choked . . . by dinner" and the constant "reaching and stooping and tramp-ing," she recognized the variety of benefits for others. Whether dealing with issues of economy, society, health, or place, she concluded that Oregon hop picking proved a democratizing and liberating occasion for men, women, and children of all classes. MacLean's only significant recommendation was in the offering of more "wholesome entertainment" for the campers. And her final remark exclaimed, "Long live the Oregon hop pickers!"[9]

MacLean's depiction of the Willamette Valley's seasonal hop harvest affirms much that has been remembered and written about Oregon's golden era of hop production. Until the mechanization of harvests in the mid–twentieth century, a wide range of the region's residents embraced the oppor-tunity to augment their regular source of income, and a festive spirit guided the harvest season. But there were also glaring omissions in her work. Most notably, MacLean failed to comment on issues of racial and ethnic diversity or poor working and living conditions. These oversights might have reflected an emphasis on women workers, or, perhaps, an experience that lacked cultural diversity and social tension. The absence of these themes at least provides an avenue to investigate the multiple meanings of the harvest, which is the most remembered and celebrated aspect of the Willamette Valley hop industry.

THE ENGLISH ORIGINS OF THE AMERICAN HOP
HARVEST

Like so many other aspects of American hop agriculture, the origins of the harvest labor story grew out of European traditions. By the early nineteenth century, when commercial hop farming expanded along with commercial breweries, growers in Bavaria, Bohemia, and England found difficulties in acquiring enough hands to pick their crops. Once able to depend on the labor of family and neighbors, growers began advertising for and attracting nearby working-class peoples.[10] The situation in England offers the best example, since it would ultimately offer a direct model for American growers to emulate.

To solve their harvest dilemma of needing many hands but only for a short picking season, hop growers in Kent, Surrey, Sussex, and Hampshire looked to the city of London. By 1800, the metropolis of approximately one million provided a labor source for growers who could not acquire enough workers in the rural areas alone. Growers posted advertisements in stores and newspapers, and recruited by word of mouth. They often preferred women and children for their perceived picking dexterity and their lower wage demands. But, because the work was unskilled, they hired a great diversity of people: old and young, men and women, married and unmarried, skilled and unskilled, lower- and middle-class. The work attracted both individuals and families seeking additional income.[11]

As Annie MacLean would only hint at later in her Willamette Valley experience, the sojourn to the hop harvest was not entirely pleasant. The labor was monotonous, occurring mostly under the hot summer sun. The work dating back centuries was simple: pick ripe cones from the bines and place them in a basket, bucket, sack, or box carefully, keeping stems and leaves from intermixing. Unlike seasonal harvests in other agricultural sectors, hop picking did not require a tremendous amount of physical stress or endurance. Compared with fruit or berry harvests there was less time spent stooped over or stretching and reaching into trees. Hop pickers mostly stood or squatted near the bines, which had been lowered for their reach. There could be uncomfortable heat, the possibility of skin rashes from contact with hop bines, and respiratory congestion that resulted from breathing in the airborne lupulin particles emanating from the hop cones. In most cases, however, the greatest difficulty facing pickers was not physical but mental. Picking cones for three weeks could be exceedingly tedious.[12]

Provisioning for the seasonal migration to the hop harvest began in London. Workers assembled clothes, tents, cookware, and foodstuffs for the extended stay in the countryside, as well as bonnets, hats, and gloves to protect them from the elements while picking. For comfort and entertainment, workers brought portable furniture, musical instruments, and games. With these items in tow, workers made the initial trip by foot and wagon in the early nineteenth century, offering a remarkable spectacle. Onlookers marveled at the flocks heading to the hop fields along with their possessions. The extension of train service from London to Kent by the mid-1800s moved much of this travel and spectacle to the interior of rail passenger cars. Still, be it on the roads or rails, it was hard to miss the yearly carnival-like exodus from city to country.[13]

Hop farmers prepared for the arrival of temporary workers by providing basic amenities. They cleared land for camping, established sleeping quarters in dormitories or barns, and set up means to distribute water and collect waste. Growers also collected wood, bedding, and some food supplies for sale to those who arrived at the hopping grounds ill equipped. Preparation was critical. If pickers did not feel the standards of pay or camp adequate, they often packed their belongings and moved to another farm. Although there were instances when London hop pickers faced detestable living and working conditions, Kentish hop farmers recognized the advantages of taking good care of their seasonal employees. By the second half of the nineteenth century, growers secured rail lines for efficient transportation, sanitary housing, access to water, and good pay. Fair and generous treatment ensured a solid reputation among workers and also proved helpful in preventing the occasional strike or riot.[14]

In the mid–nineteenth century one English grower, Edward Albert White, became legendary among the hop pickers. In addition to paying fair wages, he sponsored and organized sporting events and concerts, striving more than any of his neighbors to generate an appealing end-of-summer occasion. The recreational additions to the harvest culture occurred simultaneously with other work rules, including assignments to specific sections of each hop garden and calculations of payment by field bosses. White's judicious business decisions set precedents for other growers seeking to promote the hop harvest as a convivial "paid vacation." His workers embraced the social and cultural opportunities and repaid his efforts with hard work and loyalty. Furthermore, his reputation as a generous boss ensured that he never feared for labor supplies even in periods of shortage.[15]

Other English growers embraced White's innovations. In the second half of the nineteenth century, the hop harvest became an increasingly inviting opportunity. Even if the work was tedious, the picking season became an occasion for Londoners to spend time with old friends or meet new acquaintances. Music, dancing, storytelling, and camaraderie across social classes concluded hard-working days, and these festivities themselves became the main attraction of the harvest. One Londoner suggested that the work appealed to those looking forward to a "month of Bohemian life."[16]

The English hop harvest drew attention from a variety of journalists and other writers curious about the nature of the late-summer country holiday. Most affirmed the idea that growers promoted: positive associations with hop picking, which was fundamentally a monotonous and unskilled task. But there were exceptions. In the late nineteenth century, novelist Charles Dickens traveled among the hop pickers with an eye trained to the downtrodden of Victorian society. Rather than seeing happy vacationers, he observed masses of city dwellers traveling to the countryside to set up squatter camps in barns or tent villages. He reported, "I have been amazed . . . by the number of miserable lean wretches, hardly able to crawl, who go hop-picking."[17] Dickens did not veil his concerns for the working classes of the hopyards. He worried that the conditions of the primitive hop camp and its labor promoted social abuses not unlike those found in the industrializing streets of London.

An 1877 account from a London magazine provided a different perspective. "Judging from appearances," wrote the author, "one would have set down the whole crowd as belonging to the lowest class—as composed of the scourings of the slums." Unlike Dickens, who did not spend much time at a hop camp, this journalist's discussion took a sharp turn after spending three weeks in the field. He continued:

> No greater mistake could have been made. Such beings were there, and too many of them, but they were far from forming a majority. I was soon to discover that it is as customary with working families of comparative respectability to go "a hopping" in September as it is for members of another section of society to go to the moors, the seaside, Switzerland, and Norway, and for similar reasons—relaxation and health.[18]

While the writing may have reflected a bourgeois glossing over of the conditions of poverty and filth, the journalist urged his readers to think of the hop harvest as offering an opportunity for a late-summer romp in the countryside. Not only did the harvest provide family entertainment, it also provided health benefits in a period when doctors stressed the importance of time spent outdoors.[19]

In 1902, another Londoner captured the spirit of the hop harvest, writing, "One of the most delightful holidays I have ever spent was in the Kentish hop-grounds." He spoke of eating well and enjoying tea and wine. He socialized and took half-day cycling tours. The writer extolled his experiences, noting, "The people in the hop-fields were quite well behaved" and included both local farmers and villagers, as well as a "larger number of Londoners."[20] In general, the descriptions in this piece and others teetered on lavishness, casting hop growers and hop picking in a favorable light while promoting the hop harvest as a spirited pre-autumnal occasion. Little wonder that American growers embraced the Kentish model of harvest labor just as they embraced English hop cultivars and associated agricultural knowledge. Promoting the hop harvest as an appealing folk occasion seemed a practical and effective way of attracting a labor supply. All played out in a similar fashion as the industry became transplanted onto North American soil.[21]

THE TRANSPLANTED LANDSCAPE AND CULTURE OF LABOR

In the early 1800s, when commercial hop cultivation emerged in North America, small growers relied on the hands of family members and neighbors for the seasonal harvest. In this way, the harvest resembled small-scale European practices over the preceding generations. With expanded acreages by midcentury, growers needed a labor pool from which to draw workers. It was at this moment that larger operations in New York adopted the European, and particularly the English, model of hop-labor recruitment and the treatment of the hired help. The labor of women and children from urban areas became important, since farms or factories in the industrializing towns mostly commanded the labor of men. Women also played essential roles in tending to camp needs of cooking, sewing, and other domestic duties.[22] Growers in the prime agricultural region of Oneida County, New York, also hired recent European immigrants and local American Indian labor. As could be expected, some middle- and upper-class onlookers shared attitudes similar to those of Charles Dickens, seeing the seasonal workers who slept in the "cow-barn, horse barn or hop-house, whichever is most available" as uncouth harvest tramps.[23] These people did not approve of the mixing of lower- and middle-class families in this unskilled-labor pool. Nevertheless, various classes of workers appreciated the wages and social atmosphere, returning yearly in substantial numbers to the hop fields.[24]

When commercial hop production spread to the Midwest in the mid- to late 1800s, the idea of a harvest cultural event in late summer enlivened the task and prospects of a laborious three weeks.[25] American growers continued to emulate Kentish traditions, providing social events for workers, with live music and dancing becoming common after days spent in the field. The most popular harvest event was barnyard dancing and live music around bonfires. For many, the harvest remained just work. But, as was the case in England, workers were drawn to a change of scenery and the sights, sounds, and smells of life in the hop camps set in the verdant countryside.[26]

By the second half of the nineteenth century, the seasonal harvest in hop-growing localities of the United States became a romanticized event. Just as they did in England, articles, folk stories, and poems published in popular newspapers and magazines captured the harvest spirit.[27] The Mennonite Publishing Company of Elkhart, Indiana, circulated one such work, entitled "Hop-Picking Time," that described the varied meanings of the hop harvest for different participants:

> "Hop-Picking is coming!" the boys shout in glee,
> "What glorious times we are going to see!
> We'll meet all the girls we have met years before,
> And have all those jolly times over once more!"
>
> "Hop-picking is coming!" the girls smiling say;
> "We've been looking ahead for this many a day,
> To the beaux we will have, and the dancing and fun,—
> We'll enjoy them so well when hop-picking's begun."
>
> "Hop-picking is coming!" the poor widow sighs,
> As she looks on her child with love-light in her eyes,
> And thinks of the comforts her earnings will buy
> For herself and her child when the winter winds sigh.
>
> "Hop picking is coming!" We'll earn what we can,
> My wife and myself," says the stout working-man;—
> "'Tis our harvest time now, but the winter will come
> When we'll need all our earnings to gladden our home."[28]

The poem revealed the cheerful nature of the season, with all classes, from farm children to city factory workers, able to engage in harvest activity. It also showed how the harvest season held different social and economic meanings for different groups, including, as Annie MacLean hinted, the presence of a sexual libertine spirit. All of this transferred to the Far West when hop fields expanded there by the 1870s. The growers and seasonal workers of Oregon, Washington, California, and British Columbia (beginning in the 1890s) drew from a harvest culture that stretched across time and two continents.

On the Pacific Coast, a central story of the hop harvest in the late nineteenth century continued to be its convivial, healthful, and socially uplifting nature. After a visit in 1893, for example, a *Cosmopolitan* reporter wrote a refrain strikingly similar to that of the London and Kent connection. "If your nerves have become supersensitive from the corrosions of city life and you are the victim of ennui," the author noted, "or your liver asserts itself to the prejudice of your digestion, your duties and your friends, in fact, if you have reached the acme of general miserableness, take a vacation among the hop-fields in the gilded early autumn of California." That vacation, the author promised, would be unlike any other, and continued, "Your days will be made up of dew-exhaled mornings dwindling to the golden point of noon, of afternoons losing their superfluous heat in sunsets flaming the evening summits, and of nights so cordial and sleep-inviting they seem but moments of oblivion."[29] The description might have been a romanticized view from a Pullman car. Yet, it appealed to the intended middle-class literary audience of the era.[30] Journalists wrote similarly of Washington and Oregon hop harvests. In 1888, for example, an *Oregonian* writer reported, "The pickers I visited I found to be full of sport. Before I entered the field I could hear roars of laughter coming from it which convinced me that life in them was not so bad as I first supposed."[31]

While publishers of these praiseworthy reports intended to address the curiosity of readers, hop farmers delighted in them. According to one historian, "Growers knew well that the actual work . . . was anything but a 'vacation.' . . . Hop pickers endured excessive dust . . . oppressive heat, a contact rash similar to poison oak . . . and even the threat of electrocution from an unexpected storm 'taking possession of the wires.'" Still, as he continues, "[g]rowers were not about to disavow 'the picturesque feature of hop picking time.'"[32] While this was surely the case from the grower's perspective, non-journalistic accounts of workers shed light on the idea that hop-picking time was a time for celebration and excitement for many. (See figure 5.)

Unpublished diaries and reminiscences from Oregon provide a broader perspective on seasonal life in the hopyards. A young girl, Iris Tarbell, for

FIGURE 5. Hop pickers in Aurora, Oregon, circa 1900. Notice the hops growing up timber posts, as opposed to the trellises that would be adopted by most growers in subsequent decades. Courtesy Aurora Colony Historical Society.

example, who picked hops in the Willamette Valley for the first time in 1895, embraced the opportunity. "Hop picking was over three weeks ago," she wrote to her cousin. "I'll just say that I liked the work much better than I had thought I would even though I started out with some pretty grand ideas about it. I couldn't make much at first but I kept gaining until I could make a little over $1 per day. I wish I could get work like that year round."[33] Robert Hamill, a Portland man, wrote nostalgically about his journey to the Independence hopyards from 1904 to 1906. He recalled a childhood adventure that began on a Portland streetcar and concluded with a southward river journey. He wrote, "The trip was beautiful and we saw many wild animals in the fields and woods. . . . We all marveled at the number of small rivers that flowed into the Willamette." He continued by noting, "Arriving at the camp site on the farm we were assigned to cabins. These were nice cabins arranged in long rows with plenty of trees for shade. We soon got acquainted with our neighbors and I found a nice boy to play with."[34] Hamill's words showed how the hop harvest offered an opportunity for growing urban populations to connect or reconnect with rural life—a reality seen as increasingly important as the century progressed.

Finally, reflecting upon her work in the Willamette Valley hopyards during the 1930s, longtime resident Amanda Grim summed up the feeling of many when interviewed about her experiences: "[H]ard work to pick hops? No, it was a picnic. It was fun."[35] In all, these sources agreed upon the buoyant spirit of the Willamette Valley hopyards. They also highlight the good pay and amenities offered by growers. Not to be lost in cheerful accounts was the fact that the central objective of the hop-picking time was indeed work.

LABORING FOR THE LUPULIN

In the Pacific Northwest, growers divided their fields into quadrants and assigned workers to specific areas. There, one could—whether man, woman, or child—work at one's own pace for as long as one wanted in a day. Before the introduction of the trellis system, growers hired stronger men (called *pole men* or *pole pullers*) to remove poles from the ground and lay them down for pickers.[36] In *Hop Culture in the United States,* Ezra Meeker noted that "[o]ne stout man to every twenty pickers is considered necessary as a helper in taking down poles, cutting the vines apart, making roads and as a general assistant."[37] After the adoption of the wire trellis system, field workers, called *wire men*, unhooked the top wires of the trellis to ready a new set of bines for picking. One scholar notes, "To the pickers, the pole pullers and the wire men were the most important figures in the field."[38] They determined the pace at which one could work and ultimately weighed the hops for pay. The relationship indicated a certain amount of politicking in the hopyards. Knowing the right people could benefit one's ability to pick faster and make more money.[39]

Throughout the Willamette Valley's golden era of hop agriculture, workers averaged around one hundred to five hundred pounds of hops picked a day. At a rate of a penny per pound, that meant average daily earnings of one dollar to five dollars.[40] To provide context, during a six-day workweek in the 1920s an efficient hop picker could earn the equivalent of as much as two weeks' or a month's pay in a factory. Exceedingly savvy workers could make more. One newspaper reported in September 1912 that Lela Murray, a Yamhill County schoolteacher, set a record in picking 1,001 pounds of hops in one day. Given such an extraordinary feat, the paper noted that "[s]he has a written statement from J. G. Morris for whom she picked, that these hops were picked clean and that there were no hops left on the ground or on the vines picked by her on that day." Furthermore, "[t]hese hops were picked entirely alone as the yard boss and Mr. Morris would not allow any one to help the young woman that day. In making this record Miss Murray used two baskets and kept up two rows alone."[41]

Not all hop pickers felt the need to labor so intensely. That individuals or families could work at their own pace contributed to the leisurely nature of the hop harvest. Growers were not always concerned if pickers arrived for socialization more than harvest work, because they paid them by the pound. While Miss Murray's earnings were extraordinary, the pay in good or bad economic times provided a quick source of extra cash for Pacific Northwesterners. It was not enough to live on, but enough to boost standard yearly incomes.

In nearly all Willamette Valley hopyards, payment occurred in the form of tickets representing a number of pounds picked. The tickets were redeemable at the end of each day, week, or harvest season, and could be turned in for cash or used in local stores, restaurants, or taverns per agreements between growers and shop owners. This, too, was modeled on the English hop harvest and designed to keep money within the community. While some workers blew through their earnings in local shops or taverns, most saved as much of their wages as they could for the purchase of items needed in the coming year. Many former hop pickers recall using the money for school supplies and winter clothing. Overall, work in the hop fields was an important part of the regional economy. Reports from the first half of the twentieth century estimated that farmers paid hop pickers millions of dollars each year.[42]

Taking a page from their Kentish counterparts and farmers in other specialty crops, Oregon hop growers recognized the advantage of a stable, productive labor force. They offered incentives so workers would stay for the entire harvest. Before the turn of the twentieth century, growers such as Krebs Brothers and the E. Clemens Horst Company recognized the importance of urban workers, too. Early on they coordinated with rail operators for the *Hop Special* train from Portland to Independence and Willamette River steamers for efficient transport to their fields.[43] (See figure 6.) Although railroads could create other issues (mostly related to the costs of return trips home), they served hop pickers even after the 1920s, when many arrived by automobile with furniture and camping equipment in tow.[44]

In addition to transportation, discerning Willamette Valley hop growers provided housing and services for picking crews. Smaller operations allowed families and single women to sleep in the house and offered single men

FIGURE 6. Hop pickers arriving in Independence, Oregon, by the *Hop Special* train, early 1900s. Courtesy Benton County Historical Society.

accommodations in the barn or hop dryer. Most set aside areas to pitch tents and supplied running water, toilets, and fuel for cooking for a small fee. Some larger operations included cafeterias, and nearly all hopyards included a commissary on the property, usually run by the female head of household or the owner's children. These small operations provided basic foodstuffs—flour, bacon, and butter—as well as treats, such as soda. It was also common for butchers or merchants from nearby towns to bring goods for sale into the camps, where women usually cooked in the evening. In both campground and fields, most growers offered water, lemonade, or soda to help with dehydration and boost worker morale. As the large hop growers became established and economically secure, they built temporary dormitories and recreational facilities, and paid for live music and entertainment.[45]

As was the case in England, the efforts of growers to promote the Pacific Northwest hop harvest as a "paid vacation" did not always meet the realities for seasonal workers. Though rarely documented in its first several decades, some hop camps failed to meet the high-quality living standards promoted by regional media outlets, with lack of sanitation being a common complaint. Contract disputes regarding pay were also not out of the ordinary. While those problems necessitated time and energy, they were not at the forefront of the growers' minds. The primary problem was simply a fact of agricultural life in the American West: the region's population was a fraction of that of

England and New York, and competition for available workers presented a yearly challenge. To ensure proper harvest, hop growers had to hire all available hands. Herein arises the distinctness of the Pacific Coast hop harvest, given the region's sparse and diverse populations. The hop harvest was a decidedly multicultural affair.[46]

THE DIVERSITY OF THE WEST, CAPTURED IN THE HOPYARDS

Early on in his tenure as a hop grower, Ezra Meeker recognized that securing enough harvest labor highlighted a critical problem not only for his farm but also for his neighboring western Washington farmers. In the 1870s, the railroad boom that encouraged the arrival of larger Pacific Northwest populations was still a decade off. The Bureau of the Census reported a modest Tacoma population of seventy-eight residents; Seattle barely topped one thousand. While Meeker and his fellow hop growers hoped to attract as many workers as possible by adopting labor-recruitment methods found in the hop regions of England and New York, there simply was not a significant enough population available to them to harvest crops. This forced Meeker and his peers to buck a common perception of the time regarding the inferiority of nonwhite workers. Hop growers fervently reached out to American Indian groups surrounding Puget Sound.[47]

Meeker's decision paid immediate dividends. Indians from Washington, British Columbia, and even Alaska seized the yearly opportunity to earn good wages in a brief time during the hop harvests. By the time he penned what was to become a hop manifesto in 1883, Meeker noted, "The bulk of the hops are picked by Indians; they come from far and near, some in wagons, some on horseback, a few on foot, but the greater number in canoes. Two thousand, five hundred Indians came into the Puyallup valley during the hop-harvest of 1882. They were of all conditions, the old and young, the blind and maimed, the workers and idlers."[48] Despite their diversity in age, gender, and health, Meeker acknowledged his good fortune that the indigenous laborers were not only available in large numbers, but also quality hop pickers. Tribes with backgrounds in gathering food and medicinal plants in the wild adapted to the work easily. Growers were most impressed with their ability to pick cleanly. Meeker suggested that they were "reliable workers" who labored until "pitch dark."[49] The decision to hire Indians spread across the Pacific Northwest and into the Willamette Valley by the mid-1870s. There, many growers agreed that American Indians were "the best of hop-pickers."[50]

While happy to have American Indian workers, growers still treated them as second-class labor. The tribal groups worked on segregated hop fields and spent their evenings and nights in segregated camps. They also earned about one-tenth less than what white workers were paid, or approximately ninety cents per one hundred pounds of hops picked. Additionally, not all members of white society were pleased with the use of indigenous labor. Many hop growers complained about the slow and intentional pace of Indian workers, with some flat-out refusing to hire them. Others derided the hop harvest along with seasonal wage labor as counterproductive to upward social mobility for American Indians. But the decision not to hire could be risky. To ensure a timely harvest, hop growers needed any available worker.[51]

The annual journey of indigenous peoples to the hop fields and their arrival there became events of note that shaped the character of the region's agricultural landscape. Particularly on Puget Sound, where families arrived by canoe from British Columbia, Alaska, and other parts of Washington to pick hops, newspapers enthusiastically reported the arrival of native peoples at the beginning of a festive harvest season. In 1885, the *Oregonian* covered the Indian arrival at Port Townsend, the first stop on the way to the Puget Sound hop fields. "Nine canoes filled with Indians . . . arrived here yesterday from New Westminster [British Columbia]," the writer noted. "Twenty-one more canoes are expected to-day," he continued. "These Indians are bound to the hop fields, and from present appearances there will be no lack of hop pickers this season. There are between 200 and 300 Indians of all ages."[52] Every harvest season, newspapers and magazines reported on these arrivals for both community interest and to prepare urban shop owners in Seattle and around Puget Sound for an influx of business from the American Indian communities.[53] One *Overland Monthly* writer described the experience as the "event of the year," explaining that "[t]hese Indians, with their boats and rush tents, their baskets and babies, their cards and gambling, and all the hoo-doo and *tamanamus,* or midnight dances, make the autumn in the Sound country a time of panoramic interest."[54] Horse racing and sweat baths also marked the Indian encampments, which assumed the status of a pleasurable cultural landscape. In sum, American Indian festivities contributed to the harvest culture, bringing various experiences, languages, foods, and events.

In the last decades of the nineteenth century, American Indians and whites recognized the entrepreneurial possibilities surrounding what was becoming the celebrated hop harvest. At that time, white Americans wanted to view native peoples in authentic clothes and in rustic settings. Various sources, including Buffalo Bill's Wild West Show and Helen Hunt Jackson's *A Century of Dishonor* (1881), drew attention to the plight and culture of American Indians.[55] The Pacific Northwest hop harvests provided an ideal window into indigenous life. One historian suggests that Indian women in particular sought to exploit the popular practice of "Indian watching." They sold traditional basketwork and took photographs with white tourists for money. Some likely engaged in prostitution. Equally important in the spectacle were the white businesspeople who jumped at the opportunity to exploit the harvest culture. They built hotels for tourists interested in viewing the hop harvest and encountering genuine Indians, and provided wagon and rail transportation to the hop fields for visitors. Although many tourists arrived from within the region, others arrived from more distant points. Famed naturalist John Muir even commented that the Puget Sound hop harvest was one of most interesting tourist destinations he had visited.[56]

Willamette Valley residents similarly took interest in the arrival of American Indian communities in the hopyards. In *The Settlers Handbook to Oregon* (1904), Wallis Nash captured the scenes of the yearly native trek when he wrote, "The Indians on the reservations enjoy it heartily. There is a regular exodus from the Grand Ronde Reservation in Polk County and the Siletz in Lincoln County. The road out from the latter passed through our ranch and the procession of the wagons, with their dusky occupants, men, women and children all bound for the hop yards, was a long one, every year."[57] (See figure 7.) Recognizing the significance was not simply a one-sided affair, however. American Indians embraced their fortune at the opportunity to work and earn money. One Siletz tribal member, Gale Evans, recalled that in the first half of the twentieth century, "as long as you could walk, they made you pick. No monkeying around."[58] Thus, while enjoyable, the hop-picking time was an important part of the indigenous economy, achieved as a family and community effort.

American Indian participation in the Pacific Northwest hopyards had multiple meanings. The most obvious was the economic benefit. Following an array of Indian wars and a federal policy of designations of reservations and confinement, native peoples and local economies sought integration of native labor into the wage system.[59] As one writer noted in 1891, "Before the introduction of hops into Washington . . . Indians did not earn a dollar in money in a year, but now, at the close of the hop-harvest, a single Indian family . . .

FIGURE 7. Hop pickers from the Grand Ronde Reservation working in an Independence, Oregon, hopyard, 1922. By this point in time, most hops were grown on trellises. Courtesy Oregon Historical Society.

will carry home with them one hundred dollars in cash. The difference to that poor family, in comfort and civilization, can easily be understood."[60] As American Indians became increasingly reliant on wage labor, hop picking occurred amid a cycle that included fishing, clamming, cannery work, farming, picking berries, and logging. Of all these sectors of work, the hop fields paid the most. In fact, for some tribes who continued traditional economic activities, money earned from hop picking was enough cash to subsist for the year. A less obvious benefit of the harvest for tribal groups was the opportunities for cultural events. Yearly gatherings of Pacific Northwest tribes had been common for centuries. The hop fields allowed for continuity and shared celebrations surrounding games and trading. One might even argue that the hop fields provided a middle or transitional ground between traditional indigenous culture and the dawning of a commercial and industrialized world.[61]

Despite the welcomed presence of Indians by most hop growers, racial tensions and exclusions often appeared. To avoid the hiring of any nonwhites, some hop growers insisted that school be delayed to allow children to pick instead of Indians.[62] At other times whites enacted antivagrancy laws to force white workers into the hop fields. In periods of national depression, out-of-work white men contested the use of nonwhite labor, thinking that it was inferior or that they had a right to such work.[63] Still others publicly longed for a day when a population increase in the Far West might allow for an all-white picking labor force.[64] The worst criticisms of American Indian labor arose simply from a common white superiority complex that saw nonwhites as inferior and backward. A writer for the *Overland Monthly* sarcastically noted that for the Siwash Indians of Puget Sound, "[t]he Hop ranch is the El Dorado. . . . [H]e is enabled to fill his pipe with tobacco, his stomach with rum, and to clothe his body with an odd conglomeration of the habiliments of civilization."[65] Another article, in an 1888 issue of the *Oregonian,* noted, "The poor Indian oft lends his untutored mind to the task of hop picking, and relieves the monopoly of the occupation by trading off his ponies and squandering, as fast as earned, the proceeds of his industry."[66] These racist terms offer a sober indication that the hop harvest must be remembered not simply as a "paid vacation." It also offered a microcosm of the complex social relations and intertwined prejudices of the era. And that is still just part of the story.

Growers had no choice but to fulfill labor demands by looking beyond white and Native American help. In 1883, a year after the infamous global hop shortage, the *West Shore,* a Portland-based promotional magazine, reported that even with white and Indian workers in the fields, much of the crop would be lost.[67] The constant pressure forced growers to follow California's lead in hiring Asian immigrants. Beginning with the employment of Chinese workers, successive waves of individuals from Japan and the Philippines arrived at the hop fields during the late nineteenth and early twentieth centuries.[68] In the Central Valley, California social critic Carey McWilliams would note that the need for agricultural labor was "made to order."[69] His remark pointed to the need for Asian workers, and emphasized that a majority of them came from agricultural backgrounds. According to one historian, "Horticulturalists held the Chinese in high regard for more than their availability, stamina, and organization. These immigrants brought with them a vast agricultural knowledge accumulated by their ancestors over centuries of experience, which they readily adapted to California's environmental conditions."[70] Another historian notes that Chinese workers in particular "earned a reputation for organization, stamina, and stoicism coupled with an unmatched willingness to toil under disagreeable conditions."[71] Specialty-crop growers of all types saw Chinese workers as "docile," "reliable," and "industrious."[72]

Although less prominent in the Willamette Valley, Chinese laborers did not escape notice. Many growers praised them for their speed and efficiency during harvest. Some were so impressed with their work that they kept them on the farm year-round. Still, because of a perceived racial inferiority, growers paid Chinese workers up to 25 percent less than whites, even lower than indigenous workers. At one point the *Oregonian* reported simply that "growers do not feel justified in offering more."[73] As was the case in their treatment of American Indians, growers assigned Chinese workers to segregated living quarters and field areas. The Chinese laborers brought their own distinct culture—language, foods, and games—which added to the multicultural atmosphere of the hop harvest. Over time, Chinese agriculturalists in the valley also turned to hop farming and hired out for temporary labor.[74]

Amid a time of anti-Chinese sentiment, many onlookers criticized the presence of Chinese workers in the Pacific Coast hopyards. Yet as early as 1880, the *Oregonian* defended their use because of the fundamental need for labor. The newspaper quoted a grower from Butteville as saying, "A great many papers on this coast are condemning the employment of Chinese in the hop fields." But, the source went on to suggest, "[t]he writers of these articles will probably view the matter in a different light when they are informed that after all the help available—whites, Chinese and Indians—has been mustered into the hop fields of Oregon and Washington, still a considerable part of the crop must spoil this season for want of pickers."[75] In later years, the paper made similar claims. "Throughout the Willamette valley the work of harvesting the hop crop is beginning," an *Oregonian* writer noted in August 1889. "About 500 Chinamen have already gone to different hop fields, and 150 more leave Ash street dock this morning for the vicinity of Independence."[76] Not all of the region's residents were pleased.

Employment of Chinese workers coincided with a deep racism that surrounded Asian immigrations in the American West. One *West Shore* article explained, "Hop picking in the 192 hop yards of Oregon ... has been in progress for several weeks. Scarcity of pickers has led many growers to employ Chinese, but it seems to be the universal opinion that their work is not as satisfactory as that of the whites and Indians. They do not accomplish as much and are not so thorough."[77] In Northern California, the *Chico Enterprise* commented that, although better than Indians, Chinese workers ranked below Japanese and whites, particularly young white women.[78] Other criticisms were not so tame. In the same 1888 article that questioned the effectiveness of American Indian labor, the *Oregonian* suggested, "The meek, quiet and sometimes despised Chinaman . . . is largely employed as a last resort in default of white labor. His dark ways and vain tricks have not served to render him a desirable article in the hop fields."[79]

In general, many whites protested Asian labor of all types, believing that Asians were both in competition with whites and un-American because they refused to integrate into what whites believed was the traditional American lifestyle. The most glaring criticism of Chinese workers was that they sent wages to their family homes across the Pacific, which was perceived as disengagement from the common society and economy. Particularly in California, but also in Oregon and Washington, growers took heed and often replaced Chinese labor with American Indians or simply advertised for non-Chinese workers.[80] Other actions were violent. Chinese workers faced discrimination, and sometimes hostilities. One of the worst of these instances occurred in September 1885, when an angry white mob descended upon a Chinese hop picker's camp in the Kittitas Valley of western Washington, killing three men. Similar violent incidents occurred in Tacoma and other cities throughout the late-nineteenth-century West. The events connected to other violent activity elsewhere—from agricultural fields in California to mining camps in Colorado—in what is a dark cloud that hangs over the history of the American West and its resource-based economies.[81]

But anti-Chinese protests and the eventual Chinese Exclusion Act, passed in 1882, only exacerbated the labor shortage for the hop harvest. By the 1890s and early 1900s, Japanese and Filipino immigrants helped fill the demand. Their arrival arose in part because of Japan's modernizing effort during the Meiji Restoration, a period when the Japanese government encouraged young men to move abroad to learn new skills and earn money. Filipinos began arriving shortly after the United States acquired the Philippines following Spanish-American War negotiations with Spain. Like many Chinese immigrants, Japanese and Filipinos often had agricultural backgrounds and were good workers.[82] At first, farmers across the West welcomed their arrival, but as was the case with Chinese labor, a backlash occurred. On American soil, whites saw the sophistication of Japanese working crews and the presence of Japanese families as a more immediate threat.[83] One historian notes, "As late as 1907, many Sacramento County growers posted notices discouraging both Chinese and Japanese from applying because they intended to supplement white pickers with native Americans."[84] White hop farmers' feelings were clear: they preferred white labor over all others.[85]

In sum, the story of harvest in the hopyards provides an important indicator about diversity found in the West, and a labor situation unique to the Pacific Coast. As one writer for *Out West* magazine, a popular journal of life and culture, noted, "Hardly elsewhere can be found so many nationalities, classes and types, busied at a common occupation. There are the office man and woman, the clerk, the professional man, the student, the teacher and the invalid, eager for the physical benefits . . . or bent on enjoying a vacation."[86] The article mentioned, more specifically, that the social makeup included white families and "bands of Japanese . . . Chinese; some negroes, the ever-present Italian, the 'Hobo,' and lastly, the Paiutes."[87] Although camps tended to be segregated, many former hop pickers recall constant interaction between different social and cultural groups as a highlight of their experience. Variations of language and foods, and of dances and games, offered a range of contexts, from the prosaic to the exotic, in the Willamette Valley hop harvest and others up and down the Pacific Coast.[88] Unfortunately, as the story of American Indian and Chinese labor indicates, there were also times when the hop harvest connected to the dark cloud of intolerance. Although harvest labor conditions would change as the twentieth century progressed, there was one thing that could not be denied: beer cultures around the world depended on hops handpicked by a diverse Pacific Coast society.

FIVE

Hop Center of the World

IN ONE OF THE EARLY SURPRISES of the Pacific Northwest hop industry, Ezra Meeker's fortunes came crashing down in the late 1890s. Whereas many credited hop-aphid infestations as the culprit of his demise, the depression of the decade was to blame.[1] Meeker's millions earned from the wolf of the willow dwindled in the face of upswings in European production, and scores of small farmers defaulted on loans that he as a middleman extended. By 1897, Meeker abandoned hops and joined the gold rush in the Yukon and Alaska, showing not only that the industry was volatile but also that it attracted those willing to take risks.[2] But the hop culture of the Pacific Northwest endured beyond the charisma and expertise of this one man. Emil Clemens Horst, the Seavey family, and other grower-dealers filled his vacuum. So, too, did brewing-industry professionals and representatives from the U.S. Department of Agriculture. Despite ongoing obstacles, Pacific Northwest hop acreage continued to expand.

By 1905, Oregon became the nation's largest hop producer, with the Willamette Valley cultivating the majority of the state's acreage.[3] Once a speculative undertaking by a handful of enterprising farmers, the plant became Oregon's most important specialty crop in the first half of the twentieth century. The hop offered an increasingly visible sign of the valley's agricultural diversity and a marker of how rural commodities linked urban centers and marketing networks near and far. Mostly, small diversified farmers and larger grower-dealers contributed to the expansion of hop farming, but they were not the only players in the industry. In Mount Angel, Oregon, for example, the monks of St. Benedict's Abbey made good money in raising hops beginning in the late nineteenth century.[4] While growers would face new challenges, the Willamette Valley's hop industry increased production year after year. In the opening decade of the twentieth century, boosters proclaimed the region the "Hop Center of the World," a title long held by Bavaria, Germany.[5] By decade's end, Oregon averaged approximately twenty thousand acres of hops harvested each year, contributing millions of pounds to local brewers and others around the world.[6] Constantly on the agenda of expansion and the search for recognition lay a series of questions: How might the regional hop industry work to better produce and market the crop, reduce the risk of market fluctuations, and overcome international stigmas?

THE GREAT EXTRAVAGANZA

In the same year that Oregon claimed the title as the nation's largest hop producer, the state's growers saw a tremendous opportunity. The city of Portland announced its debut as an international economic and cultural center by hosting its first and only world's fair, marking the one-hundredth anniversary of the Lewis and Clark expedition to the Pacific Northwest. The "Great Extravaganza," as many dubbed it, provided an invaluable opportunity to display Oregon's bounty—goods produced and processed in the state and shipped from Portland. Organizers sought to highlight the state's diverse products and market Oregon goods as a brand name, taking a page from California agriculturalists and entrepreneurs.[7]

When Oregon business leaders considered the world's fair, they thought deeply about how to promote the products of their state. Decades prior to the turn of the century, Oregonians contributed to the global marketplace, with timber, wheat, and salmon leading the way.[8] But the state's leaders felt slighted during the previous year's fair in St. Louis, where organizers allotted little room for Oregon's display. Topping the list of complaints, one representative noted that "[t]he Oregon exhibits were placed in various exposition buildings at the Louisiana Purchase Exposition" and "the State did not receive from these exhibits the amount of good that it should." To rectify this slight, organizers of the Portland event decided to feature the wide range of the state's accomplishments "under the roof of Oregon's own State building."[9] Situated near the fair entrance, the Oregon Building would house exhibits from each of the state's counties to showcase their exceptional contributions to the world. More pointedly, as one historian notes, "[t]he exhibits and activities elaborated on the theme: Oregon had grown by harvesting its natural bounty from forests, fields, and rivers for national and international markets. In the coming decades, Oregonians hoped to produce more and more and sell its products more and more widely."[10] The historical timing of these efforts, for hop growers and all other Oregon producers, could not have been better.

In the years leading up to Portland's 1905 world's fair, the United States emerged as an international military and economic power. The Spanish-American War of 1898 and its aftermath created an American empire with outposts in the Caribbean and across the Pacific Ocean. It opened new marketplaces among colonized peoples.[11] More pressing was the pervasive industrialism that spawned growing cities with expanding populations in the United States and Europe. Millions of people, primarily from Europe, immigrated to the United States from the end of the American Civil War in 1865 to the outbreak of World War I.[12] The world's fair organizers recognized opportunities for advertising Oregon goods to these expanding markets inside and outside the United States proper. The exhibition's official title, "The Lewis and Clark Centennial American Pacific Exposition and Oriental Fair," pointed to expansion into the Pacific world while at the same time emphasizing an American rediscovery of Oregon a century after Lewis and Clark trekked across the region.

At the fair, the Oregon Building displayed samples of the state's abundant natural resources. Neatly stacked grains stood arranged in pyramid fashion in the Umatilla County exhibit. Josephine County featured giant pinecones from its forests and a wide assortment of fruits and vegetables from its agricultural lowlands. In the Washington County exhibit, hop bines saturated with cones hung heavy from the columns and arches—not unlike the star-spangled bunting and flags that draped countless others on fairgrounds. Far from simple adornments, the display of all of these products was calculated and purposeful. New technologies and expanded transportation networks by rail, ship, and road opened markets across the world. With innovative marketing and exposure, Oregonians associated with the world's fair knew that they could become global leaders in the production and sale of the wide range of products from the state.[13]

The fair's timing was propitious for Willamette Valley hop growers. Not only did Oregon achieve the U.S. lead in hop production ahead of New York and California in 1905, but a year later hops also became the fourteenth most lucrative crop in the country. Oregon contributed 40 percent of the U.S. crop, an astounding amount and enough to provide for one in ten beers in the world at the time. Topping sixteen million pounds of hops annually on over twenty thousand cultivated acres by the end of the decade, the crop attracted increasingly larger investments. The presence of hop plants and cones hanging in the Oregon Building signaled the continued importance of the Willamette Valley to the national and international hop and brewing community in a period when beer consumption grew dramatically. The displays also established the tone of how leaders of the regional hop industry hoped to improve upon its success and better connect to the global marketplace with their new standing as national leader and "Hop Center of the World." What remained to be seen was how a still largely unorganized body of small growers might respond to surrounding forces of modernization, ranging across new technologies in the industrial age, corporate organization of the economy, and government response in the Progressive Era of the early twentieth century.[14]

TRANSFORMATIONS, TECHNOLOGY, AND TENSION IN THE EARLY 1900S

In the first decade of the twentieth century, over a thousand farmers cultivated hops in the Willamette Valley annually. The mere presence of hop acreage and hop dryers left an imprint on the landscape.[15] Independence, Oregon—southeast of Salem—held the highest concentration of hop raisers. Growers believed the location on Willamette River bottomlands made it ideal because nutrients from intermittent flooding often replenished the soil. For this reason, Independence residents were perhaps the first to claim the title "Hop Center of the World."[16] Nevertheless, the expansion of crops to uplands instigated debates about the best place to grow hops. The key difference from the bottomlands, many farmers suggested, lay in the fact that higher ground stood immune from flooding and subsequent outbreaks of insects, mold, and disease. In *The Settler's Handbook to Oregon* of 1904, Wallace Nash reported that the issue became quite a controversy, noting, "Indeed today it is a matter in dispute which class of hop yards do the best. The upland hop man claims that his hops are the richer and more fully ripened in quality and produced at less expense because needing no spraying to defend the hops from mold and lice—the lowland hop man says that his hops are earlier ... and so run less danger of rain."[17] Regardless of the debates, it was clear that hopyards expanded to all areas of the Willamette Valley, with Marion, Polk, and Lane counties leading the way.[18]

FIGURE 8. Willamette Valley hopyard, circa 1932. Courtesy Special Collections and Archives Research Center, Oregon State University.

Beyond spreading to uplands from the flood plains, the continued success of hop farming depended on the introduction of a critical new technology. Advancements in wire trellis systems, which some English and American farmers began implementing as early as the 1860s, inspired growers to abandon timber poles.[19] An industry report explained what the new method of cultivation entailed:

> In the high trellis system posts are set every fourth or fifth hill and extend 12 or 15 feet above the ground. The bottom of the posts may be creosoted to make them last longer. Wires are stretched over the tops of the posts across the yard each way at right angles. Above the hills not covered by the main cross wires, extra wires are strung from the wires supported by the posts. Anchors keep the system of wire taut.[20]

The benefits were ample:

> Growers claim the following advantages for the trellis: (1) The hops remain healthier, (2) spraying against the aphis [aphids] and other pests is easier, (3) the hops mature earlier, (4) the cones have better color, (5) the cones are easier to pick and can be picked cleaner, (6) vines do not require cutting

which would weaken the stock by loss of material that otherwise returns from the vine to the root.[21]

As the century progressed, all hop growers in the Willamette Valley adopted the wire trellis system.[22] (See figure 8.) The technology saved on costs and benefitted from reducing the labor needed to set and remove timber poles each spring and fall. The trellis also continued the process of connecting other natural-resource commodity industries to the Willamette Valley—specifically, twine, cotton, and hemp for the hops to climb on trellises.[23]

In the ongoing effort to save on labor costs, the E. Clemens Horst Company patented a mechanical hop harvester in 1909. The company advertised it as a machine "fifty feet long, fifteen feet high, and ten feet wide" that stripped bines of their cones. Horst touted it as the fix for labor problems that surfaced nearly every year on the Pacific Coast, noting how the machine could do the work of 450 individual hand pickers and pick hops "absolutely clean and much better than human beings." In an effort to sell the contraptions, the company also suggested that "[t]he worry on account of scarcity of labor is eliminated, and the consumer is assured of receiving hops vastly superior in quality over hand-picked hops."[24] The dream of a mechanical picker had been a long time coming, with patents for the technology dating to nearly fifty years prior. But Horst's machine presented the first genuine option. In the 1910s, as one scholar notes, the mechanical harvester reduced the amount of workers he needed.[25] Few other growers followed in Horst's footsteps, because the early models were expensive and unreliable and growers felt apprehensive about the amount of hop cones wasted. For the next three decades, farmers continued to rely on manual harvests.

Aside from the trellis system and the introduction of Horst's mechanical harvester, other technological advances did not arrive so rapidly. While the U.S. patent office approved dozens of patents related to the hop industry throughout the late nineteenth and early twentieth centuries, most dealt with enhancements to systems already in place.[26] Emerging technologies of motorized tractors and plows could not operate effectively in narrow hopyards and therefore proved impractical. Even when farming families possessed automobiles by the 1920s and 1930s, the internal combustion engine did not appear in the hopyards because there was no vehicle that could navigate between the rows. The Oregon hop industry thus continued to operate on preindustrial terms, tied to the labor of humans and animals.[27]

The lack of widespread technological advancement was in part related to one of the central attributes of Willamette Valley hop growers in the early twentieth century: the majority remained diversified family grain and livestock farms that grew the specialty crops for extra cash income. Small farmers did not have money to invest in new machines, nor did most want to take on more risk. While some hop growers entered the business with the sole purpose of speculating to get rich in high-priced years, the well-documented gambles of the industry prevented most from getting too excited. The small-scale nature of the region's hop farming also brought about another set of issues. Many growers did not even consider the destination or uses of their crops.[28] Harvey Kaser, a grower from Silverton, Oregon, observed that some "growers were completely ignorant of any trends in the brewing trade. . . . All they had to rely on was what the handlers told them."[29] Similarly, hop grower L. D. Wood noted later in the century that "[t]he growers of hops know less of their use than of any other farm product. . . . The average grower does not even know the names of the brewers who finally buy and use his hops."[30]

Kaser's and Wood's points underscored a disconcerting trend for leaders of the industry. As American hop expert Herbert Myrick feared in 1904, increasing numbers of farmers entered the business with no previous expertise and had little desire to innovate or organize collectively for advantages in the marketplace. It was a precarious position given the opportunity with the opening of industrial and urban markets across the globe.[31] Industry leaders were further discouraged because California specialty-crop growers had laid a blueprint for marketing in the previous decades. In agricultural cooperatives, California fruit and nut growers created Sunkist, Diamond Brand Walnuts, and Fancy Brand Almonds, among other trademarks. Success followed these brands. It showed that farming cooperatives could be a major economic boon in bringing crops to market and obtaining good prices for products. But California events made little impact upon Willamette Valley hop growers at that time.[32] For the most part, Oregon hop-grower organizations failed to educate growers on the advantages of collective marketing.[33]

What did affect Oregon hop growers was the larger historical movement of temperance and prohibition. At the same time that the state became the national leader in hop production, a tide of Progressive Era moralism and health consciousness rolled from coast to coast. Farmers tied to the back-to-the-land and agricultural-cooperative movements evoked Thomas Jefferson's views of the American farmer as the sturdy, moralistic citizen of the republic.

Hop farming, for its connection to vice and agricultural speculation, did not fit this vision, particularly within the context of Pacific Coast specialty crops. In simple terms, one historian suggests, "Fruits and nuts . . . made California synonymous with health and prosperity," but hops stood outside of this image.[34]

By the mid-1890s, members of the Oregon State Board of Horticulture shunned hop growers. They likened hop growing to the ill effects of alcoholic drinks, consistent with the views of some Progressive Era reformers who would support the prohibition movement. In one report from 1894, the board outlined how fruit growers were not of the same ilk as hop and grape growers. "When we consider the grape and hopvine, we enter directly into a consideration of great sources of revenue," they noted, "but yet they are in one sense two of the greatest products of evil, for both are the indirect cause of most of the leverages that make wrecks of men."[35] The writers turned then to passages of the Bible warning against abuses of alcohol. Later reports made similar claims while promoting fruit orchards as healthful and moral, in contrast to hops and grapes.[36]

Willamette Valley hop growers had no such qualms about their enterprise. Moralist rhetoric either fell upon unreceptive ears or never reached the hundreds of families who continued to cultivate the crop. More likely, the diversified growers simply recognized the potential for cash income as acceptable regardless of the moral rhetoric aimed at them. They happened to live in one of the best environments in the world to grow the crop and in an era that witnessed an expanding brewing industry because of growing populations and the growing popularity of lager beer. As a historian of beer suggests, the era "spawned the saloon," a period in which millions of Americans and immigrants in the industrializing and urbanizing world sought respite in malted and hopped concoctions.[37] Whether or not individual farmers recognized the significance of the expansion of national breweries, as Harvey Kaser noted, they seized the opportunity to sell more hops.

At the very least, hop growers knowledgeable about their product took comfort when breweries stood up to moralist reformers with their own marketing efforts. Catering to Progressive Era audiences, Anheuser-Busch issued a series of advertisements proclaiming that their product was the "natural drink of America" and inspired "[a]ppetite, health, and vigor." One ad noted, "In every glass is health; and what is health but efficiency and power? It comes to your table a delicious, sparkling food—a wholesome malt beverage, exhaling the aroma of hop gardens and fragrant scent of new-mown northern barley fields." The advertisement concluded by suggesting, "Nothing better for the system than good beer, and Budweiser is the best."[38] Pabst marketed its beer similarly in the early twentieth century as a "Clean, Pure, Wholesome" product. Mirroring Anheuser-Busch, Pabst also had a series of advertisements that noted, "When you drink beer you want Quality and Purity, because that means health and strength."[39] Willamette Valley beer makers and hop growers tried their hand at similar advertisements in local newspapers, explaining how hops were wholesome and aided in digestion. But growers were better served by the nation's large brewers, who had extensive advertising campaigns.[40]

For the time being at least, beer sales rose continuously and spurred domestic hop production. Leaders of the hop and beer industries used their resources to improve marketing, growing, and breeding from their collective expertise. These modernizing efforts set professional and scientific standards for the rest of the century. Professional businesspeople and scientists also worked energetically to achieve ever-greater levels of production.[41]

THE COMMERCIALIZATION OF PACIFIC COAST HOPS

On the other side of the globe from Oregon, the continued expansion of two German companies and one from England signaled a new corporate era in the hop industry by the turn of the twentieth century. Johann Barth and Son of Nuremberg, Simon H. Steiner of Laupheim, and Wiggins, Richardson & Co. of London did for the hop what Andrew Carnegie did for steel and Gustave Swift did for the meatpacking industry in the United States. Started in the late eighteenth and early nineteenth centuries as small trading firms, the three companies vertically integrated over time to include new crops, storage facilities, and access to brewers. They utilized the tactics of industrial-era big business to monopolize much of the European hop market, streamlining international hop markets and better connecting brewers to the central hop-growing regions in Germany, Belgium, Bohemia, and England.[42] By the early twentieth century, Barth, Steiner, and Wiggins, Richardson & Co. stood out as the world's largest hop dealers. They provided breweries with the varieties, quantities, and qualities of hops needed to keep pace with industrial-era beer production. Given that global consumption of beer doubled from 125 million barrels per year to 250 million barrels between 1880

FIGURE 9. Emil Clemons Horst (1867–1940), one of the most influential hop growers and dealers of the North American West and inventor of one of the first mechanical harvesters.

and 1910, the monopolization efforts of the European companies were substantial. A modernizing and corporatizing beer industry required a modernizing and corporatizing hop industry.[43]

During the early twentieth century, however, Barth, Steiner, and Wiggins, Richardson & Co. did not have a hand in the Pacific Coast hop market. In the United States, their agenda centered on the importation of European hops for American brewers, not exportation of American hops. The door lay open for other entrepreneurs to fill a role that these leading hop dealers had carved internationally. Primed to assume that role was Emil Clemens Horst, the German immigrant with a penchant for the business who superseded Ezra Meeker as a Pacific Coast "hop king." (See figure 9.) From the late 1890s to his death in 1940, Horst borrowed business practices from corporate hop dealers and regional leaders. He became the Pacific Coast's quintessential grower-dealer, acquiring land and planting three thousand acres of hops in California, Oregon, and British Columbia. Furthermore, he established marketing and sales offices in Sacramento, San Francisco, Portland, Salem, Chicago, New York, and London. Not one to shy from self-praise, Horst advertised himself as the "largest hop grower in the world." The large European dealers may have disagreed with him, but the fact remained that his new standing in the Pacific Coast hop community indicated changes in the nature of the industry.[44]

Horst arrived in the United States from Germany in 1874 at the age of seven, coming from an educated and industrious family. In his early twenties, after schooling in New York, he managed a hop-dealing corporation with his brothers. The business folded in 1896. Soon after, Horst moved to San Francisco and established the E. Clemens Horst Company. He initially planted acreage in Yuba County, California, and made such a remarkable impact on the community and economy that the postal service opened an office at a site they named Horstville in 1898.[45] When Ezra Meeker exited the hop business, Horst swiftly expanded holdings. By the second decade of the twentieth century, the company grew nearly 10 percent of all hops grown on the Pacific Coast, including about 20 percent of the California crop and 10 percent of the Oregon crop, including a five-hundred-acre hopyard in Polk County. The company netted nearly a million dollars a year and employed thousands.[46]

Horst contributed to a new professionalism of the West Coast hop industry. Like Ezra Meeker before him, he acted as an agricultural supplier and offered loans to small farmers. Horst also became an invaluable source of knowledge, contributing to print media on hops at the time and introducing ideas from outside of the region. In the Willamette Valley, Horst's influence eventually surpassed the levels that Meeker attained. The key lay in his ability not only to sell his own hops and negotiate contracts with smaller growers, but also to understand the demands of international markets.[47]

Horst's success occurred in large part because he understood that by the turn of the twentieth century, hop cultivation in Great Britain was in decline. As one English journal commented, "The very risky character of hop cultivation has not only prevented the extension of hop cultivation in England, it is slowly contracting it." The author highlighted what Horst and other growers already knew, suggesting, "[F]or some years past there has been a steady increase in the growth and production of hops in Oregon, which has now become ... the principal hop-growing State in the Union."[48] The observations captured the fact that the center of hop cultivation in the world had shifted. Horst knew that he and his fellow growers of the American West were at the new center of hop growing. And he sought to exploit it.

Horst solidified a lasting influence on the Pacific Coast and the global hop marketplace in 1904 when he negotiated an exclusive contract between Ireland's Guinness Brewery and a large group of Oregon hop growers. Offering hops at a cost "appreciably less than the prices being charged by the English merchants" but "on par" in quality, the deal set in motion a long-term commitment with the famous brewery. While his methods in outbidding well-known English suppliers may have been cutthroat, Horst cemented early his legacy as a champion of the Pacific Coast hop industry. Growers in Oregon may not have been savvy to the details of the negotiations, but they understood the value of working with the E. Clemens Horst Company.[49]

The Horst-Guinness deal underscored several trends, not the least being Horst's new standing in global hop trading or the reality that Guinness drinkers from 1904 onward consumed Oregon hops in a blend with other English varieties.[50] First, the deal demonstrated how English production had faltered by the first decade of the new century. With hop production down in Great Britain and beer production and consumption rising, the country's brewers had to look elsewhere for hops. The Pacific Coast offered English varieties, including the Fuggle hop, in great quantity and soon took over as the largest global exporter to British (and Canadian) breweries.[51] Second, the deal demonstrated that Pacific Coast hops were in the process of overcoming their perceived stigma in European brewing traditions. Many brewers in Europe and European colonies continued to view American hops as second-rate and refused to use them, but others became attracted to the Oregon product. To a certain degree, Horst's efforts began in earnest the ability to market Willamette Valley hops as an international brand. Third, Horst's negotiations with Guinness revealed a new collective power for Oregon growers if they could organize together with the assistance of business leadership. Such organizing activities on a smaller level—which had stopped and started from the 1870s onward—failed without such a skilled leader. Though a substantial number of farmers were still unaware of the destination of their hops, Horst helped weave their livelihoods into the international culture of hops and beer production.

Ultimately, the deal that Horst struck with Guinness also instigated a new level of global competition between hop traders. Although buying and selling the commodity had always been fiercely competitive, with buyers and sellers engaged in unending legal battles over prices and contracts, a new era was born when European companies finally set up shop on American soil. This is exactly what happened in August 1906, two years after the Horst-Guinness arrangement, when the London firm Wiggins, Richardson & Co. established an office in Salem. The arrangement, as the *Oregonian* reported, marked a new period in the industry as European players, who did have offices in New York by this time, planted themselves firmly in the Pacific Northwest. This new presence was significant. For starters, the London firm had a near exclusive deal to ship hops to England's Bass Brewery, the largest in the world by the late nineteenth century. Wiggins, Richardson & Co. soon purchased land in the vicinity of Independence, Oregon, and began growing and exporting the wolf of the willow from its own Wigrich Ranch.[52]

While Horst was the most recognizable leader of the Pacific Coast's grower-dealers in the early twentieth century, he was not alone in the efforts to better commercialize and streamline an industry that depended heavily on small farmers. Several other individuals emerged to fill the vacuum that Ezra Meeker left upon his departure for the Yukon. In Northern California, the Durst family joined Horst in owning substantial acreage and marketing American hops, and in spending significant amounts of time in Europe to improve their standing in the industry.[53] To the north, Herman Klaber, the purported "hop king" of Washington after Ezra Meeker, joined Horst in connecting regional growers to the international marketplace. Klaber spent considerable time lobbying Congress for changes to hop tariffs and vehemently fought against smaller growers who wanted to form cooperatives. Klaber might have achieved more if not for perishing in 1912 on the *Titanic* following a European hop-selling venture.[54]

In the Willamette Valley, several residents and companies engaged in professional leadership roles. Salem's T. A. Livesley Company and Independence's Krebs Hop Company both organized and contracted smaller growers to compete in the international marketplace. Independence's Willard "Arch" Sloper also became well connected to the regional hop industry not only in his efforts at growing and expanding his business, but also by introducing a number of labor-saving technologies. His "wire-trellis dropper," patented in 1909, revolutionized the process of picking hops. He also invented a plow specific to hop cultivation, and in the early 1930s, his "automatic baler," which operated like an elevator to process hops into standard bales, also saved on labor costs.[55]

The most visible Willamette Valley grower-dealer was none other than James Seavey, the son of Oregon pioneer Alexander Seavey. The family had a much different relationship to hops than did Emil Clemens Horst. After immigrating from England and living in Maine for three generations, the Seavey family arrived in Oregon in 1850 with Alexander acquiring land under the Donation Land Act. An early focus on grains proved profitable for his family, but like many others they turned to specialty crops in the closing decades of the century. In 1877, the elder Seavey planted twenty-five acres of hops near the McKenzie River, and the crop showed promise. From that point onward, he spread his land holdings throughout the Willamette Valley, from Lane County in the south to Washington County in the north. He trained three of his sons, James, Jesse, and John, in the arts of agriculture and commerce.[56]

Upon their father's death in 1908, the Seavey sons divided his land, and expanded their hop acreage. They became one of the largest and most successful hop-growing families in the Pacific Northwest. James, in particular, took control of 150 acres near his father's original McKenzie River site and expanded from there. By 1912, he owned over 500 total acres in three counties of the Willamette Valley. Equally important, Seavey opened a Portland office for the marketing and distribution of Oregon hops: the J. W. Seavey Hop Company. For several decades, the business competed with the E. Clemens Horst Company and others in providing a major connection from Oregon growers to brewers around the world. His relationship with Willamette Valley growers, however, was more intimate since he had grown up in the farming community. Historian Joseph Gaston noted that, for other farmers, James Seavey's "opinions concerning hop growing are accepted as authority."[57]

Together in the opening decades of the twentieth century, Emil Clemens Horst, James Seavey, the operators of the Wigrich Ranch, and a handful of other grower-dealers became the new faces of an increasingly market-wise and commercialized Oregon hop industry. Their achievements stemmed from longtime experiences with hops, West Coast agriculture more broadly, and global markets. But there were other forces at play. The professionalizing Pacific Coast hop industry started to benefit from scientific programs concurrently emerging in the private world of brewing and the public world of federally supported agricultural research.

FROM "STUDIES IN BEER" TO "HOP INVESTIGATIONS"

As much as hop growers and dealers sought to take control of their own destinies at the turn of the century, specialists outside of the industry contributed to advancements in science and technology. In this period of modernization, the developments were a calculated matter of staying competitive in the global marketplace. Thomas Edison and his colleagues in their Menlo Park, New Jersey, facilities set precedents for privatized research and development in the closing years of the nineteenth century. At the same time university departments across the nation became more specialized as the educated public embraced positivism, or the foundational belief that science held solutions to the world's problems. The goals for both private and public sectors were to advance technology, industry, and the American quality of life. The American beer industry followed suit, and the U.S. Department of Agriculture also added its expertise to the development and marketing of specialty crops, including hops.[58]

The leading American brewers in the late nineteenth and early twentieth centuries—including Anheuser-Busch, Pabst, and Schlitz—relied on the innovations of professional scientists on issues varying from chemistry to marketing, and in locations inside and out of their own brewing facilities. Louis Pasteur contributed one of the most important developments in the late nineteenth century when his 1871 *Studies on Beer* provided the foundations of understanding yeast and fermentation. The pasteurization process, along with expanded railroad networks, allowed national breweries to extend their marketing footprint at unprecedented rates.[59] But not all advancements came from widely recognized scientists. For example, among the dozens of inventions that William Painter of Maryland developed was a superior seal for bottles in his crown bottle top of 1892. In general, according to one historian, the new industrial brewing leaders of the late nineteenth and early twentieth centuries "dived into the age, gambling on new and untested technologies: artificial refrigeration, pressurized carbon injection, pasteurization, and automated bottling machines." They created sophisticated laboratories and employed leading scientists from around the world while at the same time expanding breweries with mechanically powered machines and labor-saving systems. The result of these initiatives in the United States was the extraordinary rise of the major American brewing companies. The businesses outcompeted smaller brewers in volume and costs—a trend that occurred throughout other American industries at the time. In 1880, there were nearly three thousand breweries in the United States. That number was reduced by nearly half by 1910. The industrialization of brewing opened a new capacity for beer making and new market shares across the country and abroad.[60]

By 1905, the *New York Times* reported of the brewing industry, "Constant progress has been made in various directions," including "important advances of the brewer into the realms of science, as related to his business. The teachings of the various brewing schools and scientific stations have been extensively adopted and turned to practical account."[61] Advancements occurred rapidly amid these innovations and in the formation of professional organizations. The American Brewers' Academy (1880) and Master Brewer's Association of the United States (1887) provided new research on beer making and related agricultural products, and published information for their members in magazines and journals. Standard German and British brewing publications remained the most widely read in the world, but the United States witnessed the beginnings of its own professional-journal print culture. The *American Brewing Industry Journal,* the *Yearbook of the United States Brewers' Association,* and *American Brewer* became the avenues through which the brewing industry, including agricultural producers of grain and hops, engaged in an interrelated knowledge exchange.

In the early part of the twentieth century, Emil Clemens Horst contributed to the new brewing societies and journals more than anyone else in the American hop industry. While most of his published works concerned the history and future of his trade, he also offered his company's land and hops for professional brewery research. In one 1904 study, for example, the E. Clemens Horst Company donated eighteen hop plants to brewing scientists for judging quality and effectiveness of fertilization. The judges rated seven of the samples "very good" in terms of their overall appearance and chemical properties. It was a good sign. The collaboration pleased both hop producers and buyers in an era when many brewers of the European tradition still discounted American hops. Horst's efforts attracted positive attention to Pacific Coast hops, and at the same time generated feedback for future improvements in his fields. As the century progressed, such alliances continued to benefit both brewer and hop grower. They collectively determined what varieties of hops were useful and how best to cultivate, process, and store them. Still, private investment by the brewing industry in research specific to hop agriculture was minimal. Growers, dealers, and brewers looked to federal assistance, another crucial step in the modernization of the hop industry.[62]

The effort of the federal government to assist in scientific and technological advancements in agriculture dated to 1862, when Congress created a non-cabinet-level Department of Agriculture and passed the Morrill Act to create land-grant universities. In both instances, the goal was to advance the nation's citizens in the agricultural and mechanical arts through research and applied knowledge. But progress moved slowly. Congress did not create a cabinet-level USDA Secretary of Agriculture until 1889. It had left research at land grant institutions underfunded until the Hatch Act of 1887. Only then did the federal government generate new funding for research at agricultural experiment stations connected to land grant universities. These stations would become the scientific backbone of American farming for generations to come.[63]

The USDA initiated the experiment stations with a specific role: to offer research and education to farmers. With better funding and vision, the agency adopted the motto of helping "people improve their lives through an education process which uses scientific knowledge focused on issues and needs."[64] The agricultural experiment (or research) stations aimed to improve crops in specific geographic and climactic regions, with each state focused on its own needs. Much of the government's research activity centered on cereals and other major crops, but it gave some attention to hops and other specialty crops. Experiment stations across western states seized upon the specialty-crop revolution to help farmers. Even in arid Nevada, the experiment station in Reno published a bulletin on the potential of hop growing in the high desert. As the author noted, "Moderate fortunes have been made with special crops, with a large market near by, and the bringing together of the right man and the right crop."[65]

At the same time, the Government Printing Office released several circulars on the international hop trade and U.S. participation. One of the first publications was an 1891 pamphlet simply entitled *Agricultural: Hops,* followed by others including *Hop Cultivation in Bohemia* (1899) and *Hop Culture in California* (1900). Though they offered only a handful of pages with statistics on production in the growing regions, the works demonstrated a collaborative ambition on the part of American producers and government to compete with older, more established growing regions in the world. Perhaps not surprisingly, the scientists employed to do so delved deeply into the literature already produced in those places.[66]

Oregon's Agricultural Experiment Station in Corvallis first planted hops in 1895, though scientists had already been working on some hop issues—namely, pest control and studies on hop varieties from Europe.[67] It seems as if that work was minimal, even though hops had clearly become a favorite specialty crop for the region's farmers. Hops drew more attention at the experiment station in the next decade. In 1909, the experiment station at the

Oregon Agricultural College in Corvallis (predecessor of Oregon State University) engaged in Oregon's first federally sponsored hop research project, simply called "Hop Investigations." The research program included fieldwork that integrated local growers, including test acreage from the hop-yards of James Seavey, E. Clemens Horst, T. A. Livesley, F. S. Bradley, Krebs Brothers, and H. Hirschberg.[68] Unfortunately, the efforts failed to produce useful conclusions within the first several years.

In his annual report to the USDA for 1911, Oregon's Agricultural Experiment Station director James Withycombe highlighted chemical studies of hop lupulin and related issues of kiln-drying hops to preserve essential resins. Recommendations for growers included standardized heats for drying, with slow turning every six hours and adding sulfur for uniformity.[69] The experiment station's first major published report was a 1913 bulletin entitled *Hop Investigations*. Its contents focused on work from the institution's experimental hopyard in Corvallis and collaboration with James Seavey. Though the report discussed research surrounding qualities of hard and soft resins and the methods of kiln drying, most conclusions were fairly innocuous. One main recommendation, for example, noted, "Of all the fertilizing materials which have been used in the Willamette Valley, ordinary barn-yard manure seems to give the best results."[70] While the comment may have seemed insignificant, it was useful. The deep root systems of *Humulus* deplete soil of nitrogen and require attention at the end of the growing season. The recommendations for manuring, and, in later times, planting cover crops such as legumes or winter grain that would return nitrogen to the soil, were important.[71] Still, the strength of the initial *Hop Investigations* report lay in its recommendation for growers to seek better quality-control standards and for the experiment station itself to continue researching.

Although slow to generate useful conclusions from its hop research, Oregon's Agricultural Experiment Station successfully established lasting relationships with similar programs across the world. The most important of these was England's Wye College in Canterbury, Kent. In 1906, under the direction of plant pathologist E. S. Salmon, researchers there initiated the world's first and most successful hop-breeding program. Primarily, Salmon sought to increase the level of alpha acids, or bitterness profiles, in hops. This would increase the commodity's value to brewers because they could use fewer hops in their beer recipes while maintaining standard bitterness levels. But the cross-breeding (or hybridizing) program would find other benefits, too, including hops that demonstrated disease resistance, higher yields, and better storability. Members of the hop and brewing industries naturally looked to these developments as an important part of their futures.[72]

In 1919, Salmon bred his first successful hop hybrids, the product of English and North American parents. After substantial testing, he officially released the two new hops in the 1930s as "Brewers Gold" and "Bullion," offering a competitive advantage to the dwindling English industry and other growers around the world—specifically, those who looked to the new hybrids as a solution to disease problems. The developments suggested potential for continued success at Wye College and other fledgling research programs across the world, including the one in Corvallis. Salmon himself immediately attained notoriety as the father of modern hop breeding and gained esteem throughout the first half of the century when he released additional hybrids, such as "Northern Brewer."[73] Yet Salmon's historical significance transcended his own lifetime, and that point cannot be emphasized enough. Though he died in 1954, hop breeders in the twenty-first century continue to study his copious, if not hieroglyphic, notebooks, now held at the Imperial College London.[74]

American advancements in hop breeding and research lagged behind the English because U.S. brewers, growers, and scientists did not have a unified agenda. Brewers wanted to learn more of hop qualities specific to brewing, while the government and hop industry wanted to find ways to improve methods of cultivation, storage, and dealing with the threats of pests and disease. This disconnect was visible in the discussions and publications of brewers, including a 1907 issue of *Transactions of the American Brewing Institute* that suggested bluntly, "There are many difficulties in the way of the government making these investigations. In the hop investigation apparently we have been unfortunate."[75] The main problem, the author noted, was the inability to get the federal government to support the American cultivation of the European noble varieties, not just the Clusters and Fuggles that dominated the American agricultural landscape. Efforts failed, even after consultations with renowned Santa Rosa horticulturalist Luther Burbank, to at least seek European-American hybrids. In 1907, Burbank claimed, "The hop should be a most promising plant on which to carry on . . . experiments in selection."[76] The inflexibility on the part of farmers rested in their propensity to growing the Cluster hops for their high-yields, as opposed to the noble varieties grown in Continental Europe, which produced significantly fewer cones per plant. Nevertheless, the Pacific Coast hop industry flourished despite a growing undercurrent of prohibition. Agricultural sciences and new technologies rapidly advanced in the twentieth century, and the Smith-Lever Act of 1914 expanded the role of cooperative extension. Later in the century, Corvallis became a major center for hop research and breeding, joining in the success that E. S. Salmon and his peers at Wye College first enjoyed.[77]

HOP CENTER OF THE WORLD

Amid the corporatization and new science connected to the broadening commercialization of Pacific Coast hops in the early twentieth century, Oregon boosters latched on to the same narrative as their predecessors and the 1905 world's fair organizers: Oregon's "natural bounty." Hops added to the crop varieties produced in the agricultural utopia of the Willamette Valley. Perhaps most important, the crop highlighted the state's economic vibrancy. Promotional literature captured these ideas and eagerly anticipated new technologies and advancements in transportation for further success.

In 1903, a *Sunset* article noted rather extravagantly, "While Oregon hops are already a factor in the markets of the world, the day is not far distant when the industry will have attained gigantic proportions, and he who owns a hop field will have his hand on the lever that moves the world's financial wheel."[78] A book published two years later, *Benton County, Oregon: Illustrated,* elaborated on the "agricultural possibilities of the country," and noted specifically, "Hops of the finest quality are grown, and the annual yields prove conclusively that this crop can be grown at a minimal cost. Four hundred acres of yards in the county produced last year 300,000 pounds of cured hops, nearly all of which rate first class."[79]

The Benton County publication also spoke to the importance of the expanded marketplace, explaining the urban role in marketing the wolf of the willow among other crops. The author noted how Portland (a city of one hundred thousand by that time) became the launching point of regional agriculture for destinations across the nation and the world. "It has lines of steamships to all Alaskan points," the text noted, and "also to San Francisco, Honolulu, Yokohama, Hong Kong and Manila." After explaining the various railroad connections via the Southern Pacific and Northern Pacific, the text finally suggested, "Through its ocean service it has large and growing trade with Alaska, Mexico, Central and South America, the Sandwich Islands, Philippine Islands, China and Japan. It does a large and lucrative business with Europe and the Atlantic Sea board which will be quadrupled when the Nicaragua Canal is completed."[80] As railroads continued to expand and quicken their pace, and the Panama Canal opened faster shipping between the American West Coast, Europe, and Africa, specialty crops reaped the benefits of the market. All played to the favor of Willamette Valley hop growers, who took a greater hold on global hop production.

Into the 1910s, the sentiments remained. Joseph Gaston, in his *Centennial History of Oregon* (1912), captured the momentum when he noted:

> Allowing a pound of hops to the barrel of beer, Oregon is now producing hops enough to produce yearly thirty million barrels of beer; or 30 barrels for each man, woman and child of the State. But only a small amount of the hop crop is converted into beer in Oregon, and not one fourth of what is produced here is consumed by people of the State. Nearly all of the Oregon hop crop is shipped to the eastern states or to foreign countries; while the Oregon breweries ship probably half their brew to the Philippines, Hawaii, Alaska and British Columbia.[81]

The world's fair organizers of 1905 and all individuals counting on the success of hops would have been proud to see such advancement by Oregon's farmers in a few short years. The attention was unparalleled in other agricultural enterprise and genuinely solidified hop culture as a distinguishing feature of the Willamette Valley's agricultural identity.

At the same time that insiders and outsiders alike praised the Oregon hop industry, however, the American public increasingly entertained the possibility of banning the production and sale of alcoholic beverages. A century-long fight for temperance and prohibition gained unprecedented political support by the 1910s. Agricultural and industrial producers feared what this might mean for their livelihoods. The Willamette Valley's new identity as the "Hop Center of the World" stood at a crossroads.[82]

SIX

The Surprise of Prohibition

WHILE RESIDENTS OF THE "Hop Center of the World" had been able to shrug off prohibition's advance in the first decade of the twentieth century, the story quickly changed. In 1912, Oregon's women achieved the right to vote via the state ballot-initiative process, and a majority supported the outlaw of booze. This should not have come as a surprise. American women had long exercised their political voices for the causes of temperance and prohibition.[1] But it still proved demoralizing for Oregon's alcohol-related industries and consumers. In a vote of 136,842 to 100,362, in November 1914 Oregon's electorate decided to go dry five years before the nation as a whole. A loophole in the law initially allowed imbibers of beer, wine, and spirits to legally consume their beverage of choice if it was brought from outside of the state. Two years later, additional legislation declared all-out prohibition.[2]

By 1916, the early onset of prohibition in Oregon resembled stories in twenty-three other states (including neighboring Washington and Idaho). Yet the women's vote was not ultimately responsible for the national law, since women did not have a vote in national elections until 1920—after national prohibition went into effect. So what brought on the Eighteenth Amendment? The answer is complex. First, the United States joined prohibition efforts along with other modernizing nations that had become urban and industrialized in the late nineteenth and early twentieth centuries. Reformers not just in the United States but also in Canada, Great Britain, Scandinavia, and Australia fought against alcohol, believing that booze of any kind contributed to the erosion of work ethics, morals, health, and productivity, particularly in the lower, and often immigrant, classes. One historian notes that the efforts triumphed because "urban capitalists believed such a ban was ... a necessary precondition of the social reform required to ensure successful and permanent transformation of American society into an industrial order characterized by political stability and labor's social quiescence."[3] In other words, business leaders approved of a ban on alcohol to secure more efficient industrial output from sober wage earners. Although a powerful narrative, particularly when linked with the social concerns of women and Christian activists, it does not answer all of the questions surrounding the advent of national prohibition. Other global forces came into play.

In 1914, with prohibition efforts ratcheting up during the first years of Woodrow Wilson's administration, Congress held an expansive public hearing on the subject. Following marches and meetings held in the streets of Washington, D.C., a range of speakers took to the capital for thirteen hours of a prohibition debate.[4] At the same time that the hearings commenced, a complicated set of alliances and events pushed Europe to war. Although it would be three years before the United States officially entered the conflict, anti-German and nativist xenophobia resounded across the country. Quickly, prohibition efforts took on an anti-German tone because the majority of American brewers were of German descent. German-American citizens cried foul in the halls of Congress, but it was clear where the law of the land was headed. It reversed the immigrant and alcohol narrative that played out prior to the Civil War. Whereas German-Americans who brewed light lagers had once helped assuage the temperance turmoil, they became the target of the Progressive Era prohibition campaigns by the 1910s.[5]

Despite the pleas of German-American communities, American consumers, and all parties associated with the alcohol industry—not to mention economists who feared what the loss of related taxes might mean for federal and state governments—the prohibition cause marched forward. More states joined Oregon by crafting their own prohibition legislation, and the nation increasingly sided with teetotalers as America's entrance into world war grew near. As Charles Stelzle, a Christian prohibition leader, suggested, "There never was a time when America so needed her sober senses as to-day—it is a time when selfishness must be subordinated to the great task of winning the war."[6] The Wilson administration seemingly pounded the final nail in 1917 when it passed a wartime measure to divert agricultural resources, particularly grain, away from the alcohol industry and to supplying efforts of American soldiers and their allies. American farmers, they argued, should be aiding war efforts, not contributing to the vice trade. The national wartime slogan, after all, proposed that "Food Will Win the War," not booze.

After examining various versions of the law, Congress ratified the Eighteenth Amendment of the Constitution in January 1919. The text noted, "After one year from the ratification of this article the manufacture, sale, or transportation of intoxicating liquors within, the importation thereof into, or the exportation thereof from the United States and all the territory subject to the jurisdiction thereof for beverage purposes is hereby prohibited."[7] Even more so than Oregon's and other individual state prohibitions, the effects of the "noble experiment" on the nation's society, economy, and culture were drastic. Would this spell the end of Willamette Valley hop farming? As it turned out, the Prohibition era in Oregon, from 1914 to repeal in 1933, proved full of surprises.

THE HOP GROWER AND BEER MAKER'S APPEAL

With Prohibition and world war looming, many farmers abandoned hops, believing the crop's heyday had come and gone. Other agriculturalists felt motivated to continue growing their specialty crop of choice. In Oregon, the prohibition fight left hop producers frustrated, particularly after their overcoming various international stigmas and creating lucrative contracts in the first decade of the new century. While some local hop farmers were of German origin, the makeup of growers was ethnically and racially diverse, making problems of nativist persecution a nonissue. The overall problem was the collective strategy to be adopted in responding to the potential loss of the domestic beer industry.

Facing a possible collapse of the domestic outlet for their crops, hop growers turned their organizational efforts away from price controls and tariffs. Their main hope rested with persuading political leaders to protect their livelihoods. As early as 1908, Willamette Valley growers penned a petition for Oregon senator Charles William Fulton, noting, "We, the undersigned growers and dealers in hops, petition you to do all in your power to oppose present hostile legislation affecting brewing interests.... Consider these, if passed, ruinous to the hop industry of Oregon."[8] Californian counterparts also wrote to their congressman Duncan McKinlay, "calling upon him to do all in his power to keep the country from going dry by opposing the Prohibition movement now before Congress."[9]

At the same time, large brewers such as Gustav Pabst called upon cultivators of corn, rye, barley, and hops to join him in the fight against Prohibition.

"The continued growth of prohibition and the destruction of the brewing and distilling industries," Pabst wrote in an address of 1908, "will result in the allied trades in all lines of manufacture being made to suffer great losses through the destroyed market for their product."[10] In the next several years, farmers across the country joined brewers in protesting state and national prohibition. Several anti-Prohibition organizations formed to dispute the anti-alcohol voices of the Temperance Party, the Women's Christian Temperance Union, and the Anti-Saloon League. In various publications, the National Wholesale Liquor Dealers Association of America informed readers not just of the economic and social consequences of the nation becoming dry, but also the historical. They emphasized that George Washington and other founding fathers distilled liquor, thus suggesting that alcohol had always been an important part of national culture. In other publications, the United States Brewers' Association tried a different strategy, suggesting that national prohibition would lead to the rise of an underground trade and a rise in crime.[11]

In the Pacific Northwest, local breweries banded together against anti-booze organizations. Portland's Henry Weinhard's, Gambrinus, Mount Hood, and other Portland breweries, along with the Salem Brewery Association, the Roseburg Brewing Company, and the Eastern Oregon Brewing Company, joined their counterparts across Washington in strategic meetings. Their fundamental objective was to work with saloon owners to prevent Prohibition—namely, through media campaigns that warned the public of the potential ill effects of an alcohol ban.[12] In some cases, Pacific Northwest saloon interests touted their roles as societal benefactors, providing social gathering places and, at times, providing for the poor and homeless during harsh winter months. Of course, members of the Anti-Saloon League and the Women's Christian Temperance Union dismissed such claims.[13]

As the largest brewer in Oregon (which had by that time changed its name to the Henry Weinhard Brewery and bought out many competitors up and down the West Coast), Weinhard's took the lead in championing the same ideas as Pabst and other industrial brewing heavyweights in local newspapers.[14] The prohibition issue became the central concern for Paul Wesserman, Henry Weinhard's son-in-law, who had taken control of the brewery's operations after the patriarch died in 1904. He faced an uphill battle in swaying public opinion, but had enough financial resources to bombard newspapers with advertisements. Efforts to promote beer as "wholesome and invigorating" remained from the late nineteenth century. As the twentieth century progressed, Weinhard's ads became more political. The company took out

large advertisements in the *Oregonian* in the decade that preceded state prohibition. "If it is wrong for Carrie Nation to smash a man's property with a
hatchet," one such ad argued, "it is wrong to ruin it with a ballot box. The
principle is identical. Think about it." Other portions of that advertisement
warned of increased taxes and honed in on the loss of personal liberty. At one
point, it even asked, "Shall the vegetarian prohibit meat markets?"[15] As convincing as the range of anti-Prohibition arguments might have been, they
failed to sway the antibooze crowd. In these regards it is important to note
that as early as 1908, Weinhard's began to manufacture "near beer," branded
as "Malt Tea," in an attempt to buffer its business from the temperance movement. The brewery marketed the low-alcohol beer by noting that consumers
"will find it in taste as well as nourishment, very superior to any of the so-
called 'soft drinks' now on the market."[16]

Even after the successful prohibition vote in Oregon, hop growers made
final pleas to the lawmakers before the "full" prohibition law went into effect
in 1916.[17] A last-minute appeal by Salem's J. L. Clark outlined the situation
best as it affected state farmers financially. "Since the beginning of the hop
industry in Oregon," he noted, "more than $65,000,000 has been returned
to local growers. In 1914 alone more than $6,000,000 was added to the
income of the state from this industry and some 50,000 men and women
participated in the income through employment offered them in picking the
crop."[18] Clark saw the collapse of the industry resulting in the decline of two
million dollars of state income in the following years. The economic consequences reverberated from hop farmers and dealers to shipping companies
and seasonal laborers. At the center of it all was recognition of the industry's
history. As Clark noted, "Through hard and consistent work Oregon has
gradually climbed to the lead in the hop industry of the United States. This
supremacy will, however, no longer be possible unless present conditions are
changed."[19] Clark's words offer candid insight from one of the region's serious hop growers who understood the hop industry's unique history and
genuinely appreciated the efforts of his predecessors.

But the prohibition fight proved futile, and many hop raisers believed
their livelihoods lay elsewhere. In 1915, the political climate led to Oregon's
demise as the national leader in hop production—a standing it had held
throughout the previous decade. California, which had not enacted state-
level prohibition and had a large population to which brewers could sell their
beer, stepped into the void to become the largest hop-producing state. Prior
to full prohibition in Oregon by the beginning of 1916, the state's growers

kept within close striking distance of California's production, but fell drastically the next year.[20] At that point, the future simply did not look promising in Oregon and the larger Pacific Northwest. In the late autumn of 1915 an *Oregonian* writer noted, "When the bells ring in the new year on the early morning of January 1, they will sound the death knell of the brewing industry in this state."[21] While that was not entirely the case, the words foretold part of the future. With state prohibition already enacted and potential for world war escalating, the fight of those on behalf of the Pacific Coast hop and beer industries faded. Prohibition had arrived in the second decade of the twentieth century, and the self-proclaimed "Hop Center of the World" seemingly stood on shaky ground.

HOP GROWING IN WARTIME AND AFTER

Across the Willamette Valley, the onset of state prohibition inspired many farmers to pull out their hops and replant their acreage with other crops. In another casualty, Oregon Agricultural College ceased funding the "Hop Investigations" program that aimed to assist growers.[22] As war approached, hop growers up and down the Pacific Coast followed suit and looked for alternative crops. These included small growers and the larger grower-dealers of the era. Speaking for all, Emil Clemens Horst summed up the situation: "Owing to war conditions, the acreage formerly devoted to hops has been greatly restricted and these lands are now used for the growing of vegetables." As it turned out, hop growers were prepared to make this conversion better than other farmers because hop dryers required very little labor to convert into vegetable dryers, a new value-added commodity industry in the country.[23] Horst captured the meaning of this boon both for the benefit of the region's agricultural economy and for war efforts when he suggested, "Dried vegetables, although not new to European countries, were practically unknown to this country until the advent of the war. There are now ten vegetable drying plants in California and Oregon, and more are under construction." Horst went on to note, "A great impetus was given to the industry when the War Department contracted $1,000,000 worth of evaporated vegetables for the use of the solider abroad."[24] Beyond the economic reasons, the changes in hop fields could be seen as patriotic. Along with the thousands of soldiers who faced military service during the war, the West Coast's farmers saw themselves engaging in what they saw as a patriotic duty to produce food.

Not all farmers abandoned the wolf of the willow. Some saw hope in the fact that a domestic market for hops continued to exist in states including California, New York, and Illinois. Additionally, even in dry states, breweries continued to produce hopped nonalcoholic "near beer" as well as "nutritive tonics," which were essentially regular beers that could be purchased with a doctor's prescription for a variety of ailments. While they provided a small market for hops, none of these alternatives could ever sustain Pacific Coast hop growing at the levels that it had achieved in previous decades. As a result, hop farming across the United States dropped in half, from approximately forty-four thousand acres on the eve of American participation in the Great War in 1916 to twenty-two thousand acres by 1919, a year after the war concluded.[25]

Yet the hop farmers who stayed the course were wise to do so. At the same time that prohibition laws took hold in Oregon and other states across the nation, the European war disrupted European hop production. The new technologies of war devastated productive agricultural lands where hops were grown in Bavaria, Bohemia, England, France, and Belgium. In Germany alone upon the onset of war, hop production declined from nearly seventy thousand productive acres to just over twenty thousand by war's end.[26] Even with global beer consumption down during those years, the reduction in European hop lands opened opportunities for American hop farmers. As if that opportunity were not enough, beer makers across the world shunned German hop imports. Willamette Valley hop growers, in particular, grasped the opportunity during the war to expand further, not only into European markets, but also to those in Latin America, Africa, and Asia that once drew their hops from Germany.[27]

And that was not all. After the armistice in 1918, European countries turned inward to address social, economic, and health costs after four years of trench warfare. As nations rebuilt, they relied on American natural-resource commodities. Oregon forests provided timber to rebuild war-torn areas in Belgium, France, and England, and the productive agricultural lands across the state contributed substantial food supplies.[28] Willamette Valley farmers also provided increasing numbers of hop bales during the peacetime era when beer consumption again rose in Europe, where farmers remained scarred and unable to produce enough hops. "The English production is scarcely ever sufficient for its needs," one writer explained, "so that Great Britain must import some and mostly takes Oregon hops, because they are especially adapted to the English ale brewer's requirements."[29] Another suggested, "The English demand for Oregon hops of the coming crop continues, and agents for London firms are offering 30 cents on contract. Only a small part of the Oregon and Washington crop remains unsold."[30] Other nations took the English lead and began purchasing Pacific Northwest hops to rebuild their brewing operations. All of these conditions allowed the Willamette Valley to reclaim its lost hop acreage, and even show an increase after the war and throughout the 1920s.[31]

In this story, the expansion of hop growing in the Willamette Valley offered the greatest surprise of the Prohibition era. Even upon the onset of the Eighteenth Amendment, the international marketplace turned toward the region to meet its brewing needs. Abstemious critics across the country chided growers for producing a beer-related crop at a time when Prohibition prevailed. But their voices fell on deaf ears, as they had in decades prior. In 1923, Oregon reemerged as the leading national hop producer, ahead of California. Throughout the decade, Willamette Valley farmers expanded hop acreage nearly on a yearly basis, with total production exceeding prewar levels. Reversing the trend from the previous decade, the state's land harvested in hops moved from eight thousand acres in 1919 to twelve thousand acres five years later, and to seventeen thousand acres by 1929.[32]

The unexpected fortune of hop growing in the 1920s allowed Emil Clemens Horst, James Seavey, Arch Sloper, and other grower-dealers to again lead the way in continued efforts to better market their crops, integrate new technologies, and solve the labor situation, just as they had when they rose to the top of the hop-growing world. Aside from the mechanical harvester and other technological developments that improved the trellis system, growers also continued the long tradition of material and knowledge exchanges from across the world. Of particular note by that point in time, nearly all Pacific Northwest hop farmers had transitioned to the wire trellis system. This inspired a new global commodity exchange as growers trained their climbing hops onto coir (a coconut string) imported from India (and, later, Sri Lanka).[33] Though a seemingly minute detail, it represented the commitment to new technologies, and the importation of new commodities underscored the confidence that hop growers had in a secure future and the continued relationship with other natural-resource industries worldwide.

The good fortune of the Willamette Valley's Prohibition-era hop growers continued. Mother Nature also aided Oregon's reemergence as a major hop producer. In 1926, in the midst of the Willamette Valley hop revival, a devastating disease called downy mildew swept across Europe's traditional hop-raising areas. The pathogenic mold attacked hop roots and leaves, leaving plants stunted and their usually brilliant green leaves yellow-spotted and wilted. *Pseudoperonospora humuli,* the causal organism of downy mildew in hops, could destroy productive plants in a matter of days, leaving a year's crop wasted and the potential for future years in peril.[34] This is exactly what happened that year in Germany's Hallertau district, still one of the most productive hop-growing regions of the world. In a word, as one German historian notes of 1926, "The year . . . went down in hop history as 'the year of downy mildew.'"[35] It was a situation that somewhat resembled the unstable environmental conditions of the global hop shortage in 1882 that first put Pacific Coast hops on the map.

For all of these reasons, from the war-torn landscapes of Europe to downy mildew outbreaks, the Oregon and Pacific Coast hop industries *thrived* during the Prohibition years. The development was particularly remarkable because hop raising gained value in a decade in which American agriculture as a whole faltered, portending ominous depression several years prior to the stock-market crash of 1929. Whereas debt brought on by investment in new industrial farming technologies such as tractors and plows, and low prices brought by lack of marketing agreements, plagued the nation's grain belt, specialty crops such as hops fared well.[36] Furthermore, although the European agricultural sector had recuperated much quicker than industry, the blight on the hop fields continued to offer possibilities for Willamette Valley farmers.[37] This is all to say that hop farming on the Pacific Coast stood outside of the major agricultural crisis of the 1920s.

At the end of the decade, American economist Herman Feldman analyzed the trends of the nation's hop growers. Looking closely at the numbers, he determined that at that time American hop production stood at levels similar to those of the prewar and pre-Prohibition periods. Although the foreign market was still volatile, Feldman's figures proved accurate. Despite a drop in production during the 1910s, the Pacific Coast hop industry fared significantly better than once feared. His evidence did not come from hard numbers alone. It also came from authoritative voices in the industry. In his final note of a published study, Feldman relayed an exchange with the region's leading grower-dealer. "E. Clemens Horst, of California, one of the world's greatest hop growers," he explained, "writes us that the hop growers have been prosperous during prohibition." Feldman concluded: "You may dry your tears for the 'poor' hop grower."[38]

If the hop industry fared surprisingly well during national Prohibition, the story unfolded differently for other Americans associated with the alcohol industry. Not only did beverage makers have to pour out millions of gallons of beer, wine, and spirits and lay off thousands of workers when the country went dry, but the nation also lost thousands of jobs in related industries.[39] Bottle and barrel makers, delivery drivers, and saloon and restaurant workers suffered. Furthermore, state and national government lost substantial income. In some cases, alcohol-related taxes made up upward of 75 percent of state revenue, leaving a crippling impact on governments during the Prohibition era. Nationally, the federal government lost out on billions of dollars in tax revenue from the onset of Prohibition in 1920 to repeal in 1933. If that was not enough, the costs for enforcement locally and nationally exceeded millions of dollars as well.[40]

Oregon obviously felt these economic consequences earlier than many other states because it had already gone dry. The Gambrinus Brewery of Portland, founded in 1875, provides a telling example. Just prior to the successful prohibition ballot initiative of 1914, the brewery planned a major expansion of facilities in Portland and Oregon City.[41] But two years later, the owners saw no future and closed down altogether. Gambrinus was far from alone, as almost every brewery in the Pacific Northwest shut down upon the onset of Prohibition. In Portland, only the Henry Weinhard Brewery and the Portland Brewing Company remained standing, joining the Salem Brewery Association (by that time operated by the Olympia Brewing Company of Tumwater, Washington) as the only breweries left open in Oregon after 1916. In addition, four hundred saloons shut down in Portland alone, along with hundreds more across the state.[42] Washington faced a similar situation, with the Seattle Brewing and Malting Company (at the time the sixth-largest beer producer in the nation) and the Olympia Brewing Company remaining as the last standing. Hundreds of saloons shut down across the Evergreen State, too.[43]

To survive Prohibition, the Pacific Northwest breweries all drew on strategies developed by the large national companies. Success in part entailed the production of near beer and nutritive tonics. Weinhard's marketed their "Columbia Brand" of low-alcohol beer as "clear and sparkling" and "the choice of men who know the joy of a carefully brewed near beer." Later, the brewery similarly sold Weinhard's Club Special and marketed it as "the Brew with the Genuine old Flavor!"[44] (See figure 10.) Like other near beers, these sold well, but far less than the company's traditional beers. The proof was in the business plan. Weinhard's sold its entire malt house in 1928 because it simply did not need it anymore.[45] Similarly, in Washington, the Seattle Brewing and Malting Company abandoned one of its brewing facilities and turned it into a feedlot. For Weinhard's, along with the other breweries in the country that remained standing, the real money in the period came from diversifying and transforming their facilities into producers of soda, juice, cheese, ice cream, baking yeast, and chocolates. Weinhard's sodas and juices sold under the names "R-Porter," "Toko," "Luxo,"and "Appo," much like Olympia's "Applju" and similarly named products from surviving breweries.[46] Established distribution networks and marketing capabilities were what really kept any brewery open during Prohibition. Big regional and national brewers had name recognition and the ability to repurpose their beer marketing.

Outside of the brewing industry itself, Oregonians experienced, with the rest of the nation, a period of immense social change during the Prohibition era. While the majority of Americans obeyed the law, the Eighteenth Amendment created a willingness to violate the law on the part of millions. Americans did not even have to frequent a speakeasy to find their drink of choice. Millions turned to home brewing, what some called the national pastime of the era, or to distilling the harder stuff. Ironically, the underground beer and liquor makers added juice or soda produced by the major breweries to make the swill palatable. If finding a speakeasy or making booze at home was not available, there were also legal options. Wine could still be acquired for religious purposes, and, besides nutritive tonics, doctors could also prescribe whiskey and other liquors. It goes without saying that in many regions of the country both church membership and the pharmacy profession grew during the dry years. Like others throughout the country, Pacific Northwest residents engaged in all of these activities.

On a national scale, Prohibition famously marked a time when organized crime seized the opportunity to engage in the exact activity the Eighteenth Amendment prohibited: the manufacture, sale, and transport of intoxicating liquors. A thriving economy emerged to traffic alcohol across national borders and to erect underground breweries and distilleries. Canadian and Mexican border towns drew upon the opportunity to generate fortunes. Monterey, in the state of Nuevo León, Mexico, for example, thrived during the period and reinvented itself as a mecca of brewing, using that success to sustain a vibrant

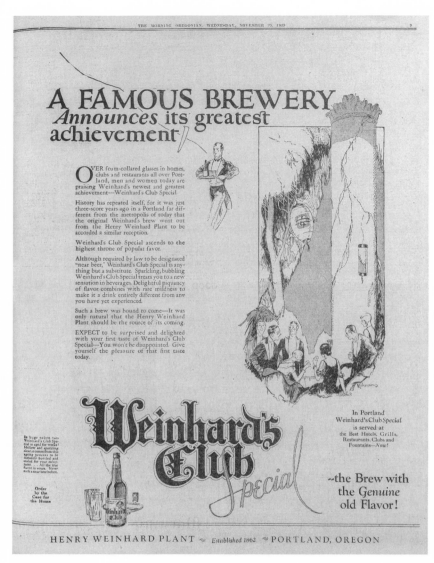

A FAMOUS BREWERY
Announces its greatest
achievement

OVER foam-collared glasses in homes, clubs and restaurants all over Portland, men and women today are praising Weinhard's newest and greatest achievement---Weinhard's Club Special.

History has repeated itself, for it was just three-score years ago in a Portland far different from the metropolis of today that the original Weinhard's brew went out from the Henry Weinhard Plant to be accorded a similar reception.

Weinhard's Club Special ascends to the highest throne of popular favor.

Although required by law to be designated "near beer," Weinhard's Club Special is anything but a substitute. Sparkling, bubbling Weinhard's Club Special treats you to a new sensation in beverages. Delightful piquancy of flavor combines with rare mildness to make it a drink entirely different from any you have yet experienced.

Such a brew was bound to come---It was only natural that the Henry Weinhard Plant should be the source of its coming.

EXPECT to be surprised and delighted with your first taste of Weinhard's Club Special---You won't be disappointed. Give yourself the pleasure of that first taste today.

In huge oaken vats Weinhard's Club Special is aged for weeks! Mellow and sparkling clear it comes from this ageing process to be instantly bottled and sealed for your enjoyment. . . . All the true flavor is retained. Never such a near-beer before.

Order by the Case for the Home

Weinhard's Club Special

In Portland Weinhard's Club Special is served at the Best Hotels, Grills, Restaurants, Clubs and Fountains---Now!

--the Brew with the *Genuine* old Flavor!

HENRY WEINHARD PLANT ~ *Established 1862* ~ PORTLAND, OREGON

FIGURE 10. Advertisement for Weinhard's Club Special "near beer," published in the *Oregonian,* November 25, 1925.

brewing industry to the present day.[47] As one historian explains about Washington's illegal booze operations, "The hundreds of miles of the Canadian border were impossible to patrol without a huge standing army, and the waters of the Pacific Ocean and Puget Sound with their adjacent wooded beaches provided a foggy and private paradise for the rumrunners. Anyone

FIGURE 11. Portland Police and Women's Christian Temperance Union representative with 108 cases of liquor to be destroyed, circa 1920s. Courtesy *Oregonian* archives.

with a small boat could enter the business."[48] In short, Prohibition in the United States opened economic opportunity for neighboring nations and members of organized crime, who helped with distribution and sales.

As was the case across the country, law enforcement in the Pacific Northwest was simply inadequate to monitor and enforce dry laws. The Prohibition Bureau had limited power of enforcement. At any given time there were never more than three thousand agents employed across the country to bust bootleggers, rumrunners, and speakeasy operators. (See figure 11.) Furthermore, many agents and other officers of the law did not take their work seriously; some even participated in the clandestine economy. Historians go so far as to speculate that most Prohibition agents supplemented their income with bribes. But perhaps one Prohibition Bureau official spoke for many in the underfunded service when he noted, "We are doing as well as we can."[49]

In Oregon, Portland became a center for illegal alcohol manufacture and trade for the entire region. Underground breweries and distilleries arose almost as soon as Prohibition came into effect, as did networks to import and distribute foreign products. Both provided booze not just for the city's speakeasies, but also for a large swath of the Pacific Northwest. As was the case around the country, police raids made for newspaper headlines. But scholars guess that these efforts involved only a tenth of all alcohol consumed in the city. Some of Portland's city officials often turned their heads away from the underground activity, and received under-the-table rewards for doing so. Few of the distributors and even fewer of the enablers were ever caught and prosecuted.[50] All in all, the clandestine liquor economy and speakeasy culture became legacies of Prohibition.

WHEATLAND AND THE WHITENING OF HARVEST LABOR

One of the final surprises of the Prohibition era in the Far West hop industry took place amid the late-summer harvest. In a story driven by the same tensions over immigration and American identity that led to national Prohibition, the character of the seasonal gathering changed. At the same time that Oregonians voted for prohibition when World War I erupted in Europe, western agricultural labor began a period of racial and ethnic change. On the national level, many white Americans developed increased leeriness toward the millions of immigrants who had arrived in the previous decades. Although many were themselves children or grandchildren of immigrants, voters believed that the newest arrivals stole jobs and failed to Americanize. The "Narrow Nationalism" after the war (from 1919 to 1924) witnessed the rebirth of the Ku Klux Klan and a strong fear of socialists and other perceived anti-American groups. This played out particularly strong in Oregon. The once progressive state that had enacted some of the first direct democracy laws—including the Oregon System of initiative, referendum, and recall—turned swiftly conservative.[51] In the hop fields, associated racial and ethnic tension inspired growers to avoid hiring non-white labor.

A prewar riot in the hop fields of Wheatland, California, played a more localized role in the sociocultural transformations of labor in the Pacific Coast hopyards. On August 3, 1913, the seasonal harvest had begun at Durst Brothers Hop Ranch, one of California's largest hop ranches at the time. But all was not well. A crowd of twenty-five hundred workers became agitated with working and camping conditions that included diseased water sources, lack of toilets, and extensive waste surrounding the living areas. Seizing the opportunity to organize for better standards in agricultural labor, a representative of the Industrial Workers of the World (I.W.W.) arrived to protest. In the searing heat of the Central Valley summer, he incited the crowds with pro-labor speeches and established a list of demands for the ranch owners. Needless to say, the Durst family was not pleased and, as a result, called upon local law enforcement. Upon their arrival, a riot broke out in which two workers, one officer, and the ranch's lawyer were left dead while many others were wounded. The following day California governor Hiram Johnson ordered two hundred national guardsmen to the ranch, though the damage was already done.[52]

Although murder trials and investigations into the role of the I.W.W. immediately grabbed the public's attention, the ramifications for capital and labor relations in western agriculture had a more lasting impact. The most obvious result was the 1914 congressional investigation that sought to understand seasonal workers and working conditions by questioning industry leaders including the Durst brothers and their neighbor Emil Clemens Horst. California created a Board of Labor to survey and enforce fair labor practices that had been lacking in Wheatland and many other hopyards. The decision reverberated across the Far West. By 1924, Oregon created a similar institution, the Oregon Seasonal Employment Commission, both to professionally recruit workers and to regulate work sites.[53] The process had broader implications, as growers now feared that hiring immigrant and nonwhite labor would draw unwanted attention to the conditions of their hop harvest.

Willamette Valley hop growers responded to the Wheatland riot by seeking more white workers. Demographic shifts aided these efforts. By 1910, Portland's population had grown to over two hundred thousand, and the surrounding towns of Oregon City and Salem also grew.[54] More peoples of northern European descent lived in the area generally, including a larger white, Protestant middle class. Hop growers seeking to minimalize the hiring of diverse labor pools turned to these new populations for their harvest. In the process, as had their predecessors in England, they realized the need to improve the overall seasonal experience. Responding to Wheatland and general Progressive Era concerns over moralism and healthfulness, growers wanted to make camps more family friendly and festive. They believed that this would be best achieved if the camps were less diverse. The result was that, as one historian has suggested, "[h]op picking, in fact, increasingly became a white, even middle-class undertaking."[55]

James Seavey's fields offer an apt portrait of changes to the seasonal hop labor camps in the Willamette Valley. In 1915, a Corvallis newspaper described one of his hop ranches outside of the town, noting, "Each year just before picking begins, water from the various wells in the camp grounds [is] tested and has always been found pure and free from typhoid or any other disease germs. Garbage barrels are placed at convenient points throughout the camp, and these are emptied daily."[56] It continued by noting, "[C]ertain rules are enforced and every care taken to keep the camp clean and sanitary. A certain camp discipline is maintained."[57] The article showed how growers enforced discipline in the workplace. It also revealed Progressive Era concerns about health and the problems associated with the Wheatland riot. As a result, the author concluded that many of the same hop pickers became loyal to the Seavey ranch, returning year after year.

Growers like Seavey understood what was at stake. So they worked hard to advertise their quality working and living conditions, hoping to turn the harvest into a wholesome family affair. A 1918 advertisement, for example, read:

> Picking will commence Sept. 4 at J. W. SEAVEY'S, one mile south of Corvallis. Good camp grounds, plenty of good water, sawed wood, one table for each camp, baskets furnished to pickers FREE. Grocery store and a meat market on the place. Will meet trains and move pickers out and in after picking is over, free. Auto truck morning and evening to accommodate those not able to camp on ground. 180 acres on high trellis and clean yards. Please register early.[58]

Seavey's yards were some of the most popular, and employed over five hundred workers alone in Corvallis, and hundreds more in Eugene and Forest Grove. His camps' reputations were useful in recruiting workers.[59]

Still, as was the common theme prior to mechanization of harvests, even the most attentive growers failed to meet their seasonal labor needs. Particularly amid Prohibition, many of Oregon's Protestant families were averse to contributing their labor to vice. With increasing acreage each year in the 1920s, growers had to up their ante. Among others, Emil Clemens Horst worked with a social organizer from Portland to determine an effective approach to meeting family needs. One historian suggests that the outcome was a program that included:

> first, housekeeping, tent pitching, sanitation, water, and fuel supply; second, primarily providing first aid and preventative medicine but also issuing a camp newspaper and supervising concessions; and third, a recreation program that included the installation of playgrounds and nurseries and assumed

the responsibility for nightly camp fire meetings, dances, religious services, and special programs on Sunday afternoons.

He continued:

> Not only were sandboxes, swings, and slides built under huge pine trees, but a playground worker and a nurse supervised over one hundred children in daily activities. Other large ranches throughout the region adopted many of these activities in one form or another.[60]

Along with these attractions, growers emphasized the benefits of fresh air and exercise, and the retreat from industrial life. Some might even argue that the ordering of public spaces appealed to white middle-class social values of the period.

Seavey and Horst were not the only two growers who were affected by the turn away from diversity in the hopyards and developed new programs to appeal to potential white pickers. Given the importance of the crop to their community, residents of Independence, Oregon, came together as a whole with new ideas to appeal to white families by making them feel comfortable during the harvest. In 1922, hop growers and shop owners pooled their resources to provide an entertainment program. Later in the decade these efforts reached greater levels with carnivals and parades; some hop farms employed recreation directors and nurses to appeal to the urban middle class. *Recreation* magazine reported in 1923 that various hop growers had installed "simple playgrounds with day nurseries attached" and conjoined efforts with the Independence post office and medical services. The changes worked. Although the number of laborers was never completely secure, growers successfully attracted white families while also passing various inspections from state and federal labor agencies.[61] This served both farmers for their labor needs, and local businesses because many of the pickers spent a great deal of their earnings in town.[62]

Throughout the 1920s, the Willamette Valley hopyards took on new meanings. By no means was nonwhite labor excluded, but the increased presence of urban white families harked back to London and Kent in the early 1800s. The hop-picking time became an occasion to cut ties with the city. Conversely, the hop harvest connected the city once again to the rural countryside. In an era before the proliferation of summer camps, children had opportunities for new friendships and entertainment. (See figure 12.) Adults, both in families and single, had similar opportunities. For all, the hopyards

FIGURE 12. Children picking hops with hop basket, circa 1915. Notice the gloves, sweaters, and hats for protection against the sun and potential hop rash. Courtesy Special Collections and Archives Research Center, Oregon State University.

presented not just an economic opportunity, but also a chance to spend time immersed in rural life surrounded by health, vitality, and community. Thus, in the midst of Prohibition and the perceived threat of alcohol to American families, the hop harvest offered some residents of the Willamette Valley and larger Pacific Coast outdoor activities in a secure, healthful environment.

The final surprise of the Prohibition era proved ominous. Despite the resurgence of hop agriculture after World War I and throughout the 1920s, the most feared disaster arrived in 1929—and it was not the stock-market crash. Instead, in that year the dreaded plant disease downy mildew appeared in Pacific Northwest hopyards for the first time. The disease occurred in Washington first. A year later, growers found it in the Willamette Valley, where the disease that rots hop rhizomes and leaves thrived in the wet climate. A scientist from Oregon Agricultural College reported of downy mildew that "[a] part or all of the shoots arising from an infected hill may be diseased. Badly infected shoots are unable to climb and are severely checked in growth." He continued to note that farmers could train unaffected shoots from the crown of the rootstock, but "[t]his practice may delay the ripening and subject cones to infection before they can be harvested. Infections on the vines kill many of the buds from which the side arms develop, and the attached leaves usually become affected also."[63] In the beginning, downy mildew did not spread rapidly because even if a hopyard was affected, the disease would not necessarily destroy the entire crop. This changed over time.[64]

But Willamette Valley hop farmers pressed on. Immediately, growers worked with one another and with USDA scientists to find solutions. Farmers employed a variety of organic topical treatments in the spring for prevention. Without a known cure, the key rested in spotting damaged plants and promptly burning them before the disease could reach others. As *Pseudoperonospora humuli* became the blight of the valley, hop growers often avoided even talking about it. Yet, the timing proved complicated. Although the disease's arrival was unsettling to hop farmers, the new decade, even with the onset of the Great Depression, brought hope that the "noble experiment" of Prohibition would end. There remained tremendous hope for the future of the wolf of the willow in the valley of the Willamette.

SEVEN

Fiesta and Famine

IN THE ELECTION YEAR OF 1932, "Happy Days Are Here Again," a song composed by Milton Ager and Jack Yellen, blared from radios across the country as Franklin D. Roosevelt's campaign anthem *and* the anthem of the repeal cause. In the midst of the Great Depression, the song optimistically declared:

> So long sad times, go 'long bad times,
> We are rid of you at last.
> Howdy, gay times. Cloudy gray times,
> You are now a thing of the past.

> Happy days are here again
> The skies above are clear again
> Let us sing a song of cheer again
> Happy days are here again.

The song reflected faith in Roosevelt's political vision and the possible renewed freedom to drink booze legally. It inspired hope for consumers across the country who had been criminalized during Prohibition and those who produced beer, wine, and spirits. Growers of grain, grapes, and hops also looked forward to the happy days of repeal. On the Pacific Coast, the pages of a newly founded agricultural hop journal, the *Oregon Hop Grower* (later to become the *Pacific Hop Grower*), expressed support for Roosevelt's election and the renewed domestic beer market.[1]

As the year progressed, the country turned its eyes to repeal developments at both the state and national levels. Come election month, Oregon even received attention from across the country. A November issue of the *New York Times* noted, "Early reports indicated that Oregon's prohibition law might be repealed." The author also suggested that "[t]he vote in Oregon was the heaviest on record."[2] Clearly, the repeal issue struck a chord with the state's populace. Although these events in the Far West were not necessarily center stage in most of the nation's mind, they signaled a sign of things to come. More important, that autumn Oregonians joined voters across the country who sought change, casting their five electoral votes for the Democratic candidate for president, who won in a landslide.[3]

Roosevelt followed through with his campaign promise early on. Though industrial and agricultural reform lay at the center of his New Deal agenda, he understood the need to once again legalize the production and sale of alcohol. In one of his first actions after he assumed the presidency in March 1933, Roosevelt helped push through the Cullen-Harrison Act to legalize beer produced at 3.2 percent alcohol. The limits in alcohol content reflected a measured approach to reintroducing beer to a country in which many still supported temperance. But the decision had an immediate impact that spring. Most notably, on April 6, the day leading up to the legal selling of 3.2 percent beer, spontaneous public celebrations erupted for what the media dubbed "New Beer's Eve." Representatives from St. Louis's Anheuser-Busch brewery took to the nation's capital, ceremoniously trucking in fresh batches of lager to the doors of the White House. Local breweries from coast to coast engaged in similar celebrations.[4]

The return of domestic brewing held a deeper meaning than the ability to legally drink beer; it also translated into more jobs and revenue created in the associated industries.[5] In May 1933 the national media reported, "The legal beer faucet has now been turned on for more than a month, and the gurgle of its noisy stream has spelled good news to industry, capital and labor." The article mentioned that nearly half a billion dollars of taxes would be collected within a year from the beer industry, and it estimated that, in New York state alone, the Cullen-Harrison Act opened the door for forty thousand jobs, including "19,000 directly in the breweries, 221,000 in cooperage, lithography, bottle making, [and] lumbering."[6] These numbers did not even reflect the immediate growth of the hotel and restaurants industries. It made sense, then, that as the year progressed, Roosevelt promoted broader repeal of Prohibition, citing economic stimulus to create new jobs and alcohol taxes. By December 5, Congress fell in line with the president and ratified the Twenty-First Amendment to repeal the Eighteenth Amendment in its entirety. In an even greater celebration than New Beer's Eve, the cheerful nation opened countless bottles of booze not limited to 3.2 percent alcohol. The American public could turn now to their beverage of choice in depressed times.[7]

In hindsight, many agreed that the Prohibition experiment was misguided. Perhaps the most famous of these voices was that of Eleanor Roosevelt, who tackled the issue in one of her popular My Day columns. While she noted that she was "one of those who was very happy when the original prohibition amendment passed," her mind had changed. "Little by little," she wrote, "it dawned upon me that this law was not making people drink any less, but it was making hypocrites and law breakers of a great number of people."[8] Her column concluded that Americans should be free to make their own choices regarding alcohol consumption and it was not up to national law to interfere. Truer words could not have been spoken for the members of the beer and hop industries who had fought Prohibition from the beginning.

Undermining the enthusiasm for legalized booze was a harsh reality seldom addressed in the immediate years following repeal. The Eighteenth Amendment (and earlier state laws) had transformed the inner workings of the American brewing industry, destroying a vibrant national brewing scene of thousands and putting strict regulations on those who remained. Additionally, the Prohibition years had severed the American public's relationship to flavorful, hand-crafted beers, setting in motion a long period in the nation's history when bland big beer reigned. This proved a critical development for Pacific Coast hop farmers, as the future of American beer lay in the hands of corporations who produced a beverage devoid of robust hoppy notes.[9] Still, growers looked eagerly to the prospects of a revived domestic beer industry.

THE POST-PROHIBITION ERA

The rebuilding and restructuring of the American beer industry in the 1930s and 1940s set the pace for American beer consumption for the rest of the twentieth century. Breweries that survived Prohibition had done so with sophisticated corporate strategies and deep enough pockets to alter and repurpose their facilities. As noted previously, major national and regional breweries stayed in business during the 1920s by manufacturing near beer, nutritive tonics, soda, juice, cheese, and other products, and they achieved success by selling these products within preexisting distribution networks. While the ad hoc business kept many companies viable, not all breweries could adapt. Of over one thousand domestic breweries in operation prior to national Prohibition, fewer than two hundred remained in 1933.[10]

Of those that stayed in business, many did so by partnering with other businesses—not just breweries, but also malting companies and marketing firms—in an effort to reduce competition and share marketing strategies.[11] In Oregon, this trend was most recognizable when the Henry Weinhard Brewery took on Arnold Blitz's Portland Brewing Company (the only other remaining brewery in Portland after the onset of state prohibition) as a corporate partner, continuing after 1928 as the Blitz-Weinhard Company. Headquarters remained at the original City Brewing facility on Burnside Street in Northwest Portland, and the company continued to sell Weinhard's famous lager. But there were changes. For the first time, the Weinhard family relinquished the presidency of the company to an outsider. Blitz, Portland Brewing's longtime leader, assumed the presidency over the younger and more inexperienced heir apparent, Henry Wessinger. The brewery also began selling its popular Blitz lager and bock beers.[12] Elsewhere in the Pacific Northwest, takeovers had already occurred, as the Salem Brewery Association had become an arm of Washington's Olympia Brewing Company, in Tumwater, by 1904. The Seattle Brewing and Malting Company had also become one of the larger beer makers in the nation, having purchased many smaller breweries along the West Coast.[13] Despite the reduced competitive field, the revival of all of these breweries in the 1930s not only meant economic opportunity for many in the region, but also transformed urban life with the return of malty aromas that once again wafted through the brewery blocks.[14] This was also the case across the country as the number of breweries surpassed seven hundred within the first year of repeal.[15]

Given the expectations and fanfare of repeal, the Blitz-Weinhard Brewery began the post-Prohibition era on somewhat shaky ground. Second-in-command Wessinger admitted that "[b]rewers were totally unprepared for the resumption of normal demand.... About all we can promise is samples—and those are for our old customers." As a result of the lack of beer, several events were cancelled, including the RKO Theater's midnight show on "New Beer's Eve." Customers also cancelled restaurant reservations in that first week of legalized beer, given the lack of the product on hand. As an *Oregonian* writer suggested, "The situation may easily be comprehended when it is realized that on the entire Pacific coast there are only as many breweries left as were formerly situated and operating in almost any one of the major coast cities."[16] Luckily for the brewers and the drinking public, the beer started to flow faster in the next several months and joined national trends with a substantial upsurge in production. Though it would take a decade to regain the levels of production that existed prior to Prohibition, it was clear that beer was back to stay, with general applause from the American public.

Rebuilding the American beer industry was not an easy task with temperance and prohibition forces persisting across the country. Following the Cullen-Harrison Act, those special interests kept a watchful eye on the mandate that limited breweries to selling beer with only 3.2 percent alcohol. After full repeal in December 1933, state regulatory commissions, including the Oregon Liquor Control Commission, emerged to play the watchdog role. This provided a buffer between brewers and the still-potent dry forces.[17] But the new oversight and limitations challenged the brewers' ability to sell their products, with the greatest challenge being the emergence of a three-tiered distribution system. In that system, alcohol producers could not sell their products directly to retails outlets, thus requiring them to use middlemen or distributors.[18] Preventing brewers from marketing their products as they had in the past presented challenges. A desire to continue selling light beer after full repeal also indicated that brewers wanted to tread lightly. As one scholar suggests, "In a delicate position after their 1933 rebirth, the brewing industry had to gauge the best way to appeal to the broadest market while not alienating key segments. Any misstep and they might provide prohibitionists with ammunition."[19] Together, the new regulations and the inability to manufacture stronger and more flavorful beer called for a creative advertising response.

Brewers immediately gathered their creative talent. First, they successfully lobbied for new laws that allowed grocery stores to sell beer. Next, they utilized the new technology of aluminum cans (with the use of cheaper electricity), which would make purchases more efficient, eliminating any obligation to return glass bottles. In 1935, grocers in Virginia first stocked cans of Kruger's Cream Ale, and Miller, Pabst, Schlitz, and Anheuser-Busch followed in the next decade with similar packaging in other states around the country.[20] Finally, beer companies changed advertising strategies. While they continued to draw upon images of cheerful beer gardens and dance halls, advertisers moved product placement into the household. Advertisers paid keen attention to female consumers, believing that women played a vital role in purchasing goods for the household. A flood of radio ads featuring women as beer buyers aided these efforts, attempting to take booze out of the social sphere and plant it in the domestic.[21] Outside of those trends, the Salem Brewery Association, for one, took advantage of its unique geography by marketing its beer as "From the Hop Center of the World." (See figure 13.)

FIGURE 13. Advertisement for the Salem Brewery Association's Salem Beer, published in the *Oregonian,* November 8, 1934. In the post-Prohibition period, the company marketed its beer as being brewed in the "Hop Center of the World."

The brewers' efforts to maintain, as one scholar notes, "their good public image and restore their industry"[22] proved successful within the first decade after repeal. By 1935, the United States overtook Germany and Great Britain as the world's largest beer maker by producing over ten million barrels annually.[23] The revival continued as domestic beer consumption jumped upward by the outbreak of World War II, in contrast to the situation during World War I. As a sign of the nation's cordial new relationship with the beverage during wartime, the Food Distribution Administration ordered American brewers to donate 15 percent of their products to the American military.[24] Officials believed that low-alcohol beers did not pose a threat to military efficiency, and, in fact, they championed beer as a morale booster. The decision had an unintended consequence. While large brewers had the capacity to adjust without problem, the requirement strained smaller breweries that already struggled in the changing world of amalgamated brewing and marketing structures. The mandate therefore accelerated the nation's movement away from smaller breweries and toward dominance by big beer. As far as Pacific Northwest hop growers were concerned, the developments were welcome. In 1944 an industry journal noted, "Hop growers are glad to know that their product is adding its bit to the supremely important task of winning the war."[25] Increased beer production, after all, meant more hop sales.

Yet a feeling of uncertainty lingered for Pacific Coast hop growers, who were not entirely clear about what the new world of beer meant for their livelihoods. They had worked diligently since the late nineteenth century to gain global market shares, and growers recognized that repeal created a renewed American marketplace for their products.[26] But in the post-Prohibition era, the mass-produced beers lacked the flavor of ales and lagers of yesteryear. Those light concoctions appeared to wean the public away from the hoppy tastes of more flavorful brews. Would this manifest itself in a decreased demand for Pacific Northwest hops? Although not an immediate concern for growers after repeal, it became an issue as the century progressed.

HOP GROWING AFTER REPEAL

In the short term, the reemergence of a domestic beer market benefited all of the Pacific Coast hop growers, but none more than those in Oregon. As it had from the mid-1920s onward, the state produced roughly 50 percent of the American hop crop and roughly 10 to 12 percent of the world crop.[27] After repeal, cultivated areas increased from approximately 15,500 acres in 1932 to 23,000 acres in 1934. By 1936, Oregonians harvested over 26,000 acres of hops, the most in the state's history.[28] At that time, the acreage devoted to the specialty crop was more than that of almost all other countries, including Germany. Only Czechoslovakia dedicated more land to hops than Oregon, and that was only by 1,000 acres.[29] To borrow a phrase from Ezra Meeker, the "palmy days of '49" seemed to have returned.[30] Still, many obstacles remained from previous eras.

In a familiar refrain from the late nineteenth century, industry experts of the 1930s feared that the rapid expansion of hop raising, particularly by inexperienced farmers, would glut the market and depress prices. More-established hop growers also worried, even more than they had in earlier periods, that low-quality hops entering the marketplace from novice farmers would undermine the reputation of quality that the Pacific Northwest had worked so hard to obtain.[31] Yet this era of growth was different from those of the 1880s or early 1900s. Having taken early steps toward modernization prior to Prohibition, Oregon hop growers were better prepared to compete in the global marketplace. First, the expanded presence of corporate hop dealerships, including transnational companies, signaled the promise of this new era. Joining the English-based Wigrich Ranch near Independence, the German companies Barth and Steiner also invested in the region's hop farms, showing that the Pacific Northwest had indeed become a global center of hop production. Along with those firms, E. Needham and his Needham-Taylor and Company emerged as an outspoken player in the Willamette Valley, joining James Seavey, Emil Clemens Horst, Arch Sloper, T. A. Livesley, and Krebs Brothers. In Washington State, John I. Haas arose as a force in the Yakima Valley, but also had farms in Oregon, California, New York, and British Columbia. All brought more stability and quality-control oversight along with the longtime grower-dealers.[32] Second, and more important in regarding to the new professionalization, Pacific Coast hop farmers joined together in grower organizations with success. Attempts dating as far back as the 1870s had seen little cohesion among hop farmers.[33] But given the economic uncertainty of the 1930s, regional hop farmers recognized an urgent need to collaborate for price controls, quality controls, marketing, and education for growers and consumers alike.

In 1932, Oregonians organized a lasting hop growers' association, with California and Washington farmers doing the same shortly thereafter. To reach their members, the three-state associations began publishing the *Pacific Hop Grower* (which replaced the *Oregon Hop Grower* after just one year in existence). Dean H. Walker, the first president of the Oregon association, captured the problems that the collective group hoped to address in one of the journal's first issues. He wrote, "The hop grower has suffered severely in the past from lack of organization. A strong individualist, the hop grower fought his own battles in his own way, when he saw that it was a losing fight."

And he continued, "A year ago the Oregon Hop Growers association was formed and what this association has accomplished in one year is enough to convince even the most sceptic [*sic*] as to the value of cooperation in fighting for growers' welfare."[34] His words resounded with growers, who had witnessed previous failed attempts at organization. Growers hoped that working together could solve long-standing concerns.

Concurrent with these developments, other longtime problems in the industry remained. As a period study suggested, even in the 1930s, after many American and European companies had turned to hops grown in Oregon and the Pacific Coast, "Continental brewers do not consider the American hops suitable for the type of beers which they brew, and therefore will use our product only under conditions of an acute shortage in Europe."[35] There were many interrelated issues that plagued growers in the previous century: the stronger taste profiles of American hops, the presence of seeded hops, and the persistence of stems and leaves in bales. Regarding the domestic market, a similar study noted that "the market for American hops has been unnecessarily limited because many American brew masters, being of German birth, believe that foreign hops are superior." This report also noted that because hops are a small percentage of finished beer, "breweries are willing to pay a higher original price . . . in order to obtain foreign hops."[36] The recently created American Hop Promotion Board aimed to educate "American brew masters concerning the merits of domestic hops" at this time, but it made little headway. Furthermore, even though scientific hop research continued at the Agricultural Experiment Station in Corvallis, there was little progress toward developing unique hybrid varieties for the region.[37]

Other early efforts of the new hop-grower associations included the continued fight for quality control. As Willamette Valley grower C. F. Noakes worried, the inclusion of stems and leaves in hop bales had persisted as a problem from the beginning of the industry in the region. In 1933, Noakes suggested, "Oregon hops received so bad a reputation from the terrible picking that was done during the few years after the war that some English brewers still refuse to have anything to do with our hops."[38] He then argued that the fifty-year-old quality-control checks were outdated and that no other agricultural industry would permit such shoddy production. He made a brash statement: "Look around you, at Hood River apples, known and bringing top price in every market of the world," he said. "Go to your local store and ask for Tillamook cheese. The merchant will ask you for four or five cents per pound more for this cheese than other good cheeses. Why? The answer is quality goods of high guaranteed uniform standards."[39] Despite advancements made by Horst and Seavey in selling Willamette Valley hops as a named commodity earlier in the century, the issue of quality brand recognition clearly remained.

Above all of these other issues, the new hop growers' associations sought solutions to the economic instability that plagued their industry. Soon after the formation of the state hop associations, plans began for price supports within a framework similar to the treatment by the New Deal's Agricultural Adjustment Act of other crops (particularly wheat) that aimed to lower production to raise or sustain commodity prices.[40] A 1934 article highlights the sentiments of the hop association leaders, noting, "Hop growers of the Pacific Coast are standing at the cross roads. . . . One road leads to production control, through control of hop root sales, which in turn will lead to years of continued prosperity to the hop grower by insuring him a price for hops that will give him a fair return on his investment." The writer continued, "The second road leads to unrestricted production, with the inevitable cycle— more hops, less money, and finally a bitter 'survival' period which will force hundreds of hop growers into bankruptcy and reduce acreage."[41] The implications of the article became clear to Pacific Coast growers. They needed to adjust the rate of production and better manage the speculative nature of the commodity's market to ensure good returns.[42]

Over the course of a few short years, West Coast growers championed a federally sponsored hop-marketing agreement. Whereas Pacific Coast hop grower associations failed to enact one in the previous decades because of a lack of unified organization, the better-organized hop raisers of the 1930s turned to Congress.[43] They used a provision in the New Deal's Agricultural Adjustment Act that sought to control prices based on reducing agricultural production. In 1937, grower-dealer Dean Walker emerged as the hop industry's spokesperson for the related congressional sessions. Although hops were a specialty crop and often overlooked in the New Deal programs compared with staple crops, the legislation was successful. Congress included the marketing agreement plan in the legislation and explained, "As hops under ordinary barn or warehouse storage deteriorate rather quickly in brewing qualities, sound marketing practices to prevent large surpluses become necessary." The language also suggested, "The hop industry is faced with a serious and discouraging outlook, but one that does lend itself to a sound marketing program."[44] As explained to the readers of the *Pacific Hop Grower* after the marketing agreement passed in Congress, an elected board "will have the

FIGURE 14. Fuggle hops infected with downy mildew, 1932. Courtesy Special Collections and Archives Research Center, Oregon State University.

power to determine [growers'] saleable tonnage, issue certificates for 'saleable tonnage,' establish minimum prices, act with the Bureau of Agricultural Economics . . . in establishing grades and standards, hold a surplus, and . . . give reports to dealers and growers."[45] The new industry board would finally achieve many of the goals that larger growers had unsuccessfully fought for in the previous several decades of failed organization.

In 1939, the *Pacific Hop Grower* explained the new marketing agreement so elusive during the previous fifty years: "The first order approved August 11, 1938, fixed 28,500,000 pounds as the quantity salable . . . without penalties . . . of 1938 hops, as against average annual production for the previous five years of some 40,000,000 pounds."[46] It was a victory for many, even if some feared that it could restrict their business in the long run. Though that marketing agreement lasted only until the outbreak of World War II, growers would again work with Congress in the postwar period on other versions. These agreements protected growers already invested in hops, while restricting potential growers from entering the business.

The one thing the new grower organizations could not find an answer to was the continued blight of downy mildew. Expanded domestic beer production and expanded hop farming in the Willamette Valley obscured the real-

FIGURE 15. Experimental spraying of a hopyard for red spider control and "other closely related problems," 1940. Courtesy Special Collections and Archives Research Center, Oregon State University.

ity that, per acre, the region produced fewer hops than ever before. There appeared to be no easy solution. In 1933, the *Benton County Herald* took on the topic, explaining what the presence of the disease meant to the region's hop growers and how even the best prevention methods of dusting the crops with various sprays of lime and copper sulfate were not entirely effective. "The dusting method," it informed its readers, "is for the prevention of the disease rather than its cure." In a final note, the article suggested ominously, "There is no known cure."[47] The article, which drew from Oregon Agricultural Experiment Station research, summed up best that growers had to be diligent about prevention and that if the spores did reach the plant, all parts of the plants should be pulled down and burned. It was a course of action no hop grower wanted to face, but one that was required as the years progressed. (See figures 14 and 15.)

Without a cure for downy mildew, the best bet for growers resided in replanting their hopyards with the more disease-resistant varieties released by E. S. Salmon of England's Wye College in the 1930s. The reasons were obvious. Early and late Cluster hop varieties grown predominantly in the Willamette Valley proved extremely susceptible to *Pseudoperonospora humuli*. And, although Fuggle hops were more disease resistant than Cluster, they did not have high yields, nor were they as appealing to brewers at the time. Thus, the Brewer's Gold and Bullion varieties, which also had the benefit of higher alpha acid (or bittering) levels, seemed to offer a natural solution to disease, as they would in Continental Europe's growing areas.[48] Emerging again in a leadership role, Emil Clemens Horst became one of the first Oregon growers to test the new varieties on his ranch near Independence. He expressed satisfaction at the experiment, putting a mark on the new direction in cultivation. It signaled to growers that planting new hop varieties would be the major strategy in confronting the devastating downy mildew plague in the Willamette Valley.[49] But it was by no means a foolproof plan.

Despite these remedies and the potential growth of new hop hybrid varieties, the threat of downy mildew remained. In 1934, Frank Needham of Salem reported that "downy mildew is the worst since that pest infested the valley," and in some yards, he added, "not 10 percent of the hills are coming."[50] While he did note that some of his non-Cluster varieties fared better, he also suggested that there might be mildew problems for the rest of the Willamette Valley. The pages of newspapers and agricultural journals throughout the decade repeated this news. With no cure and limited prevention methods, Oregon hop growers stood perplexed. Still, they pressed on. The sheer number of acres in production and millions of pounds harvested allowed the Willamette Valley to maintain an identity as the "Hop Center of the World" for the time being.[51]

THE HOP FIESTA

Overshadowing the doom and gloom of the downy mildew situation in the post-Prohibition era in the Willamette Valley was the ongoing story of seasonal harvest events. The expansion of hop growing, even in the face of a decline in production per acre, required more harvest workers than at any other moment in the region's history. Across the state that number reached approximately seventy thousand in peak years, or the rough equivalent of 7.5 percent of Oregon's population at the time. The Independence growers alone needed thirty thousand pairs of harvest hands. No other agricultural industry came even close to these transient-labor demands.[52]

By this time the regional harvest was more than a curiosity in the local and national media. While not appreciated by commentators in the same way as the Portland Rose Festival or the Pendleton Round-Up, the hop harvest rivaled, if not in some ways exceeded, these other community gatherings in

Oregon given the three-week duration and number of participants.[53] Many hop pickers held that it was the cultural event of the year. Others agreed. Popular writers took on the hop harvest as a subject matter in a variety of literary works. H. L. Davis's *Honey in the Horn,* published in 1936 and often called one of Oregon's greatest novels, used the harvest landscape as a backdrop for one of its important scenes. So, too, did Edwin B. Self's *Limbo City,* a mystery novel. The famous poet and printer Loyd Haberly outdid even these authors, setting his 1942 work *Almost a Minister* entirely upon a Willamette Valley hop farm; intricate and colorful block prints offered homage to the state's most important specialty crop. Finally, much as they did during England's heyday, the hop harvests in Oregon showed up across the pages of the *Oregonian* and the local newspapers of the Willamette Valley. This is all to say that the harvest season had taken on a revered atmosphere reinforced by contemporary literature and media—a welcomed reality given the growers' ongoing desire to promote the event as a healthy and convivial occasion.[54]

Beginning in 1934, growers, local merchants, and the Independence Chamber of Commerce combined efforts to promote the hop harvest as the cultural event of the year by introducing an official Hop Fiesta. This initially occurred because, even in the midst of the Great Depression, when many Americans sought work, the expanded post-Prohibition harvests continued to lack enough workers to handpick the crops. As the organizers noted in one of their programs, "The yearly hop-harvest celebration is held for the dual purpose of focusing attention on the hop industry of the state, and to provide a gay four-days of merrymaking for the thousands of pickers and visitors who crowd this community at this season of the year."[55] If there was one thing that was clear, it was that the event certainly held multiple meanings. Growers continued to depend on selling the harvest as a festive affair; shop owners wanted workers to spend money earned from hop picking in town; workers wanted to earn money and enjoy themselves in the process.[56]

In the last week of August 1934, the inaugural Hop Fiesta played out as planned by organizers. Thousands flocked to the mid–Willamette Valley on the banks of the namesake river in Independence. A program that included air shows, vaudeville, daredevils, and a parade showed that the three-day event to kick off the harvest season surpassed anything in the past. The coronation of young Marjorie Plant as the first Hop Queen offered the highlight of the fiesta, a celebration that would be replicated in the years ahead.[57] (See figure 16.) For outsiders, the staging of these events in

FIGURE 16. Cover of *The Hopper: The Hop Grower's Magazine,* October 1948. Beginning in 1934, growers in Independence, Oregon, joined the chamber of commerce in sponsoring its Hop Fiesta, where a Hop Queen was anointed annually. Courtesy Special Collections and Archives Research Center, Oregon State University.

a small town of barely a thousand people may have appeared out of place. But townspeople and the hop growers knew what was at stake. Successful harvests required pickers, and the Hop Fiesta attracted a work force and encouraged it to stay. Of central significance, growers used the opportunity to offer open houses to potential workers so they could peruse specific working and living conditions. All in all, the event demonstrated to what length growers were determined to go to attract enough hands for harvest.[58]

Throughout the next two decades, the Hop Fiesta became what the *Oregonian* called "one of the greatest celebrations ever staged . . . in the Willamette Valley."[59] Along with the highlight of crowning a yearly hop queen (and hop king beginning in 1941), Independence growers and businesses invested in a three-thousand-seat Hop Bowl to host various entertainments, ranging from live music to boxing matches. Programs offered something for everyone, including elaborate decorations across the town, children's dances, bathing-beauty contests, and themed events, such as the 1940 "'Over the Rainbow' Coronation Ceremonial."[60] The entertainment could also be far more intense. For the more adventurous, the program from 1939 featured an "[e]xhibition of log rolling and speed-boat racing on Willamette river at Independence ferry landing." It also advertised a "[t]hrilling and daring exhibition of motorcycle riding given by the Portland Police Motorcycle Corps, a crack group of riders, headlined by Bob (Suicide) Dillion who crashes a flaming wall."[61] In a nod to the health and safety improvements made since the 1910s, the Hop Fiesta programs also advertised a commitment to first aid and community policing. In all, the vast array of entertainments brought alive the vibrant spirit that hop growers had sought to cultivate across the world at harvest time for over one hundred years. And they did so in a way that no other sector of agricultural labor replicated before or since.[62]

The Hop Fiesta gamble paid off for Independence growers and the chamber of Commerce. The important reality behind the festivities was the fact that thousands of acres of hops had to be harvested, and it brought workers to the hopyards at the exact moment when they were needed.[63] Yet in an agricultural sector that required the largest number of workers per acre during the harvest, even the Hop Fiesta could not attract enough labor in some years. The other half of the harvest story following the post-Prohibition expansion of Willamette Valley hop acreage offers perhaps a more typically dystopian remembered story of West Coast agricultural labor during the Great Depression.

Despite the success of the Hop Fiesta and improvements implemented by growers and the government since the Wheatland riot of 1913, the harvest labor situation did not always unfold in an idyllic fashion. In 1935, one industry representative noted, "So acute is the [labor] shortage ... that several yards were unable to start picking today. Practically all of the yards are short handed."[64] As a result, those growers who had seized the opportunity to hire mostly middle-class, white family pickers in the previous two decades, and who aimed to continue that tradition via the Hop Fiesta, had to abandon their preference. To supplement harvest hands, growers turned to migrant workers—a group identified unceremoniously at the time as hoboes, bindle stiffs, and fruit tramps, who traveled up and down the Pacific Coast following the various seasonal harvests, from berries patches to hopyards to fruit and nut orchards. Although white families continued to contribute to the harvest, government reports indicated that across the Far West migrant workers of various racial and ethnic backgrounds constituted 50 percent of hop-harvest labor—including a number of American Indian, Latino, Asian, and African-American workers. Some growers feared that the demographic shift moved the image of the hop harvest away from the convivial paid vacation that growers had fought hard to promote in the previous decades.[65]

Progressive and later New Deal reformers also complicated the image that hop growers advertised for the harvest. Governmental labor and immigration boards authorized by legislatures after the Wheatland incident to check on camps and working conditions expanded into the 1920s and the 1930s.[66] After Prohibition's repeal, the boards struggled to keep up with the expanding number of camps. Many camps resembled the poverty of the so-called Hoovervilles composed of out-of-luck farmers, workers, and the unemployed. Transient agricultural labor was a particular challenge for state and local governments, and later the federal government. Sanitation was a key concern, but so too were the comforts and quality of life for individuals and families across the West. For the most part, Pacific Northwest farms were up to code, as growers got into the habit of providing clean camps and water as part of their overall approach to recruiting. But this was not always the case.[67]

New Deal reformers also targeted child labor abuses. Oregon hop growers worried as government programs, including those instigated by the National Recovery Act, investigated the presence of children at the hop harvest. Hop growers were relieved in 1933 when the *Oregon Hop Grower* reported to its readers that the "employment of children will not be effected [*sic*] under the program as farm labor is specifically exempted from its provisions." The growers interpreted the report as suggesting that "[t]he NRA 'child labor' provision was meant to keep children out of the 'sweat shops' and 'factories.' The average child will be benefitted by rather than injured by working in the open, fresh air, of the hop yards."[68] While memories captured in oral histories suggest that this was the case for many child laborers, there was also a large percentage who could have done without the yearly heat and monotony of the so-called paid vacation.

Migratory labor, however, presented the more acute problem. Workers often traveled from the Central Valley of California and through Oregon, Washington, and British Columbia following the agricultural harvest of seasonal crops. While the migrant workers themselves worried about living paycheck to paycheck in a depressed economy, social critics and labor-rights advocates voiced concerns about their way of life. Specifically, they were concerned with the inability of migrant workers to become stable members of the American society. Marion Hathway, a Chicago professor and colleague of Annie MacLean (who had visited the Oregon hopyards in the first decade of the century), captured these sentiments in a 1934 study. She feared that single workers were "deprived of the opportunities for personal development afforded by membership in a stable family group and by participation in community life" and that married workers were "making a less satisfactory adjustment." Hathway noted, "He [i.e., the single, itinerant worker] is not only deprived of participation in family and community life, but he is denied the constructive experience of sharing responsibility for other members of his immediate group."[69] She then summed up her study by suggesting that migrant laborer was simply a scourge on the progression of modern American society. In other mediums, Woody Guthrie's folk songs and the photographs of Dorothea Lange etched the plight of migrant workers into the American consciousness. Both artists, sponsored by New Deal funding, offered raw and often stark portrayals of the hop harvest and other conditions of manual labor that were in contrast to the images portrayed by the Hop Fiesta. Growers had little choice but to hire all available workers, and those depictions provided gritty, disturbing snapshots of seasonal life in the world of western agriculture harvests.[70]

In the worst cases, strikes and riots became part of the seasonal situation in the Willamette Valley. Labor historian Stuart Jamieson suggested that, compared with those in California and Washington, the disputes in Oregon were mild. But physical altercations with guns drawn did occur. Jamieson related the following:

> The only large strikes in Oregon's agriculture occurred during 1933 and 1934 in the hop industry of Polk, Benton, and Marion Counties. Seasonal laborers employed in this crop undoubtedly were influenced by the current wave of farm strikes led by the Cannery and Agricultural Workers Industrial Union in California. It is not unlikely that many of them had taken part in strikes in California earlier in the season. The union may even have sent organizers from California to follow itinerant agricultural laborers in their seasonal migration north to the Oregon hop fields.[71]

Still, Jamieson went on to note, "[n]o union of hop pickers developed among the strikers, . . . as their period of employment in the crop was too brief. The strikes were characteristically sudden and brief."[72] This was perhaps a detriment to pickers and growers alike as tensions erupted in the following two years, as well when growers reduced pay and insisted upon cleaner picking. Rather than resorting to violence in the 1930s, thousands of pickers simply organized to walk off of the McLaughlin and Horst ranches near Independence. After recognizing that the crop would remain unpicked if not for their help, the growers agreed to increase pay. Overall, as Jamieson suggested, "[s]trikes in the hop-growing areas of the Willamette Valley were illustrative of labor relations in a crop in which neither workers nor employers were strongly organized."[73] Nevertheless, the prospect of labor strife haunted the hop harvest in the hard times of the depression decade.[74]

In a series of folksy watercolors from 1933, the Pacific Northwest painter Ronald Debs Ginther (named in part after the labor leader and 1912 Socialist Party presidential candidate Eugene V. Debs) best captured the tension of the Independence strikes. The titles of the pieces themselves framed a differing opinion of the "paid vacation": *Hop Pickers Threatened by Drunken Foreman and His Pals at Beginning of Strike; Tense Moment (Guns Drawn) in Hop Yard Strike; Independence, Oregon. State Police Intervene in Hopyard Strike.* The series was among the most intense of Ginther's paintings of the lives of migrants and homeless eking out their livings in the Great Depression. Ginther's other paintings also portrayed the harvest as an affair of society's lowest classes—those that, one scholar suggests, were "shut out of the American Dream."[75]

On the heels of all of this tension, the federal government directed funds to address migratory labor problems of the Pacific Coast hop industry. The main issue was the social standing of migratory laborers. In 1936, an economic geographer suggested, "The transients belong to the group called 'fruit tramps,' familiar to Westerners. A decrepit car or truck carries a family and their camping equipment from one temporary job to another. . . . The nondescript appearance of . . . a camp resembles that of a group of refugees." He noted additionally the concerns for migrant children: "The children early learn to dodge the truant officer and grow up with a minimum of education and home life." In sum, he suggested, "Such transient labor proves of use to fruit and hop growers, but constitutes a social problem of real concern."[76]

Later in the decade, another geographer followed up on the concerns of labor in the Pacific Coast hop industry. Considering the thousands of workers needed across the West Coast, he noted, "Obviously no community can expect to have a sufficient number of resident workers to meet a demand of this character. Twenty to thirty thousand workers must be drawn from great distances to help in the work of the hops." The scholar then suggested, "Even though they live in the democratic West, hop pickers are beginning to acquire the stigma of a lower class. . . . Few people other than professional migratory workers . . . are willing to participate regularly in such work."[77] The words offered a much different perspective than what hop growers had wanted to cultivate in the preceding decades. In fact, they harked back more to Charles Dickens, who described hop pickers as "miserable lean wretches."[78] Although the geographer explained that growers tried to maintain comfortable and safe camps, he suggested that the sheer number of laborers needed in a camp created substandard living conditions. In particular, he noted that "[f]lies and mosquitoes are almost certain to become a pest and filth tends to accumulate. Dysentery and typhoid are a constant threat to the workers and to the community as well." The biggest problem? He explained: "The hop industry is a social liability to the community in which it is located in that it attracts a large number of people whom no community wants."[79] He noted that the general migrant-labor problem on the Pacific Coast needed attention. Workers needed new skills, permanency, and better integration into American society.

Together, these geographers' writings, Ginther's folk paintings, and Dorothea Lange's photographs offered different perspectives of Pacific Coast hop labor. New Dealers contested the image of a "paid vacation" that growers had cultivated in Europe and North America in the previous century. They cast the hop harvest as unsanitary, sometimes violent, and populated by the downtrodden of society. Yet in doing so they overlooked half of the labor equation: middle-class and urban families who continued to seek the hop harvest for its lively atmosphere and social opportunities. The Hop Fiesta of Independence and its association with this group of people provides the other half of the complicated harvest-labor story. If one thing was clear, it was that from the Hop Queen to the hobo, the Willamette Valley harvest contributed a vital layer to the Willamette Valley hop industry's story.

THE END OF THE GOLDEN ERA

In 1940, three and a half decades after the Willamette Valley became the nation's largest hop producer and the Lewis and Clark Exposition in Portland broadcast Oregon's "natural bounty" to the world, the title of an article in the May issue of the *Pacific Hop Grower* humbly forecast an end of an era: "E. Clemens Horst Called by Death." Horst's life story was familiar to many in the hop industry. He was a German immigrant reared in New York. In the 1890s, he started his own hop company in Northern California and soon after invented a range of labor-saving technologies—most important, the mechanical hop picker and improvements to hop dryers. Furthermore, for decades Horst's company represented smaller growers in domestic and international contracts, typically negotiating with Guinness and other brewers for the body of Oregon growers and others up and down the Pacific Coast who might not have even known where their hops ended up. He also became a central liaison between growers and the new agricultural science occurring in governmental programs and the private sector. Horst's story encompassed Oregon's reign as the Hop Center of the World, even amid its struggles in the first part of the century.[80]

In Emil Clemens Horst's lifetime, except for a brief period following state prohibition and World War I, Oregon was the national leader in hop production, rivaling at the time the largest producers in the world. As the Hop Center of the World, the area and its associated industry underwent substantial transformations, engaging in a process of modernization essential for continued success. While downy mildew brought many farmers to their knees by the 1940s, continued efforts in research and dissemination of new agricultural information confirmed Oregon's reputation as a major player in the global hop industry. The state's leadership in corporatization, marketing, and science allowed regional hop growers to achieve success later in the century.

At the time of Horst's death, however, the downy mildew crisis was impossible to ignore, serving as a central antagonist for the rest of this story. In the second half of the 1930s and the early 1940s, Oregon hop acreage remained substantial, but per-acre production continued to decline. Over time, the botanical disease, along with other changes in the specialty-crop industry, such as the expensive mechanization of the harvest, proved too much for a majority of Willamette Valley farmers to endure. The decline of the wolf of the willow in the valley, along with the death of one of its most vibrant leaders by the mid-1940s, marked an end to the golden era of Oregon hops.[81]

EIGHT

After the Hop Rush

THE WILLAMETTE VALLEY HOP INDUSTRY FACED its greatest crisis in the mid–twentieth century. After an extended period of success for hop growers that included global praise and one of the state's most vibrant folk occasions, downy mildew threatened to take it all away. While growers increased acreage upon the repeal of Prohibition, the quality of the crops slowly began to deteriorate along with reduced yields. Adding further uncertainty were the various marketing agreements that limited farmers' abilities to expand production or enter into the business, as well as the expense of mechanized and industrial agriculture. In 1936, at the height of the industry in Oregon, farmers cultivated approximately twenty-six thousand acres of hops. That number dropped below six thousand acres by 1954 and to three thousand acres seven years later, a time when average hop yields dipped to some of their lowest levels ever. A thriving community of over one thousand individual hop growers dwindled to fewer than four hundred by the late 1940s. The downward spiral continued thereafter as farmers either planted other crops or simply retired to the city.[1]

But the blight did not deter the most dedicated hop growers. Despite widespread uncertainty, the professional and scientific foundations established in previous generations allowed the Willamette Valley to remain an important part of the global hop industry. Growers continued to use their new organizations and trade journals to implement grading procedures and marketing agreements in an effort to achieve better prices. They also benefited from another global downturn of hop production that accompanied World War II to improve their standing in the global marketplace.[2] In the fields, they tested any remedy for pests and diseases that arrived at their doorstep, and some signaled the future of growing when they slowly integrated the new English disease-resistant hops into their farms.[3] Yet, in the post–World War II era, continued success required much more. By the 1950s, hop farmers had to fundamentally adapt to broader changes in American farming. The growers who remained planted in hops had no choice but to embrace the practices of industrial agriculture and to create monocultures on their farms. Aside from the introduction of the wire-trellis system earlier in the century, those changes probably represented the greatest single period of innovation in hop agriculture since it began in Bavaria nearly twelve hundred years prior.

At the heart of this transitional era lay a determined group of farmers, many with familial roots in hop growing dating to the late nineteenth and early twentieth centuries.[4] Alongside them corporate breweries and crop scientists at the Oregon Agricultural Experiment Station in Corvallis continued to help streamline agricultural and business practices. For the individual and family farmers, there was financial risk. But it was worth taking. Some farmers still believed that the crop could bring them great wealth; most by that time simply claimed that hop growing was a way of life and in their blood. As was the case with hop growing during the World War I and Prohibition eras, their persistence proved out. With high risk came reward, as these remaining Willamette Valley hop growers eventually helped usher in the craft beer revolution by the 1980s. That historical moment, however, would have been inconceivable when these farmers were negotiating a new world of hop agriculture just after World War II. At least they were not alone, since their hop-growing counterparts across the world also had reasons for concern.

A MID–TWENTIETH CENTURY
ENGLISH ASSESSMENT

Five years after World War II ended, British agricultural experts and economists addressed their long-standing inability to compete with American hop imports. Since the end of the nineteenth century, the topic had concerned Britain's hop industry, and it intensified in the next as Oregon gained a global market share and earned a reputation among brewers for a quality product. By the late 1940s, the British government recognized that prices for importing American hops undercut even the average cost of producing the English crop. It was a disheartening reality for brewers who wanted local hops and for English growers who had been a world center of hop production in the two centuries prior.[5]

What was it about the Pacific Coast hop industry that offered the competitive advantage? In 1950, to seek answers to that question, British officials allocated Marshall Plan funds to send delegates on a North American investigative tour. Throughout July and August of that year, the members of a "Hop Industry Productivity Team"—including two representatives from the Guinness brewery, two members of Great Britain's Hops Marketing Board, and an agricultural economist from Wye College—traveled the Pacific Coast states and British Columbia studying the competition. They came to an uneasy conclusion: the temperate climates, well-draining soils, and level terrains along the Pacific Coast were simply more conducive to hop cultivation than England. Although they recognized that Germany, Czechoslovakia, and a handful of other countries including Poland, France, and Russia remained integral players in the global hop marketplace, the British tour highlighted the new American prominence.[6]

Yet the British team's visit came at a time when the Pacific Coast hop industry itself faced immense changes. Naturally, the inability to combat downy mildew effectively in Oregon provoked alarm. But that marked only part of the story. Competition with Washington's Yakima Valley contributed equally to the demise of Willamette Valley growers and acreage of their hop farms at midcentury. Beginning in the late nineteenth century, growers planted rootstock in the Yakima Valley's good volcanic soil and benefited from a railroad infrastructure that connected distant markets. The construction of federal irrigation projects in the 1910s allowed Yakima farmers to expand in the following decades. While lacking the natural rainfall of the Puyallup and Willamette Valleys, the more arid climate in the rain shadow east of the Cascades proved beneficial by the 1930s. The drier environment offered a natural defense against the downy mildew destroying Oregon crops, and virgin soils helped stave off other pests, including hop aphids that attacked Puyallup Valley hops. Upon the explosion of federal dam building in the New Deal period, the Yakima Valley had the best of both worlds in a dry climate with plenty of irrigated water resources.[7]

Oregon farmers could not compete. In eastern Washington, acreage nearly quintupled from 1929 to 1948 from just under three thousand to over thirteen thousand acres. More important, the region's hop yields averaged well above fifteen hundred pounds per acre, while Oregon growers struggled to average above nine hundred pounds upon the onset of disease problems. Along with these changes, Yakima Valley farmers favored growing seedless hops by eliminating male plants from their fields well before their counterparts along the West Coast considered the option. While most American growers preferred seeded hops for the added weight, brewers increasingly demanded the seedless product. As a contemporary industry report remarked, "It is generally conceded that seeds in hops are extraneous and undesirable and play no favorable role in brewing. . . . [T]he brewing trade considers seeded hops inferior in quality to the seedless type."[8] Thus, the Yakima Valley farmers responded to the market demands of the period rather than holding on to older practices. For all of these reasons, Washington hop farmers supplanted Oregonians as the largest producers in the nation beginning in 1943, a position they have not relinquished to the present day.[9]

If it was any consolation to Oregon farmers in 1950, the visiting British team still assessed the Willamette Valley as productive and viable. But the report did not quite grasp the gravity of changes Pacific Coast hop growers faced. First, by the mid–twentieth century hop agriculture had transitioned away from the small family farmsteads that cultivated less than thirty acres of hops as part of larger, diversified operations. Instead, growers began dedicating hundreds of acres to the crop, often abandoning almost all other agricultural activity. Second, the remaining growers took risks by investing in new industrial technologies and synthetic disease and pest remedies; this included expensive mechanical harvesters that reduced the need for manual laborers. In other words, hop growers entered the industrial age of American farming. Given the Pacific Coast growers' history of innovation, this should not have come as a surprise. Just as their forebears had adapted to new technologies and markets, farmers in the mid–twentieth century faced adaptations to the changing circumstances of agriculture and the world around them. Before getting into the heart of that discussion, however, attention must be paid to how the new world of big beer unfolded. The nuts and bolts of business history underpin an important layer in the story of the Willamette Valley hop industry.[10]

THE AGE OF BIG BEER

Aside from a brief decline in domestic beer sales in the late 1940s, big beer was well positioned for continued success in the 1950s, the first full decade in the postwar economic upswing.[11] Expanded industrial facilities and new advertising campaigns contributed to brand recognition for big beer.[12] At the center of it all, the larger brewers fundamentally sought horizontal integration, with the Pabst, Anheuser-Busch, Schlitz, Ballantine, Falstaff, and similar companies energetically buying out smaller brewers and the distribution networks that belonged to them.[13] One historian argues that mergers became the most important part of the expanding brewing industry, and that the limited "space on a distributor's truck" (not to mention grocery-store shelves) became the prime arena of competition. The same reality occurred in the Pacific Northwest. "When Blitz-Weinhard of Oregon, for example, bought Great Falls Brewing of Montana," the scholar notes, "Blitz bought not just a brewery, but access to its distributors and so to another market."[14] This was simply the name of the game moving onward as national brands usurped control from local breweries across the country and across the world. By the 1970s, Pabst would sweep up even the Blitz-Weinhard Brewery in Portland, leaving, for a time, no locally owned breweries remaining in Oregon.

Amid all of the developments following Prohibition and World War II, there lay an unsettling problem for beer purists and Pacific Coast hop growers. American beer had changed. Robust conglomerations of beer colors and flavors no longer filled saloons and beer gardens. The new age of big beer produced predominantly light-colored and easily imbibed lagers that had been modified from traditional recipes. American brewers increasingly added rice and corn to their malts instead of the traditional barley and wheat to achieve economy and to conform to the initial blandness of post-Prohibition beer. Less malt in the kettle saved brewers money since barley was the most expensive ingredient in beer making; the decision also naturally lessened the need for the bittering influence of hops. In a telling statistic, by World War II brewers used 75 percent less hops than they had earlier in the century.[15] Two social scientists investigating the American hop industry at the time provided global context, noting, "In the United States, each thirty-one gallon barrel of beer requires the addition of about a half pound of hops. In foreign countries the amount used may be up to one and a quarter pounds per barrel."[16]

The falling hopping ratio naturally concerned American hop farmers. During World War II, they had fared adequately in large part because of the federal government's support of the beer industry. Afterward, growers worried that decreased hopping rates threatened their livelihood. In 1948 the nation's remaining hop farmers, who had found value in organization in previous decades, reconfigured collectively as the United States Hop Growers' Association. While they shared many concerns, the hopping-ratio issue took center stage. In one of their first efforts, the organization successfully called upon the USDA's Bureau of Agricultural Economics for an in-depth study. This resulted in a report, entitled *Outlook for Hops from the Pacific Coast,* that surmised the following:

> Beer consumption in the United States increased by more than 100 percent between 1935 and 1948, but the consumption of hops in the manufacture of beer increased about 33 percent. This disproportionate increase of the manufacture of beer in relation to hops used in its manufacture was due to a change in the hops-beer ratio from 0.702 pound of hops per barrel of beer to .455 pound in 1947–48.[17]

As if this information did not cause hop growers enough anxiety, the study further noted that most brewers were "inclined to believe that the consumer preference for the so-called light beers will continue in the postwar era."[18] The research raised fears particularly for farmers who had turned almost exclusively to growing hops on their land and relied on the domestic market to consume their produce.

From a twenty-first century perspective, hop growers need not have worried. Light beers maintained the Pacific Coast hop industry. The new directions in American big beer reflected a larger mid-twentieth-century consumer trend toward conformity, accommodation, and the "silent generation."[19] Food culture also became homogenous, marked by a general blanding of the American palette. Miller High Life (first sold in 1957) and subsequently released light beers complemented an era of TV dinners, McDonald's hamburgers, and Campbell's soup. Like these other manufactured goods of mass production, the American lager became the staple of backyard barbecues and football parties, appealing across socioeconomic categories of gender, race, and class during the 1950s and 1960s.[20]

The popularity of macrobrews expanded from the midcentury onward, leading to a steady demand for American hops (in addition to increased quantities of European-aroma hops still sought by large brewers such as Anheuser-Busch and Coors as a nod to their traditional recipes). In sum, the sheer quantity of bland beer produced in the mid–twentieth century kept up demand for Willamette Valley and other Pacific Coast hops. Furthermore, for economic and nationalistic reasons in the early Cold War, brewers joined others in the food industry to seek American agricultural products as opposed to importations from other countries. The situation reinforced the major breweries' commitment to forging relationships with Pacific Coast hop farmers and funding hop research programs to promote their success. Still, it might have taken some hop growers time to warm to the idea. The product of their labor was no longer recognizable in glasses of American beer.[21]

In the 1940s, the Master Brewers Association of America (originally the Master Brewers' Association of the United States), United States Brewers Association, American Brewing Industry, and American Society of Brewing Chemists established the Brewers' Hop Research Institute. An initial meeting in Chicago showcased the new influence of American corporate brewers, who at the time continued to look to Europe for 50 percent of their hops but wanted American products. The institute brought big beer, small farmers, and scientists together to discuss the needs of the domestic hop farmer for better hop-drying techniques, storability, and baling methods, and, perhaps most important, for increased market demands for seedless hops and higher content of lupulin—that valuable yellow resin inside of hop cones containing the bittering agents and essential oils.[22] The event revealed corporate brewing's new position of power and their interest in investing in hop research where the federal government once took the initiative via USDA funding. Despite the obvious concerns about hopping ratios, the ubiquity of big beer stood primed to carry American hop growers into the second half of the century.

At the same time, the Brewers' Hop Research Institute revealed a continuing connection with Willamette Valley hop growers, who had now relinquished their claim as the nation's largest hop producers and the Hop Center of the World. G. R. Hoerner, a crop scientist at Oregon Agricultural College, offered the meeting's opening comments while serving as secretary in charge of research. After reminding the attendees that his home region still grew nearly half of the hops in the United States, Hoerner explained how he and his peers envisioned collaboration among the various groups attending. The consensus among all lay in the fact that the American hop industry needed to invest more in research programs. While disappointing to many of the attendees for a lack of immediate solutions, the efforts signaled the new directions of collaboration as well as Oregon's continued role in planning for the future of the American hop industry.[23] Subsequent efforts of the Brewers' Hop Research Institute seemed to help. After its first year the major effort was in sending questionnaires on hop quality to brewers and seeking answers about how to grade hops so as to provide standards. That work took place at the Oregon Agricultural Experiment Station.[24]

Aside from research funding and collaboration, big beer made its presence felt in the Pacific Coast hop industry by cultivating relationships with farmers. Hop growers commonly entered into multiple-year contracts with brewers, offering the financial stability that had been chronically difficult for specialty-crop farmers.[25] As part of the arrangement, brewing-company representatives became frequent visitors to Willamette Valley hop farms throughout the growing season and during harvest. The watchful eyes of these representatives ensured that the breweries were getting a quality product. But it also inspired growers to modernize their agricultural operations because of both the scrutiny and the fact that they could be offered a more secure financial situation. It was apparently a winning situation for all parties.[26]

Over time, the scope of relationships between big beer and hop growers expanded. Hop-growing families welcomed the familiar faces of brewing-industry representatives to their farms as part of their own, often hosting picnic meals and inviting others from the community. Employees of the Pabst, Schlitz, Miller, Anheuser-Busch, and, later, Coors breweries all made their presence known through their frequent visits to hop country. On a larger scale, big beer companies also sponsored hop farmers' visits to agricultural and beer conferences across the globe in an effort to improve and standardize hop production. Many hop growers from the post–World War II period explain that the new connections helped them integrate into a worldwide hop and beer culture. Farmers and representatives from the hop and beer industries also visited the Pacific Northwest from England, Germany, Poland, France, China, and other major hop-growing regions of the world, encouraging the networking of a global community of hop growers. Big beer's efforts in this sphere re-created a cosmopolitan outlook to the community of Willamette Valley hop growers that Ezra Meeker and Emil Clemens Horst at one time experienced.[27]

While the vibrant pre-Prohibition flavors and aromas did not tingle American beer palettes, evidence of the success of the U.S. brewing industry and therefore Pacific Coast hop farming was everywhere. The future boded well for those growers remaining in the Willamette Valley. Changes that growers embraced in a new era of American agriculture and corporate beer production may be better understood through a closer look at the work of individual farmers and their organizations. Growers took bold financial risks to meet new standards of industrial farming. Also, although dismaying to some hop producers, the public at large continued to be unacquainted with hops on a beer-to-beer basis.

In February 1953, the United States Hop Growers' Association, which had supplanted the regionally named Pacific Hop Growers' Association, convened in Portland for its seventh annual meeting. The presence of Oregon governor Paul L. Patterson and Portland mayor Fred Peterson indicated that, despite ongoing struggles with downy mildew and competition from eastern Washington, hop agriculture remained an important aspect of the state's economy and social identity. Keynote speakers G. R. Hoerner and Dean Walker, both of the Willamette Valley, reminded the attendees of that fact, offering their expertise to colleagues from California, Washington, and Idaho. More important, the group came together to confront a variety of topics related to the changing hop industry. The agenda included how to best integrate into the expanding world of corporate brewing, particularly the threat of the growing market for minimally hopped lagers. Discussions also addressed a new marketing agreement that limited production per farm, the dire need for mechanization of harvests to meet the chronic labor issue, and general research goals to improve crop productivity and commercial appeal.[28]

As a whole, members of the United States Hop Growers' Association adjusted to market changes just as had their predecessors. Still, in the post–World War II era, they remained uncertain about their future. A "Twilight Beer Party" on the first evening of the convention may have eased the nerves of some. A tour of the Blitz-Weinhard brewery the following day prompted the association's new journal, the *Hopper*, to report, "Generous samples of Blitz Bock beer and a special brew of light beer kept the visitors in such a cheerful glow during the rest of the morning that most of them were willing to overlook the low hopping ratio—at least temporarily."[29]

Of course, the glow could last only so long. There were other agenda items. Although not voiced in specific terms, one overarching concern for the members of the United States Hop Growers' Association was the modernization and industrialization of their operations to stay in business. The period marked a dramatic and rapid integration of new science and technologies, including irrigation projects from the Bureau of Reclamation and adopting new machines and chemicals produced in wartime and repurposed for commercial purposes afterward.[30] The process, as social critics and historians have suggested, meant that American farms had become "factories in the field." That is to say, American farmers focused on efficiency and productivity—terms often reserved for Progressive Era Taylorism and Fordism—while minimizing the role of individualized farm labor.[31]

Historian R. Douglas Hurt provides more context for these changes in farming by noting, "[B]y 1960 farming had become more than a way of life; it was a business where only the most efficient survived." He goes on to note, "The days of the small-scale, diversified farmer were gone. Government policy in the forms of loans, price supports, and acreage reductions favored larger-scale farmers who could produce more for less cost than small-scale landowners."[32] Statistics complete the story. From 1940 to 1960 the U.S. farming population dropped from 23.3 to 8.7 percent of the American population as a whole. At the same time, the average farm size increased from 170 to 300 acres.[33]

While Pacific Coast hop farming defied some of the postwar agricultural trends, including a general lack of attention from the USDA, the Willamette Valley hop industry could not escape the larger trends of American agriculture. A demand for efficient and modern production and larger farms, together with expanded influence by big beer and larger transnational hop-trading firms, proved central. The major expansion of one firm in particular, the John I. Haas Company, demonstrated the changed nature of hop production and marketing. In 1953, an article from the *Hopper* noted, "Haas is the biggest hop grower in the world, the foremost exporter, and as a dealer handles a third of the crop sold in Washington." The article then explained the owner's migration from England to Washington, D.C., in 1907 and the subsequent purchasing of hops in British Columbia and the Yakima Valley. The final note was the most telling, "Today the Haas Company . . . operates a series of farms around Yakima of 180 to 700 acres each, plus more acreage in Oregon and California and holdings in Chilliwack and Lillooet, British Columbia."[34] The American hop industry had changed amid a world of big beer that demanded efficient and reliable hop production overseen by powerful transnational firms. In other words, larger world markets with a booming domestic market provided by big beer demanded more efficient, modern, and uniform hop production.

The most immediate and dramatic modernizing change for all Pacific Coast hop growers in the mid–twentieth century came with the mechanized harvest machine. Upon American entry into World War II and its associated effects on domestic labor, hop growers across the West Coast had to become even more creative in seasonal hiring or leave hops on the bine unpicked. Many sought out more women than ever before, believing that their agility and availability made for the best source of labor anyway.[35] Some growers turned to prison labor camps to meet their needs, while others embraced the

FIGURE 17. Harvesting hops by portable picking machine, circa late 1940s. Courtesy Special Collections and Archives Research Center, Oregon State University.

national Bracero Program that opened the doors for Mexican nationals to harvest crops in the United States. All growers continued to advertise across urban and rural areas for families interested in the late-summer harvest. Unfortunately, there were often insufficient numbers of workers.[36]

Growers required a mechanized solution. Mechanical harvesters, much improved upon since Emil Clemens Horst's machine of 1909, became a viable option. Although expensive, the machines of the 1940s reduced the work of one thousand hand pickers to less than fifty. (See figure 17.) As early as 1944, the *Hopper* declared that the future of the harvest rested with machines.[37] The result was a fast, efficient harvest, but with far fewer workers, spelling the end of the "hop-picking time" in Oregon's rural communities. Labor crews now cut bines, transported them, and fed them into mechanical pickers. There, a series of belts, sorters, and shakers separated hop cones for drying. While it would take five to ten years to implement across the valley, the popularity of the mechanical harvester and the reduction in need for seasonal workers sealed the end of the Hop Fiesta and all that it represented in Oregon's culture of work and play.

FIGURE 18. Feeding bines into a stationary hop-picking machine near Independence, Oregon, 1947. Courtesy Special Collections and Archives Research Center, Oregon State University.

In 1947, the Goschie family of Silverton became one of the first Willamette Valley hop growers to purchase an expensive mechanical harvester. For the father, Carl (who first planted hops outside of Silverton in 1904), his son, Herman, and his wife, Vernice (the latter two of whom took over part of Carl's land in 1941 and later expanded the business), the investment represented dedication to the industry in an uncertain time. Their first mechanical harvester was portable and moved about the trellises. After recognizing its success, the Goschies upgraded to a larger, stationary model a year later. (See figure 18.) At a cost of forty thousand dollars, this large contraption, which took up an entire building, changed the landscape of the hop farm and caused fundamental changes in the hop industry. Mechanical pickers marked the most visible transition from preindustrial to industrial harvest methods. For the Goschies, the new technology meant investment and debt, but signaled that the family was willing to take risks—even if the future of Willamette Valley hop agriculture seemed uncertain.[38]

The Goschies' gamble paid off, as did those of other hop growers willing to invest in mechanical harvesters throughout the 1950s. Not only did the transition to industrial harvest benefit their own operations, but they were able to contract the machines' use to other, smaller hop growers—an arrangement that became commonplace in the 1950s. By 1951, the British Hop Productivity Team estimated that 85 percent of Americans used the technology to harvest their crops. American hop farmers led the way to mechanized harvesting—specifically, because individual growers devoted much greater acreage to hops on their farms than others across the world and could justify the investment. European centers of production would not fully embrace mechanical harvesters for another decade or two.[39]

The move toward mechanized harvests represented one more step in eliminating smaller-scale hop growers, who could not afford the elaborate machines. But this was not the only transformation of industrial agriculture that smaller growers faced to stay competitive. Other technologies certainly compounded the issue. Industrial tractors, plows, and sprayers, which wheat, corn, and cotton growers had adopted by the 1920s, became available for hop agriculture that had smaller rows and high-climbing bines. Improvements in hop dryers with concrete floors and modernized diesel or natural-gas heaters signaled a positive shift away from wood-fuel dryers that often caught fire and burned the drying houses to the ground. By the end of the 1950s, Pacific Coast hop growers needed all of these new technologies to remain competitive. Those who failed to mechanize soon left the business.[40]

In addition to mechanization of operations, successful hop growers changed methods of cultivation and tending their crops. From the Meekers' operations in the 1860s to World War II, Pacific Northwest hop growers followed practices dating back to Thomas Jefferson's era. They applied manure, potash, and other organic fertilizers to improve soil quality, and they used nicotine and other organic compounds to control pests and disease. Willamette Valley hop growers also continued to utilize the arsenal of late-nineteenth-century science when they imported whale oil and quassia chips for topical sprays. These methods persisted even after the outbreaks of downy mildew in the 1930s until shortages of those products during World War II. For some growers it was simply not economical to apply inorganic fertilizer, such as ammonium nitrate, when they could achieve similar results via manure or cover crops, such as legumes, that return nitrogen to the soil.[41]

After that time, some growers turned to creative solutions for pest control, such as the use of ladybugs to combat aphids.[42] But most Oregon growers increasingly relied on planting Fuggle hops and the Brewer's Gold and Bullion hybrid hop varieties released by England's Wye College. Again, though these hops had been bred specifically to increase alpha acid content (i.e., bitterness levels) and thus offer brewers more value per pound, farmers eventually sought these hops for both their resistance to plant diseases and their high yields. The adoption demonstrated the value of breeding hop hybrids for an organic solution to hop growing.[43] It also highlighted the importance of collaboration between big beer, small farmers, and government scientists, as E. S. Salmon of Wye College established the first experimental plots and germplasm exchange programs with assistance from the British government and private brewers—a model adopted not only by the United States by midcentury but also by those who struggled in Continental Europe. Still, the English advances made it clear that the science on the Pacific Coast lagged behind.[44]

Even with some success from these natural methods of disease and pest prevention, chemical byproducts of the war effort contributed to fundamental transformations in hop agriculture. Chemical companies repurposed wartime products used in tropical areas for combating foliage and insects for tactical and health reasons. In peacetime the chemicals were repurposed for use in American agriculture and broader insect control. DDT became the most common of these products, but other chemical pesticides and herbicides—particularly the herbicide 2,4-dichlorophenoxyacetic acid (or 2,4-D)—also proliferated. Farmers across the country, including Pacific Coast hop growers, turned to chemical solutions. They also turned to synthetic nitrogen for fertilizer. One historian has noted, "Although some people questioned the effects of chemical pesticides and herbicides on public health, most farmers believed the postwar chemical industry had ushered in a 'golden age.'"[45] The Willamette Valley also signed on to the chemical parade.[46]

There were threats beyond downy mildew. Other common troubles ranged from verticillium wilt to the insect threats of hop aphids and red spider mites.[47] What would the solutions be? By the mid-1940s, the *Hopper* ran cheerful and heroic ads from chemical companies proclaiming the effectiveness of their products against a broad spectrum of pests. While some growers worried about the applications, the chemicals rapidly worked their way onto and into the Willamette Valley hop-growing landscape. As early as 1945, the *Hopper* contained an advertisement for DDT that mentioned its benefits, even if the ad noted that further study on its impact was needed. Despite some uncertainty, Willamette Valley growers embraced the new chemical.[48] But DDT was far from the only chemical advertised to and adopted by Willamette Valley hop growers. A 1946 issue of the *Hopper* included an article entitled "A Substitute for Nicotine." It stated, "The Monsanto Chemical Company recently announced the availability of a new insecticide that carries the name Hexaethyl Tetraphosphate.... Used in conjunction with DDT, the new product tends to maintain the balance of nature, rather than upset the balance as DDT sometimes does when used alone."[49] This specific ad referred to Vapatone, the commercial name for hexa-ethyl tetraphosphate (developed in Germany during World War II to combat mosquitoes), and sold by the Ortho brand to hop growers as insect spray for aphids and red spiders.[50] Another ad, from 1947, elaborated on the impor-tance of this new spray. Frederick J. Haas, vice president of John I. Haas, Inc., suggested that Vapatone "has promising possibilities in checking hop aphids especially when it is applied by means of fog-generating machines."[51]

DDT and Vapatone marked just the beginning of the transformation of hop growers from using organic pest controls and fertilizers to using indus-trial chemicals. In 1947, the *Hopper* announced the Besler Corporation's "DEATH –DEALING FOG FOR INSECTS . . . the GENUINE fogger."[52] In 1950, chemicals to combat downy mildew strengthened, particularly Dithane, a fungicide produced by Rohm & Haas of Philadelphia.[53] A year later, the American Cyanamid Company advertised for Parathion, which "Gives Outstanding Control of Aphids, Mites, With No Effect of Hops or Brews."[54] Charles R. Joshston, a Willamette Valley hop grower, purported the usefulness of that chemical, noting in a following edition of the *Hopper,* "Cyanamid is applied in early February at the rate of 2 ounces per hill, spread evenly over the entire crown to kill volunteer weed growth and also as a fer-tilizer and soil sterilant."[55] In the same article, however, Joshston expressed the importance of natural fertilization and cultivation techniques such as planting rye cover crops and spreading manure. His note provided a reminder that organic agricultural methods remained intact even as farmers turned to industrial chemical solutions.

While effective, industrial chemicals like DDT and Dithane, often mar-keted as harmless herbicides and insecticides, caused substantial damage outside of the hopyard. They polluted watersheds and disturbed ecosystems, contributing to health problems in animal and human populations.[56] Some farmers might have been hesitant about the rapid transformation, but it was not until the publication of Rachel Carson's *Silent Spring* in 1962 that American society at large became aware of the dangers of DDT and other chemicals of industrial agriculture. But chemical dangers were not the immediate concern of Pacific Coast hop growers. They embraced industrial agriculture, seeking the same goals as always: to produce hops efficiently and reach as many markets as possible. A sustained industry and global growth confirmed their instincts.[57]

Throughout the midcentury changes to the Oregon hop industry, there loomed a strong possibility that the community of mostly small farmers were compromising their identity through rapid modernization. In actuality, they had no real choice to stay competitive in the global marketplace. But this did not mean the growers lost a sense of identity. In fact, the industrializing and professionalizing farmers turned to one another even more than in previous eras by joining several new organizations. First, to ensure a better future in 1955, Willamette Valley growers created the Oregon Hop Growers' Association. The voluntary organization aimed to assist with the long-term agenda items of marketing and grower education. According to its articles of incorporation, the association's goal was to "disseminate knowledge concerning cultural practices in the growing of hops, conduct research in the growing, marketing and consumption of hops, facilitate the marketing of hops by members of the corporation and in general to do all things which tend to promote a healthy hop industry within the State of Oregon."[58] Recognizing the need for broader support a year later, many also joined the Hop Growers of America, a national organization that shared similar aims. But it was not until the summer of 1964 that members of the Oregon Hop Growers' Association came together to seek the support of the state in the form of the Oregon Hop Commission. This new commission required growers to pay dues in return for state- and grower-sponsored research programs.[59] Remarkably, given the fickle nature of previous organizations, all of those created during that period have lasted to the present day, not just by turning their attention to marketing and outreach, but also by providing a community for growers via monthly meetings, frequent meals, and yearly field days. For Willamette Valley growers, the Oregon Hop Commission has clearly proved the most significant.

On the afternoon of June 3, 1964, the first official meeting of the Oregon Hop Commission occurred in the state's Department of Agriculture building in Salem. The crowd included individuals with deep connections to the industry—most, if not all, being second- and third-generation hop growers.

Vic Annen, Robert Coleman, Harvey Kaser, Ed Crosby, Jr., Roger Kerr, Charles Lathrop, John Smith, R. R. Troxel, Phil Wolf, and Herman Goschie represented the Willamette Valley growers, who had diverged from the previous generations by investing almost their entire acreage in hops. As with the Portland meeting of the United States Hop Growers' Association ten years prior, Oregon governor, at this time Mark Hatfield, was also supposed to be in attendance but, "due to illness, was unable to be present."[60] Although by that time the number of Willamette Valley hop growers had dwindled to its lowest number since the early 1880s, the governor's intent to participate provided a nod to the ongoing importance of the state's hop industry.

In a turn of events from the previous decades, the concern of hopping ratios in big beer no longer existed, because growers had come to terms with the fact that the popularity of blander, industrially crafted beers still ensured their livelihoods. Answers to the pressing issues of disease and reduced market share dating back to the Great Depression and World War II remained elusive. As had happened in the growers' already nearly one-hundred-year history (and, of course, a history that dated back to its European origins), success required a recipe of hard work and collaboration with government officials, scientists, businesspeople, and agriculturalists. Behind it all, the Oregon Hop Commission helped to create a community of growers and their families. Although not in the records of the commission, it is vital to note that while the men attended organizational meetings, their mothers, wives, sisters, and children played crucial roles on farms. Vernice Goschie, Maureen Coleman, and countless others who do not necessarily show up in the record books not only participated in the transforming hop industry, but were intimately involved in the fields and engaged in a range of other activities, from hosting brewery representatives and hop-trading organizations to working on accounts. As the Coleman family of St. Paul jokes, matriarch Maureen at times spent longer hours in the fields than the rest of the family.[61] The most prevalent woman in the industry might have been Elaine Annen, who was one of the few to show up in the records of the Oregon Hop Commission. There was a good reason: she held the position of executive secretary from 1964 to 1991, and the official notes she took and signed after every meeting represent a substantial investment in the industry. Her work also included record keeping, bookkeeping, and acting as a liaison for other organizations such as the Hop Research Council, founded in 1979.[62]

In the months that followed their first meeting, the members of the Oregon Hop Commission elected officers and established a funding program based on pounds of hops sold from each farm. By the end of the year, they also adopted the following guidelines tied to the thriving research program at the USDA's Agricultural Experiment Station in Corvallis:

1. Sponsor research studies including but not limited to:
   - Adaptability of Oregon-grown varieties to new or improved methods of preparation or processing for market.
   - New or improved varieties for improved quality and yields.
   - Market preferences and how best to meet them.

2. Apply research findings and other available information to educational and promotion program for broadening of market demand and returns to Oregon hop producers.

3. Study legislation, state and federal, with respect to tariffs, duties, reciprocal trade agreements, import quotas and other matters concerning the effect on the commodity industry, and represent and protect the interests of the commodity industry with respect to any legislation or proposed legislation or executive action which may affect that industry.

4. Cooperate with any local, state or national organization or agencies, whether created by law or voluntary, engaged in work or activities similar to that of the commission, and enter into contracts with such organizations or agencies for carrying on joint programs.[63]

All of these points had been pursued in years prior and reflected ongoing concerns that grower organizations had from years past. Nevertheless, the list of goals demonstrated that problems persisted for Oregon and other Pacific Coast hop growers. The goals also revealed that, despite advances in marketing and corporate sponsorship, scientific research had not developed satisfactorily. That is to say, the Oregon Hop Commission believed that American hop hybrids had the potential to save their industry, and they were willing to fund breeding programs and offer test acreage. More than anything, the Willamette Valley growers wanted better hops, genetically matched with the regional environment to be high yielding, disease resistant, and appealing to brewers. To provide context, E. S. Salmon's hops had been bred in England in the first couple of decades of the century and released in the 1930s. There was concern that USDA science at the agricultural research center not only in Corvallis, but also at Washington State University and the University of California at Davis, lagged behind.

Little did Vic Annen, Robert Coleman, Harvey Kaser, Ed Crosby, Roger Kerr, Herman Goschie, and their peers on the Oregon Hop Commission know that change was just around the corner. In response to Rachel Carson's works and rising popular concern about industrial toxins in the 1960s, the USDA expanded funding for agricultural breeding programs across the nation, including hops, in the hopes of reducing the use of inorganic herbicides and pesticides. In Corvallis, the funding created a new USDA position for an Austrian-born plant geneticist named Alfred Haunold. Whereas his initial hiring in 1965 dictated general work on disease and pest resistance, extenuating circumstances thrust him into the central role of hop breeder soon after he moved to Oregon. The necessity for this rapid change could not have come at a better time. By the time he attended his first meeting of the Oregon Hop Commission the same year that he was hired, Haunold had become deeply immersed in the horticultural work that would redefine not only the region's hop industry and culture, but also the world's. His story, along with the story of his predecessors and contemporaries in Corvallis, provides the vital link to the revival of Willamette Valley hop growing and ultimately the craft beer revolution.[64]

NINE

Cascade

AMID THE VICISSITUDES OF WILLAMETTE VALLEY hop agriculture in the mid–twentieth century, the research program at the Oregon Agricultural Experiment Station in Corvallis offered bedrock for growers to stand upon. State-sponsored studies on *Humulus lupulus L.* dated to the "hop fever" of the late nineteenth century. By 1895, scientists housed at Oregon Agricultural College planted hops in small plots for a variety of studies. In the following decade, government and brewing-industry dollars funded a program called "Hop Investigations."[1] But the onset of Prohibition forced researchers to abandon projects related to alcoholic beverages, and this included the wolf of the willow's contribution to beer. Only after Oregon's hop crops thrived during the Prohibition era did the federal government once again support further hop research. The major item on the research agenda was the downy mildew disease that had made its appearance in the hop fields west of the Cascades in the spring of 1929.

Throughout the 1930s and 1940s, scientists and farmers worked to defend the Willamette Valley against the threats posed by this disease. They tested new sprays and technologies and studied soils and climate in hopes of finding answers that would help them curb the spread of downy mildew. Early efforts yielded little. Cluster hops, the predominant variety grown in the Pacific Northwest, proved extremely susceptible. As the variety's weakness became apparent, growers first looked to experimental field trials with Fuggle, the early-maturing English aroma hop originally selected in England before the turn of the twentieth century. Fuggle was fairly resistant to downy mildew, and it matured early, but it produced lower yields than Clusters. Thus the search for a mildew-control chemical continued until the 1950s and 1960s, when growers turned evermore to the disease-resistant and higher alpha (bittering) Bullion and Brewer's Gold varieties bred by E. S. Salmon of Wye College in England.[2]

Could researchers in Oregon emulate the English work and improve upon hop stock in further breeding experiments? From 1930 onward, the USDA funded a hop scientist position at the experiment station in Corvallis to answer this question. These developments offered further confirmation that the Willamette Valley was the center of hop agriculture in the country. Researchers hoped to develop not only superior disease-resistant hop varieties, but also bines that could produce large quantities of cones that would store well and fare favorably in the brewer's kettle. By that time, advances in plant breeding boded well for successes at the experiment station.

Yet progress was slow. In fact, it took so long to release an American hybrid that farmers and brewing-industry representatives, who still benefited from other studies at the experiment station, must have written hop breeding off as a fantasy. Astonishingly, it was not until the early 1970s that the USDA released the first American-bred hybrid, much to the rejoicing of the region's crop scientists, farmers, and brewers. It may have been a small victory relegated to a specialty crop in the nation's far corner, but the breakthrough transformed Pacific Northwest hop farming and, eventually, beer cultures throughout the world. The story of this development charts an all-important connection between the Hop Center of the World that had been and the Craft Beer Capital of the World that would arise.

THE EARLY YEARS OF OREGON HOP BREEDING

After the USDA made its selection of Oregon as the center of hop breeding in the country, it chose a rising star for program director: Earl N. Bressman. As an agronomist, Bressman had already made a name for himself in the 1920s, first acting as assistant director of the Farm Crops Department at Iowa State College and then filling similar positions at New Mexico A & M and Montana State College.[3] His most notable achievement was a book cowritten with Henry A. Wallace, a leading agricultural journalist who was interested in scientific agriculture and later became the Secretary of Agriculture during the New Deal. The book, innocuously published in 1923 as *Corn and Corn Growing,* helped set the tone for a revolution in plant breeding.[4] Namely, the authors highlighted a rediscovery of the theories of Gregor Mendel, the Augustinian monk who established the foundation of modern genetics by identifying the role of heredity, or the passing down of genetic material, including dominant and recessive genes, to offspring. Though Mendel's work centered on pea plants and unfolded in relative obscurity in a nineteenth-century Czech monastery, his theory of inheritance became a foundation of agricultural science after biologists rediscovered it in the early twentieth century. Mendel's work suggested that parents determine specific traits in their offspring based on their own genetic structure. Plant breeders Bressman and Wallace experimented with Mendel's methods to select promising parents to make scores of crosses, or hybrids.[5]

Bressman's work with Wallace put him at the forefront of agricultural plant-breeding science when he took the position as hop breeder for the new program in Corvallis. While it is impossible to know exactly what he thought of his new life in the Willamette Valley, he took seriously the challenge as he transitioned from work on staple crops to specialty-crop research. Initially, he faced the necessity of reading a significant amount of scientific and agricultural literature on *Humulus* and acquainting himself with leaders in the field of hop breeding. Not surprisingly, Bressman sought guidance and expertise from E. S. Salmon and his colleagues at Wye College.[6] From these relationships he gained information and rootstock and set a precedent in the Oregon research program for similar collaboration with other scientists from around the world.

Bressman also kept company with local farmers. He made a point of noting in his yearly reports that he spent a great deal of time visiting hopyards all across the Pacific Northwest.[7] Some of his earliest writings mention collaboration with Dean Walker of Independence; the Ireland Hop Yard east of Corvallis, owned by a Portland doctor and run by the Seavey family; the McLoughlin Yard of Salem, run by Louis Lachmund; and the Horst Ranch of Independence.[8] Bressman also visited farmers across Washington and British Columbia in an effort to learn about hop culture on the various agricultural landscapes of the Pacific Northwest and to build relationships for future collaboration. All of these activities came together to establish a strong foundation for the USDA hop-breeding program.

In Corvallis, day-to-day hop breeding entailed identifying promising male and female plants to hybridize, and then selecting promising offspring to grow in the greenhouse and fields. It was a labor-intensive process that first required swabbing pollen from mature male plants and then applying it to female flowers, which had to be protected from accidental pollen carried by the wind. In his first year alone, Bressman repeated this process thousands of times,

FIGURE 19. Grafting hop varieties at Oregon State Agricultural College, 1931. Courtesy Special Collections and Archives Research Center, Oregon State University.

creating seedlings to inspect for beneficial traits. He used twenty-three different hop plants that he had gathered from local farmers and colleagues across the world, ranging from the European noble varieties to native North American varieties, and even *Humulus scandens,* the subspecies found only in Asia.[9] As he noted in one yearly report, the objectives were straightforward. "The general plan of this hop breeding project," he suggested, "is to grow seedlings from superior plants noted in various yards, seedlings from hybrids between varieties which show indication of either mildew resistance or yield and quality, make selections of superior plants, and obtain introductions from foreign sources."[10] It is important to emphasize that solving the downy mildew issue stood as the number-one goal, but he knew that a successful hybrid would also require a good yield, hold up to brewing trials, and store and ship well. In subsequent years, Bressman tested more-advanced breeding techniques such as grafting, and using artificial light in the greenhouse to promote rapid growth of seedlings.[11] (See figure 19.) But success did not materialize quickly.

As much as he set a precedent for outreach, collaboration, and using cutting-edge agricultural science, Bressman also set a precedent in recognizing that the hard work of hop breeding was in the waiting. It could take several years to cultivate mature hybrids, and several more to test for desired

.qualities. To help explain this situation in 1932, Bressman invoked E. S. Salmon, noting, "[T]he work of raising new varieties of hops is an arduous and expensive one and is necessarily very slow, since, the seedling plants do not bear a crop until the third year, and cannot be judged for character such as aroma, richness in resins, and cropping powers until the fifth year, at the earliest."[12] The lengthy time requirements explain the lack of immediate success not only under Bressman but also under his successors in the following decades. This chapter of the Willamette Valley hop industry's story, then, involves abiding patience together with trial and error. While Bressman and his successors did not revolutionize the American hop industry during their careers, their work provided the foundations.

In addition to the breeding research, Bressman worked with other scientists to solve a variety of issues pertaining to the American hop industry. In 1931, he benefited from the USDA's hiring of G. R. Hoerner, a plant pathologist. While the two collaborated in many aspects of the continued "Hop Investigations" and published useful bulletins, they also publicized the program via outreach. They regularly met with local farmers and traveled around the Pacific Northwest to attend meetings with the hop growers and brewers who were intently interested in the developments in Corvallis. Additionally, the two offered radio programs, published scientific papers, and contributed popular articles to the *Pacific Hop Grower,* all of which kept hundreds of regional farmers and brewing-industry professionals in the loop and abreast of the hop research agenda at the experiment station.[13]

Because the hop-breeding program emerged around the same time as the repeal of Prohibition and the revival of American brewing, the Willamette Valley drew national attention from an audience that ranged from New Deal agricultural reformers to farmers looking to get into the hop business. In 1933, Rexford Tugwell, Assistant Secretary of Agriculture, visited Corvallis and nearby farms, where Bressman and Hoerner gave him a tour and explained in detail the breeding program.[14] In the same year, Bressman answered letters from farmers—not just in Oregon, Washington, and California but also in seventeen other states and Mexico—who were curious about the possibilities of this specialty crop in their locales.[15] Despite little progress in the release of new hop varieties and the persistent problem of downy mildew, the USDA hop-breeding program began to make a name for itself soon after its new start following the end of Prohibition.

The flurry of attention was not enough to keep Bressman at his post. In 1934, he left for what he viewed as greener pastures in Washington, D.C., as a USDA scientific adviser under Henry A. Wallace. Despite his brief tenure, Bressman helped establish a strong foundation for the nation's first hop-breeding program. Subsequent directors, however, faced problems similar to those that Bressman confronted. D. C. Smith (1934–1936), R. E. Fore (1936–1944, 1947–1948), J. D. Sather (1944–1947), and K. E. Keller (1948–1955) engaged in the familiar work of hand-pollinating crosses in greenhouse and field, with patient inspection of their results over the years.[16] They also kept up the work that Bressman initiated in outreach, whether meeting with the hop-grower organizations and local brewers, offering radio programs and demonstrations to farmers, or taking experiment station research to the national stage and publishing the results in scientific journals. These directors also engaged in the work of exchanging plant materials with others around the globe, finding, at times, surprising collaborators. In 1937, Fore sent plant materials to Punjab, India, and Tientsin (Tianjin), China; other years witnessed exchanges between the Oregon program and scientists in England, Belgium, Germany, Yugoslavia, New Zealand, Australia, Mexico, and the Soviet Union.[17] The directors hoped that, by the application of Mendelian inheritance science, the greater variety of plant material might yield greater diversity and possibilities in the breeding program. Despite this promise, the words of Fore in the late 1930s summed up the reality of the situation when he noted, "Since this project was started . . . sufficient time has not elapsed to make possible the development of new varieties to the point where they can be distributed."[18]

During the waiting period, G. E. Hoerner, who held his position as plant pathologist from 1931 to 1954, took the lead as the public face of the hop research program. Throughout the 1940s, he distributed a monthly publication called the *Hop Press: A Memorandum of What's Brewin'*. The work kept growers up-to-date with information that ranged from the latest topical sprays, to the individual costs of materials and labor needed to grow hops, to the developments in mechanical pickers, and to the local and international market for hops. In a folksy writing style—including one article that reminded farmers that "lupulin is lucrative"—Hoerner helped keep the growers connected to their trade. He also explained his work on behalf of the hop industry, such as his recommendation that Portland roofers use hops in roofing felt—an idea that never caught on.[19] All the while Hoerner kept growers attuned to the "master plan" of developing American hybrids and explained how that work entailed collaboration with partners across the globe. A section of his memorandum called "Notes from Abroad" also made sure to inform local growers that the work was a product of not only the traditional hop-growing regions in Europe, but also temperate and subtropical regions across the world. Unfortunately, the updates on the breeding program proved limited, reflecting the underlying fact that there was little progress to report.[20]

While the USDA and big brewing companies might have been discouraged by the lack of return on the dollars invested, hop growers, too, were disappointed. The high hopes for the breeding program at Corvallis faltered, making for uncertainty about its future. And the problem intensified as downy mildew worsened by the late 1930s and 1940s. Because of their disease resistance, Fuggle, Brewer's Gold, and Bullion hops increasingly replaced Cluster in the Willamette Valley as the preferred hop, but the program and the growers yearned for better varieties. They wanted a superior, disease-resistant American-bred hybrid.

In 1955, Stanley N. Brooks took the reins of the breeding program in Corvallis and immediately introduced some new ideas. In an effort to find new genetic material that would offer diversified stock for breeding, Brooks sent his colleague Chester E. (Jack) Horner (not to be confused with G. R. Hoerner, whom he replaced as pathologist) to collect native wild American hops in Utah, Arizona, New Mexico, and Colorado. (See figure 20.) Additionally, as Brooks noted in a 1958 report, "the breeding program has been expanded to include a back crossing scheme involving several commercial varieties and selected male parents."[21] Brooks obtained additional hop-breeding materials from Wye College and other parts of the world including Japan, while sending materials to other new places including the Dominican Republic and Israel. By this point there was also a significant exchange of germplasm, or plant material, in the United States among the state agricultural experiment stations in Washington, California, and Idaho.[22] Of no small consequence, in the spring and summer of 1963 Ray A. Neve, Salmon's successor at the program at Wye College, spent his sabbatical with Brooks in the Willamette Valley. All of this activity offered continued hope for the Corvallis team.[23]

While appearing to gain some momentum, Brooks, the most promising director of the hop-breeding program in some time, unexpectedly quit in 1967 to become a USDA administrator in Washington, D.C. Like his predecessors, he might have departed in frustration with the lack of progress. Nonetheless, he left behind a wealth of horticultural knowledge and thousands of hop plants obtained from crosses bred in the previous three decades.

FIGURE 20. Hop seedlings in a greenhouse of Oregon State Agricultural College, 1932. Courtesy Special Collections and Archives Research Center, Oregon State University.

He also left behind a record of publications and work with the Willamette Valley agricultural and brewing communities. There was, however, a major difference between Brooks and those who came before him. His legacy included a promising hop hybrid simply named USDA #56013, the offspring of an older variety called Serebrianka (sometimes called "Silver Hop") found in the germplasm collection in Corvallis and an unknown male pollinator. Brooks and his team of investigators did not know it at the time of his departure, but this hop was soon to become the program's first success story.[24]

DR. HOPS

At the same time that researchers at the Oregon experiment station struggled to meet the demands for a disease-resistant hybrid hop, an Austrian-born plant geneticist made his way to the United States and to a new career in the post–World War II period. Alfred (Al) Haunold, born in Retz, Austria, in 1929 (the same year as the downy mildew outbreak in the Pacific Northwest), was the first son of a secondary school teacher and a musically inclined homemaker. Living in an agricultural region, Haunold grew up fond of plants and botany, and, partly inspired by an uncle who had graduated from the Agricultural University of Vienna, he knew by his teenage years that he would follow a similar path. He attended prep school on a track for agricultural studies, and eventually earned his college degrees from the same university as his uncle. This included a doctorate in plant production and wheat breeding by the time he was twenty-two. The whole time, even though he lived within hours of the most prolific noble-hop-producing areas in the world, he never once encountered the plant in his studies. The agricultural landscape of his childhood featured viticulture, or grape growing.[25]

Haunold's early career, in fact, gave no indication that hops were in his life's plan. At first he appeared destined for work in wheat and corn genetics. In 1953, soon after his graduate work in Austria, he obtained a postdoctoral Fulbright scholarship to work at the University of Nebraska in wheat breeding and genetics. A year later, upon completion of his Fulbright duties, he returned to Austria and was assigned to a desk job far removed from active research. Haunold's heart lay in fieldwork and plant breeding. He wanted more than a desk job working for the Austrian government. When his contacts at the University of Nebraska requested his return, he happily accepted. Once back in the States, Haunold turned again to wheat research. The projects allowed him time to take more classes at the University of Nebraska and finish an American Ph.D., as well as get married and start a family. His life appeared on track toward a career as a plant researcher in the staple crops of wheat and corn. Unfortunately, after several promising years working in Great Plains agricultural science, funding for his program disappeared in 1964.

Following another yearlong stint working behind a desk, this time for the Smithsonian Institution in Washington, D.C., Haunold applied for a newly opened position by the USDA in hop research at Oregon State University. Despite little or no familiarity with the crop, he obtained the position, but not as a hop breeder—an occupation that would later define his career. Following the publication of Rachel Carson's *Silent Spring* in 1962, the USDA began investigating widespread pesticide use in American farming. The hop industry was not exempt, for it relied on several chemicals under scrutiny. Seeking environmentally sound solutions to hop pests and pathologies, the USDA funded Haunold's position for disease-resistance research related to improved agricultural methods and breeding. It was only when Stanley Brooks unexpectedly quit as head of breeding program in Corvallis that Haunold had an opportunity to take on much more. Even with a minimal background in hops, he extended his expertise to become the head of the

FIGURE 21. Alfred Haunold (1929–), plant geneticist who bred two dozen new hop varieties during his career as a USDA researcher and professor by courtesy at Oregon State University, 1980. Courtesy Special Collections and Archives Research Center, Oregon State University.

breeding program. (See figure 21.) From the beginning, he knew patience was the key in this work.[26]

Once Haunold took to breeding hops, he kept a close eye on all promising crosses. At first he found the work challenging and less rewarding than working in wheat and corn genetics. Yet, like his predecessors, he remained committed, corresponding with scientists from around the world, making thousands of crosses more, and taking copious notes to document the efforts of his team. He also connected early and often with the hop-growing families of the Willamette Valley. His early reports made frequent mention of collaboration with the Annen brothers, Robert Coleman & Sons, Frank Fobert, Herman Goschie, Ray Kerr, Dick Kirk, the Krebs brothers, the Pokorny brothers, Joe Serres, the Stauffer brothers, and Carl and Don Weathers; reports also mentioned collaboration with Father Dominic, who headed the hop-growing activities at the Mt. Angel Abbey.[27] As Haunold came to know these growers via Oregon Hop Commission meetings and visits to their farms, he detected a certain amount of dismay because the Corvallis hop-breeding program had not once released a new variety to the public in its thirty-five years.[28]

In 1968, USDA scientists in Idaho upped the stakes. A team led by Robert Romanko in Parma, Idaho, released a hop called Talisman, an improved Cluster variety. The hop was not a hybrid like those that the Corvallis team had been working on, but an open-pollinated seedling found in the field (or landrace). It did not offer more resistance to downy mildew, but it did have an improved yield and grew well, particularly in western Idaho's Treasure Valley. While the Idaho experiment station was not officially sponsored by the USDA for hop breeding, progress there nonetheless put pressure on the Corvallis researchers to deliver. Shortly after the release, Pacific Northwest farmers turned to the Talisman for nearly 4 percent of their crops. The numbers may not have been striking, but with scientists at the Washington State experiment station also making progress in releasing superior Cluster varieties, Haunold and his colleagues in Corvallis hoped to bring home some results.[29]

They were close. Around the time of the Talisman release, Haunold began advanced testing on USDA #56013, the hop first bred by Stanley Brooks in 1956. As Haunold and his team put the plant through more field and brewing trials, they noticed its potential. First and foremost, the hop was noticeably resistant to downy mildew, especially in the root stage, which marked the greatest problem with Cluster hops. USDA #56013 also yielded well and seemed sturdy enough for harvesting and shipping. It drew Haunold's particular interest, however, because Brooks had earlier stated that this new hop resembled Hallertauer-Mittelfrüh, the noble variety long sought by brewers in the German tradition with its balanced aroma and bittering potential. From 1968 to 1970, the USDA #56013 crops harvested in expanded field trials excited all invested parties. Brewers believed that the new hybrid resembled a noble German aroma hop, and growers appreciated the high yield of cones per plant.

It appeared that change was coming after nearly forty years of frustrating results in Corvallis. The new hop seemed primed for release to the public.[30]

Prior to release, Haunold and the program received some attention that they did not expect. In the midst of a Vietnam-era counterculture that derided conformity and embraced sex, drugs, and rock and roll, word spread about hop research unfolding in Corvallis. Hopeful amateur horticultural-ists (or, perhaps, enterprising hippies) recognized *Humulus* as a genetic cousin to *Cannabis,* and sent inquiries about the potential of another cross-breeding venture. In his yearly reports, Haunold took notice of the trend. In 1970, he noted, "During the past year we received an unusually large number of requests for hop seed, as well as for rhizomes. . . . It appears now that the close botanical relationship between *Humulus* and *Cannabis* has been pub-licized . . . which apparently showed that the active ingredients in hemp (Tetrahydrocannabinol, etc.) can also be found in hop leaves when a hop scion is grafted onto hemp rootstock."[31] Although Haunold explained that the crosses would not work, he received more mail in the early 1970s on this topic than any other.

To explain the sudden interest in his research, Haunold pointed to a recently published book by Bill Drake entitled *The Cultivator's Handbook of Marijuana*. The book contained a chapter entitled "Producing an Unrecognizable Hybrid," in which the author stated that marijuana and hop plants could be grafted together, with the end result being a larger plant that would also better hide traditional marijuana buds. Drake even suggested that scientific research sub-stantiated the claim. Haunold and his team did not buy into it, nor did they appreciate the many requests that flooded their offices. As the group wrote together in a yearly report, "The hop research staff concluded that most of the requests were being made by persons wanting to grow secret grass." After that determination, they agreed to reply to requests but that hop "stock could be supplied only to persons or institutions doing bona-fide research in hops."[32] Despite the desire to separate themselves from the myth, increased requests for seeds and plants continued into the early 1970s, most likely for the same reason. It was not the attention that Haunold and his team desired.

HAUNOLD'S HYBRIDS

In 1972, the wait was over. After over forty years of collective work in hop breeding and more than a decade of attention on USDA #56013, Jack Horner,

Haunold, and his colleagues at the Oregon Agricultural Experiment Station released the hop for commercial use.[33] They named it Cascade after the mountain range looming on the eastern side of the Willamette Valley. In retrospect, the name had even more meaning. It also invoked the long cascade of knowledge of hop cultivation and breeding, a vast domain that connected Bavarian monks, English immigrants, and agricultural pioneers across oceans and ages. In any case, the Cascade arose as a potential savior for Willamette Valley hop agriculture. Its resistance to disease was unmatched, and it yielded and stored reasonably well, especially when refrigerated. Equally important, the Cascade's release coincided with a major international hop shortage and pleas from American brewers for an increase in domestic hop supplies.[34]

The Coors Brewing Company of Golden, Colorado, made the first major inquiry about the Cascade. It was just emerging as a national power in the beer industry, and its supply of Hallertauer-Mittelfrüh hops fell short. The company faced an uncertain future. To maintain their competitive momentum with Anheuser-Busch, Miller, Schlitz, and Pabst, the brewmasters and management at Coors recognized the need to seek out additional hop supplies. They believed, like other leaders in the business, that the Cascade hop was the closest match to a noble European variety in its balanced bittering and aroma qualities.[35]

In the same year that the USDA released Cascade to the public, Coors offered Oregon and Washington growers, for whom the hop had been designed and released, a momentous deal that changed the nature of the industry. If the farmers planted new acreages of the recently introduced hop, they would earn one dollar a pound. The price was unheard of compared with the approximately fifty cents that growers earned for other varieties. The offer transformed not just Oregon but the entire Pacific Northwest hop-raising landscape. Growers moved quickly from producing Cluster, Fuggle, and the Wye hops to the Cascade on an industrial scale. Coors signed multiple-year contracts with Cascade growers in Oregon and Washington and within a couple of seasons had turned to Haunold's hop for a significant amount of its brewing needs. By 1975, the Cascade made up roughly 10 percent of American hop acreage.[36] It was an apparent win for Coors, Pacific Northwest farmers, the USDA, and the Department of Crop Science at Oregon State University.[37]

There was an important drawback, though. After Coors turned exclusively to the Cascade, its loyal imbibers began to notice a difference in the lager. Haunold recalls complaints that after two beers the taste proved too strong.[38] Brewing chemists later discovered that although there was a balance between bittering potential and aroma profile that resembled the Hallertauer-Mittelfrüh in the Coors product, the aroma qualities of the Cascade were actually quite different. The Cascade, as it turned out, harked back to the nineteenth century, when European brewers complained that American hops were too rank, pungent, and strong. This is to say that the new hybrid did not integrate well with the easy-drinking lagers produced by big beer. Because of its contracts, Coors continued to use Cascades hops in its beers, but it incorporated them in blends with other hop varieties. The brewing companies began to look elsewhere for a new source of hops, and Pacific Northwest growers once again searched for a new hop variety. Haunold and his colleagues came under renewed pressure to develop varieties to better serve American brewing needs.[39]

Despite this initial outcome, the Coors-Cascade connection underscored a notable transition in American and international brewing recipes and hop agriculture. Following three-quarters of a century as a world-leading producer of hops in the Pacific Northwest, growers had for the first time commercially planted a variety created specifically for them. Prior to the release of the Cascade, nearly all varieties grown were of English and North American origin (Cluster), or English–North American hybrids bred in England. While not ideal for Coors and big beer in the long term, the Cascade proved to have lasting power because of its relative quality and cheapness as compared with imported hops. It also showed, after decades of unsuccessful breeding attempts at Oregon State University, that American hybrids could meet the high standards of traditional hop varieties and could be used with good commercial results. The widespread excitement of Cascade faded somewhat later in the 1970s. But by no means was its presence in the brewing world at an end.

What lay next for the breeding program?

The release of the Cascade undoubtedly relieved the anxiety of Haunold and others affiliated with the Corvallis breeding program, which dated to Earl Bressman's first crosses over forty years prior. Other important developments also arose as the newly anointed "Dr. Hops" and his wife, Mary, determined that their family (now with four children) should remain in Oregon for his work on the wolf of the willow.[40] Aside from the favorable recognition that Cascade brought him and the program, his career was bolstered by a convivial and cooperative team of researchers with whom he enjoyed spending time. This included chemist Stan Likens, pathologist Jack Horner, and agronomist

Chuck Zimmermann, who would later take charge of the hop research program in Washington.[41] The cohort also included Gail Nickerson, a bright and lively individual who engaged not just in hop breeding but in research of all sorts, ranging from chemical properties to traits such as pickability (what the team called "pluckability"), or the measure of resistance expressed from pulling hops off the bine. Nickerson was an anomaly for her work in the late 1960s into the 1980s because there were few women employed as scientists in university settings at the time and because she did not even have an undergraduate degree until she returned to school later in her career. Her intelligence, dedication, and personality transcended of all of this. She played a vital role not only in the testing, but also in the publishing of articles and reports, and she participated at conferences in the United States and abroad.[42] Along with Nickerson and the rest of his research team, Haunold also strengthened connections with representatives of the brewing industry following the release of the Cascade, as well as with the region's hop growers. At the gatherings of the Oregon Hop Commission or in one-on-one meetings with hop farmers in their own fields, he became a part of the hop-growing community. He had found his home as well as his career in the Willamette Valley.[43]

For the rest of the 1970s, Haunold kept up the day-to-day work of breeding and carried on the legacy of former directors. He continued to engage in an extensive exchange of biological specimens and knowledge with scientists in many countries around the world. As an Austrian by birth, however, he was ridiculed by other European scientists and growers for his assistance to the U.S. hop industry. This became evident at the 1973 International Hop Congress in Munich, Germany, when he was showing off the Cascade to his Continental colleagues. They questioned his new allegiances and statements of the Cascade's similarities to Hallertauer-Mittelfrüh, their flagship noble aroma hop. Nevertheless, Haunold maintained global connections with the world's breeding programs, particularly with R. A. Neve and then his successor, Peter Darby, at the Wye College program.[44] Haunold worked extensively with Pacific Northwest growers and their professional organizations, and continued to familiarize himself with the landscape and climates of the Willamette Valley in Oregon and the Yakima Valley in Washington. Haunold also worked with representatives of the United States Brewers' Association, who supported ongoing hop research activities in Corvallis as well as at the agricultural experiment stations in Prosser, Washington, and Parma, Idaho.[45]

Prodded by large brewers and regional growers, Haunold's breeding program expanded in the early 1970s. After the Cascade, the USDA released several additional varieties to the public. Zimmermann at Prosser, in collaboration with Haunold and other colleagues at the agricultural experiment stations in Oregon and Washington and the United States Brewers' Association, first released Comet in 1974. The result of a 1961 cross between a "Sunshine" hop and a wild hop from Utah, Comet had higher alpha acid content than other varieties grown at the time, and it tested well against several plant diseases. Mostly, it was designed for the Yakima Valley, where, as Haunold's research team noted, Bullion and Brewer's Gold hops were "poorly adapted."[46] The Comet did not find the quick success of the Cascade because its aroma profile was judged to be too harsh and reminiscent of wild American hops to gain significant followers.

Even following the lukewarm reception for the Comet hop, American brewers continued to invest and believe in American hybrids. Recognizing that its ongoing persistence and dedication to these projects paid off in the early 1970s, the Oregon Hop Commission optimistically planned for the future. In a January 1974 meeting the organization reported, "[T]he fastest growing breweries are all using the 'Specialty' hops grown in Oregon." Upon that note, the text continued, "Beer consumption is increasing 3 to 5% annually. We must concentrate on meeting the supply demands."[47] A year later, the organization noted that Anheuser-Busch, in particular, made a commitment to the use of American hops. Meeting minutes from April 1975 reported, "The situation is the best in the U.S. it has been in many years, especially Oregon because of the varieties preferred by Breweries."[48] After a long period of decline in production and uncertainty about the future, the Willamette Valley's hop-growing prospects were on the rise. Attention to the breeding program also flourished, indicated in part by the creation of a Hop Research Council that brought together scientists from Oregon State University, Washington State University, the University of Idaho, and the U.S. Department of Agriculture. Although federal funding ran dry in the early 1980s, the program persisted, largely from funding from the brewing industry.[49]

Amid this optimism, Haunold remained focused on still-larger goals. Upon successful release of the first American hop hybrids, Haunold turned his attention to another pressing demand of the hop industry: the creation of a seedless hop that met standards of disease resistance, high yields, storability, and quality in brewing. Seedless hops drew a premium in the global marketplace because brewers preferred not to pay for the extra weight of the seeds, which held no alpha acids, no essential oils, and were also thought to add undesirable flavor constituents to beer. Whereas growers in Continental Europe had long moved in this direction by banning male plants from hop fields, Haunold envisioned a different solution: he would breed a hop that could not breed, and thus would not carry extra seed weight.[50]

In these tasks Haunold's background in wheat genetics dovetailed with his new research. Since the 1940s, scientists working on wheat, rye, tobacco, cotton, and scores of other valuable crops had introduced a revolutionary method of breeding called *polyploidy,* or the introduction of multiple sets of chromosomes to diploid plants (those with normally two sets of chromosomes). Whether done naturally or artificially, introducing multiple sets of chromosomes allows for future evolutionary diversity of traits, including potential seedlessness. Haunold began such tests soon after his arrival in Oregon.[51] His polyploidy experiments centered on Fuggle, the English aroma hop. It took time and hard work to double the chromosome number, but the results proved worth it.[52]

In 1976, Haunold and his colleagues released the first American hybrid triploids, or seedless hops, naming them Willamette and Columbia, for the rivers that framed his adopted valley.[53] Almost immediately brewers took to the Willamette for its "exceptionally desirable aroma characteristics," and growers appreciated its higher yields and disease resistance. More important, the hop appealed to growers and brewers who had desired an American seedless hop for decades. As indicated in the official release, this resulted in a "very low seed set when pollinated by fertile male plants. Therefore, growers receive the customary premium for seedless hops, regardless of pollination by fertile male hop plants."[54] Pacific Northwest growers now had an aroma hop that could compete with the seedless European noble varieties.

With the costs of importing European hops continuing to rise throughout the 1970s, the release of the two new seedless hops could not have occurred at a better time. Macrobrewers such as Anheuser-Busch, a large Fuggle customer, embraced the Willamette as their saving grace at a time when the market for European aroma hops was volatile. It was also another saving grace for a dwindling Oregon hop industry that had been helped only somewhat by the introduction of the Cascade. By the early 1980s, Pacific Northwest hop acreage moved toward Willamette hops, though growers continued to grow Cascades and the Wye varieties. Domestic and international brewers also sought the new hop varieties. Methodically, Haunold had solved several of the long-standing problems of West Coast hop production in his first fifteen years on the job.

Throughout the late 1970s and 1980s, Haunold continued the tradition of publishing, reaching out to the media, and working with growers and brewers. After the success of the Cascade and Willamette varieties, he was able to acquire more funding from the USDA, the United States Brewers' Association, and Pacific Northwest hop-grower organizations. He used it in part for researching North American wild hops, and for travels around the world.[55] His goals were to acquire new scientific information and make contacts, as well as express American interest in expanding its presence in the international hop community. Haunold visited Mexico, Kenya, Japan, Australia, and New Zealand, along with West Germany, Yugoslavia, and England. Because of his success, competence, and charisma, Haunold became one of the leading ambassadors of the American specialty crop, assuring the Pacific Northwest a key role in both agricultural and brewing circles.[56] His visits underscored the continued global nature of hop growing. For at the same time that Haunold helped inject new life into the Pacific Northwest industry, regions in Australia, South Africa, and China also emerged as new centers of hop cultivation.[57]

All the while, Haunold kept up a rigorous and successful breeding program in Corvallis that sought to address new trends. In the 1980s, the hop industry looked toward high-alpha varieties. More than ever, brewers demanded these types of hops, knowing that increased bitterness per pound of hops could be cost-effective. In large part the movement stemmed from the introduction of a hop-extract industry, in which alpha acids could be converted into liquid form by using carbon dioxide as a solvent. Researchers around the world sought to breed high-alpha hops to meet the needs of this new technology. Haunold's release of the Nugget variety in 1984 became one of the most successful of his high alphas. In fact, Oregon and Washington growers turned to the Nugget and other high alphas as a significant portion of their acreage, a trend that carried into the twenty-first century.[58]

Still, Haunold was not finished. In the successful aftermath of the Cascade, Willamette, and Nugget, he became determined to address a final long-lasting problem in American hop agriculture: the inability of American growers to produce types similar to the noble aroma-hop varieties grown in Europe. Since the rise of industrial brewing in the mid–nineteenth century, brewers coveted Hallertauer-Mittelfrüh, Tettnanger, Saazer, and Spalter hops, but American growers had limited success in raising them. The environmental conditions of the North American growing regions—or, as viticulturists suggest, the *terroir*—could produce those hops, but they did not take on the same balanced attributes as their counterparts grown in Europe. Nevertheless, Haunold maintained his commitment to raising the American hop industry onto a competitive level with the European aroma-hop producers.

After multiple-year trials that followed similar methods used throughout the entirety of the breeding program in Corvallis, Haunold finally achieved success. In 1990, he released the Mount Hood, the closest U.S. hop ever bred to a Hallertauer-Mittelfrüh. It was a revelation, a success on so many levels because he achieved what the U.S. brewers had long wanted: an American hybrid that could replace the noble varieties at a lower cost. It was also a sign of things to come. In the next three years, Haunold and the USDA released the related Liberty, Crystal, and Ultra varieties, which also acted as worthy replacements for the Hallertauer-Mittelfrüh. Soon after, Haunold played a major role in releasing Sterling, a replacement for the Czech Saazer hop, and Santiam, a replacement for the German Tettnanger.[59] Although these varieties would not take off as quickly as Willamette, Cascade, and Nugget, growers and brewers embraced all of these new hybrid hops into the twenty-first century. Dr. Hops achieved what few had thought possible.

HAUNOLD'S LEGACY

In 1995, Al Haunold retired from his position with the USDA and as professor by courtesy of crop science at Oregon State University. A range of articles in local newspapers and agricultural and brewing magazines preserve his legacy. They praised him for bringing high-yielding, disease-resistant, and desirable hops to the domestic marketplace. As a representative for Anheuser-Busch noted, "He has probably done more for the U.S. hop industry in terms of breeding than any other single person."[60] Statistics back up that claim, particularly in the Willamette Valley. As a whole, from the 1960s to the 1990s, Willamette Valley hop acreage expanded from approximately three thousand to over six thousand acres. Underscoring that transformation was the replacement of Cluster hops with Haunold's hybrids. The *Barth Reports,* which provide yearly numbers on world hop production, show that Willamette Valley growers moved first toward increased Cascade plantings in the 1970s, then Willamette and Nugget in the 1980s, and, finally, the assortment of aroma hops (including Mount Hood, Liberty, Crystal, and Ultra) by the 1990s and into the turn of the twentieth-first century. During that same time period, Yakima Valley growers moved toward the high-alpha varieties, but also incorporated Cascade and other varieties developed by both Haunold and the research teams in Oregon, Washington, and Idaho. Idaho growers grew mostly Cluster and Galena, a high-alpha variety developed by Robert Romanko at the Parma experiment station.

Haunold's legacy in hop history encompasses much more than the release of twenty-four hop varieties. Although he has noted that he is "just a common guy" and that his life has been "just a series of events,"[61] this is not entirely the case. In the hop and brewing circles of the Pacific Northwest, he is recognized for systematically addressing the many and varied hop-breeding needs throughout his career. Haunold, who resides in Corvallis with his wife, Mary, continues to be a dedicated scientist, but also one with a keen historical sense. His career rests upon nearly one hundred published articles and on the thousands upon thousands of plants cultivated, many now in the National Clonal Germplasm Repository–Corvallis, established by Congress in 1981.[62] His meticulous unpublished notes, documented in yearly USDA reports, often exceeded two hundred pages. The material he left behind reveals an understanding of historical and scientific processes, locally and globally connected over time through the exchange of biological specimens and information from Oregon to New Zealand.

For all of his work, Haunold has been revered by big beer companies that hoped for improved domestic hops, local growers with whom he worked to revive the industry, and scientists who recognized his ingenuity and tact. This is to say that the work of Al Haunold provided a vital link between USDA scientists, small farms, big beer, and beer drinkers from around the world. But, as the next and final chapter of this story explains, his role in the world of hops and beer had other substantial repercussions. Haunold's hybrids would not only be central to reviving Pacific Northwest hop production, but also be at the center of a craft beer revolution.

TEN

Hop Wars

AT THE SAME TIME THAT AL HAUNOLD and his colleagues introduced new hop varieties to Pacific Northwest farmers, a revolution in beer swept the nation. The craft beer revolt against blander, mass-marketed lagers began during the 1960s and 1970s when a small band of rabble-rousers had their senses awakened to the diversity of Ninkasi's hymn (the famous ode to the Sumerian goddess of beer). Some joined the movement after encountering exciting beer flavors during overseas travels and came home wanting more; others rejected the homogenization of taste prevalent in the 1950s and sought more exciting options. These craft beer pioneers included war veterans, engineers, winemakers, and people from all walks of life. But they all had a common goal: the desire for quality ingredients in beer making and a complexity of taste—something unknown to two generations of American consumers since the onset of Prohibition. While big beer companies and their lagers thrived among American consumers during the postwar period, these beer nuts could not stomach any more.

The craft beer revolution arose in relative obscurity on the Pacific Coast, and gained steam when small-scale brewers worked together to break the monopoly of big beer. Since the end of Prohibition, federal and state laws limited the production and sale of alcohol outside of the larger breweries. Challenging those regulations proved as crucial as making beer. Finding financial backers also proved difficult, as did acquiring building space and brewing equipment. In almost every instance, the craft brewers purchased old kettles and other implements from industrial-scale breweries, bottling factories, and dairies. They built their businesses with personal effort and sacrifice. After meeting these challenges, the new brewing pioneers still had to negotiate for shelf space in barrooms and grocery stores and convert a consumer base generations removed from quality beer. To make progress, the craft brewers depended on one another. While they knew that they offered competition among themselves, they also knew that big beer stood as the more formidable opponent. Not unlike the cooperation among small-scale hop growers of the late-nineteenth-century Pacific Northwest, small brewers worked together, exchanging recipes, ideas, ingredients, and equipment in an effort to forge a new identity for American beer.[1]

Historians and popular writers celebrate craft beer's revolution.[2] They have described how from the 1970s to the early twenty-first century the movement spread from California to the rest of the nation, with New England, Colorado, the upper Midwest, and the Pacific Northwest all staking claim to a burgeoning new beer culture. Craft brewers, too, have taken to publishing their own histories.[3] All of these accounts explain the larger meaning of craft beer to American consumers by providing intimate biographies and stories with casts of characters that span the social spectrum and the nation's geography. At the center of what was a revolution of the senses, however, one character shone brightest: the wolf of the willow. At its heart, the craft beer revolution represented a cultural awakening to hops.

STEAM BEER AND LIBERTY

The craft beer movement began inauspiciously. In the spring of 1965 a young Stanford graduate named Fritz Maytag purchased a majority share in the Anchor Brewing Company of San Francisco, one of just a handful of remaining independent breweries in the country. The business dated to the late nineteenth century, but since Prohibition various owners struggled with the company. While not outstanding by most accounts, the signature brew—a steam beer, or a malty lager made in an ale style—offered local imbibers something out of the ordinary. The owners sold the beer mostly to area restaurants, including the famous Old Spaghetti Factory in San Francisco (a funky countercultural joint that closed in the early 1980s, not to be confused with the chain that operates today), of which Maytag was a regular patron. Still, the brewery appeared destined to fail prior to Maytag's purchase. The Iowa transplant was a true lover of quality beer and local food culture. More important, he was an heir of the Maytag appliance and blue-cheese family fortunes. Although he had no prior experience in beer making when he purchased the dilapidated brewery, Maytag had the drive and resources to right a sinking ship.[4]

In Maytag's first years, the brewery still struggled to appeal to consumers and keep its accounts with local restaurants. He worked diligently, seeking to perfect his brewing skills and find the traditional recipe for steam beer that had been long lost. In that process, Maytag abandoned the use of rice, corn, and other additives used commonly in the era by large beer makers. Instead, he added all-grain mash and significantly more hops than the era's standard lagers to infuse rich flavors in his vats. Maytag's beer was not perfect in the immediate aftermath of his taking over the Anchor Brewing Company, but it commanded attention, if for nothing else than dedication to traditional brewing with quality ingredients.

As Maytag's product improved in the late 1960s, a small group of home brewers and beer lovers in California and along the West Coast took notice. And there were reasons beyond just the idea of making good beer. As one historian notes, "Maytag grasped what was lost on mainstream brewing: Out there was an audience eager for authenticity."[5] In this way the early craft brewing crowd was not alone in its endeavors. Shunning the conformity of the long 1950s, Maytag joined farmers and restaurateurs across the country who championed a consumer revolution built on novel ingredients and foods deemed whole, local, and seasonal. Many participants drew inspiration from the writings of J. I. Rodale, Wendell Berry, and others who railed against the industrial agricultural system. For these individuals, the revolution in food connected to a broadening environmental consciousness. But this was not the case for everyone. Some consumers simply wanted more choice outside of the normal food options.[6] This desire arose in large part from the new affluence of the postwar period in the United States, leaving the populace with more leisure time and disposable income.[7] American consumers became swept up with European travels and cuisine, not to mention fine wine, which joined the larger food revolution centered at first in Northern California, just like craft beer.[8]

Upon this backdrop in the early 1970s, Maytag had upgraded his steam beer, introduced a porter, and developed a bottling line.[9] Sales increased. So, too, did attention from like-minded individuals. As it turned out, several small brewers had been keeping a close eye on Maytag with the notion of following a similar path. These included home brewer (and former naval engineer) Jack McAuliffe and his business partners, Suzy Stern and Jane Zimmerman. In Sonoma, California, they opened the first new craft brewery of the post-Prohibition era. While short lasting, their New Albion Brewery (1976–82) was also the first "microbrewery," distinguished as such because it produced a fraction of the volume of national breweries. Running the first business of its kind since Prohibition was not easy. But the owners drew respect for their efforts. McAuliffe aimed for high-quality beer made in traditional styles without additives, and the brewery stood fiercely independent from the grasp of big beer. These became the defining attributes of the craft beer revolution.[10]

With the movement for quality beer gaining momentum in the mid-1970s, Maytag concocted a brew at Anchor that has been called "unforgettable" by the early craft beer crowd.[11] He called it Liberty Ale as a nod to the bicentennial of Paul Revere's ride. But *liberty* had another meaning: freedom from bland lagers. The most noticeable attribute of the new beer was its hoppiness. Liberty Ale was much more bitter than any other beers of its time, and it also held floral and citrusy aromatic notes unlike any other beer in the world. What was it about this ale that gave it such a distinct flavor and aroma? Al Haunold knew. It was the Cascade hop.[12] Whereas Coors and other larger brewers had ultimately rejected the new American hybrid because of its powerful punch, Maytag embraced it. In doing so, he immediately connected the craft beer revolution to the Willamette Valley's rich history of hop farming, labor, business, and science that had unfolded in the previous century. Moreover, this was a sign of things to come. Liberty Ale signaled that craft brewers would not shy away from using the new American hybrids and that they would feature aromatic hops in great quantity. These trends emerged as the road maps to the future when McAuliffe also used the Cascade as the signature hop in his New Albion Ale.[13]

Several other characters and events surrounded Maytag's and McAuliffe's efforts, helping to grow the craft beer movement and its hop connections. Of particular note, a unified band of home brewers, led by Oregon's Fred Eckhardt and Colorado's Charlie Papazian, helped to increase the awareness of craft beer culture in the 1970s and 1980s. Eckhardt's *Treatise on Lager Beers* (1969) and Papazian's *Complete Joy of Home Brewing* (1984) inspired many amateur beer makers. As more people familiarized themselves with quality beer, they began holding home-brewer meetings and festivals. In 1978, under Papazian's lead, American home brewers came together to form the American Homebrewers Association. Shortly thereafter, together with his friend Charles Matzen, Papazian began publishing *Zymurgy,* a home-brewing newsletter that would later become the most important magazine in the genre. That publication, along with popular writings on the craft beer scene by international beer expert and writer Michael Jackson, gave the American public a new access to brewing knowledge and beer culture.[14]

As the movement widened, along with its rising popularity it confronted significant legal and tax hurdles that craft brewers had to overcome. By 1978, there remained only eighty-nine total breweries in the United States, owned by half as many companies.[15] Big beer had a firm grasp on the American beer industry as it continued to conglomerate and take advantage of a legally mandated marketing and distribution system that dated to the repeal of Prohibition. Federal law determined that alcohol production and sale occur within a three-tiered system. This meant that brewers had to sell their beer to distributors, and the distributors then sold that beer to retail outlets—namely, bars, restaurants, and grocery stores. It was a system in which small brewers could not sell their own product without a middleman, and one that safeguarded big beer and its control of large distribution companies. Furthermore, small brewers were taxed the same amount per barrel as larger brewers, thus suffocating their ability to make profits. Finally, post-Prohibition law made it illegal for individuals to home-brew beer, which was essentially the backbone of the craft beer revolution outside of Maytag, who had taken over an existing brewery. Altogether, this framework of regulations made it difficult for small brewers to get off the ground, let alone compete.[16]

The emerging craft beer revolution pushed lawmakers to change the rules. A 1976 tax equity law stood as the first barrier that Maytag, McAuliffe, Papazian, and Eckhardt helped to overcome. In looking at the distribution differences between big beer and small brewers, Congress declared that small brewers would not have to pay as much per barrel in taxes as the Anheuser-Busch, Miller, Pabst, and Coors breweries of the world. Second, in a direct response to his Californian constituency in 1978, Congressman Alan Cranston helped push HR 1337 through Congress, a bill that legalized home brewing for the first time since Prohibition. While the Cranston Law, as it came to be known, was a seemingly small win that did not affect brewing companies seeking to make thousands of barrels of beer each year, it provided opportunity for a craft beer culture. The legislation inspired the new members of a craft beer community and opened the path for states to remove restrictions on beer makers that overwhelmingly favored large breweries. States began providing licenses for small brewers to sell their beer at regional restaurants and bars and eventually their own establishments.[17]

The craft beer–friendly legislation enabled several more small brewers to open shop. Some came and went—for example, the California Steam Beer Company (1979–81) of San Raphael and River City Brewing (1980–84) of Sacramento. Others had staying power. In 1980, the Boulder Brewing

Company of Longmont, Colorado, became the first craft brewery to open east of California. Two years later the Redhook Ale Brewery opened in Seattle, Washington (cofounded by Starbucks founder Gary Nash), as did the Horseshoe Bay Brewery in British Columbia—the first in Canada. All three of those new players had tremendous success, with the Horseshoe Bay Brewery lasting until 1999 and the others doing business into the twenty-first century.[18]

Ken Grossman's Sierra Nevada Brewing Company of Chico, California, however, stole the show after it opened in 1980. Whereas some of the other Northern California breweries found difficulties integrating into the market and winning over customers, Grossman succeeded. As is now the stuff of legend, the Southern California native had a knack for tinkering, fixing, and otherwise figuring out how the world works. His first professional love was working on bicycles, but, upon discovering home brewing and the Anchor and New Albion breweries, Grossman set his sights on making good beer. In 1976, he opened a home-brewing shop in Chico that he ran with his wife, Katie. Four years later, after acquiring financial support from family and friends, he opened what would become the most successful brewery of the revolution, the Sierra Nevada Brewing Company.[19]

At the center of Grossman's success lay his signature beer, Sierra Nevada Pale Ale. Other brewers and beer writers have described the beverage as their "go-to beer" for its consistent quality and complex, balanced flavor.[20] Like Liberty Ale, the beer's hoppy profiles stood out. Deciding on its flavor and aroma was not something Grossman took lightly. He had gone to the source—that is to say, gone north—to meet with Pacific Northwest hop growers to try their wares and tinker with his recipes. Eventually, he settled on three hops for his pale ale: Cluster, Tettnanger, and, perhaps not surprisingly given its role in the craft beer revolution to that point, Cascade.[21]

By deciding to feature the Cascade as the primary aroma hop for his pale ale, Grossman joined Maytag, McAuliffe, and many others in turning the world of brewing on its head. From that point onward, the wolf of the willow was no longer buried in the bland lagers of the post-Prohibition era. Rather, along with traditional malts, hops became the key source of flavor and unique aromas. By embracing the Cascade in particular, the early craft brewers paid homage to an American hop industry that was long seen as inferior to Europe's. Since the rise of commercial hop farming in North America, experts had suggested that the landscape and environmental conditions produced cones that were too bitter and overpowering. Featuring the Cascade twisted that narrative around because consumers began to crave American hops. As Grossman noted in his autobiography, "[I]t took a few years to start gaining acceptance, but in part due to our prominent use of this hop, it is now the number one American hop used by craft brewers in the United States."[22] This was all high praise for the plant, once known simply as USDA #56013, that grew humbly in the greenhouses and fields run by Al Haunold and his colleagues in Corvallis.

During the craft beer revolution as a whole, the hop assumed a new importance. Rather than marking an unrecognizable component of a beer's taste profile, thanks to brewers, hops were becoming the star of the brew. Like Grossman, the brewers studied the ingredient intently to devise novel recipes, and they added more to their kettles than had ever been seen on the North American continent. Many brewers continued to use noble varieties from Europe, but, increasingly, they featured hops selected from the American hybrid varieties bred and released by the agricultural experiment stations of the Pacific Northwest. Many of the new hops—such as Nugget—were the high-alpha varieties, used in the earlier stages of brewing for bitterness. But it was the aroma hops, those added at the end of the wort's boil or later, that drew the interest of the new brewing pioneers. Cascade, along with Willamette, Mount Hood, Sterling, Santiam, Crystal, Ultra, and Liberty, became the hallmarks of the hoppier craft beers that emanated smells and tastes unfamiliar to most Americans.[23]

Throughout the 1980s and into the 1990s, the craft beer revolution blossomed across the country, with the hop as its icon. In Massachusetts, Boston Beer Company's founder, Jim Koch, like Grossman, became one of the most recognizable figures in the movement. Koch's success stemmed from business as a contract brewer, enlisting larger beer makers to produce Samuel Adams Boston Lager from an old family recipe. A well-balanced lager, Koch's signature beer differed from the hoppy ales concocted on the West Coast. Nevertheless, the success of the Boston Beer Company (1984) along with Alaska Brewing in Juneau (1986), Brooklyn Brewing in New York (1987), Boulevard Brewing Company in Kansas City (1988), Great Lakes Brewing Company in Cleveland (1988), and New Belgium Brewing in Fort Collins, Colorado (1991) showed that the movement had spread from coast to coast. The simultaneous growth of the Great American Beer Festival, hosted by Colorado's Boulder Brewery and Charlie Papazian's Association of Brewers, further emphasized the popularity of craft beer. Collectively, the breweries introduced a host of novel beer styles to the public, including pale ales, amber ales, cream ales, pilsners, porters, and stouts. A common denominator in almost every new brew was the elevated status of the hop.[24]

THE RISE OF BEERVANA

In the early 1980s, beer culture in the Pacific Northwest was fairly dismal. The revolution had not yet taken hold in the same way that it did in Northern California with the Anchor, New Albion, and Sierra Nevada breweries. In contrast, the public of the Pacific Northwest was left with a choice of lagers from the country's big beer companies or one of the four big remaining breweries of Washington: Olympia (Tumwater), Lucky (Vancouver), Heidelberg (Tacoma), and Rainier (Seattle). Even Blitz-Weinhard, the longtime standard of Portland brewing, had changed. In 1979, Pabst Brewing Company took over the popular brewery with roots to the 1850s. This left Oregon without a single independent brewery.[25] While Pabst continued to distribute Blitz Lager and the premium Henry Weinhard's Private Reserve, the brewery operated out of the Midwest and no longer drew the praise that it had a century prior. Ironically, the residents of the former Hop Center of the World lived in a place seemingly devoid of beers with hoppy profiles.

Portland-area winemakers Charles and Shirley Coury stood in the vanguard of invigorating the craft beer movement in the Pacific Northwest when they opened the Cartwright Brewing Company in 1980. For a variety of reasons, mainly stemming from difficulties in bottling and distribution, the operation closed just a year later. From that point onward Washington brewers took the reins of regional craft beer. The Redhook Ale Brewery in Seattle continued to attract attention, as did Hart Brewing of Kalama (later Pyramid Brewing in Seattle) when it opened in 1983. Yet it was what was happening in Yakima, the new hop center of the United States, that drew the most comment. There, the irascible Scottish-born Bert Grant opened the Yakima Brewing and Malting Company in 1982, which doubled as the nation's first brewpub, or pub that serves beers made on-site.[26]

More so than any other of the regional craft brewers of the time, the kilt-wearing Grant introduced his customers to dark, rich beers with strong, hoppy profiles. His signature Grant's Scottish Ale was unlike anything else on the market and drew a cult following. (Beer writer Michael Jackson singled out Grant for his innovation, offering a nod at a new global recognition for American craft beer.) His other lines of beers, including an India pale

ale, also drew fanfare for their intensely hopped flavor that followed Grant's fundamental belief that "all beers should be hoppier."[27] Because of his passion for hops, many brewers credit Grant with reintroducing the India pale ale style to American brewing. It should be noted, however, that his recipe differed in two substantial ways from the beers familiar to late-eighteenth-century British sailors embarking for India. First, there were more hops in each sip. Second, Grant's beers featured the American-born hops grown in the Pacific Northwest. And why not? At that moment, Pacific Northwest brewers had access to more hop varieties than at any point in history, with many coming from nearby farmers.

Back in Portland, residents were ready for a craft beer revival. The city had a strong bar and pub culture that dated to its rough-and-tumble nineteenth-century origins as a seaport, and beer writers have remarked that the long rainy season drove residents indoors to drink and socialize. Others suggest that local access to quality water, grain, and hops always made Portland's beer high quality and attractive to consumers.[28] Certainly, the Henry Weinhard Brewery (later Blitz-Weinhard) acted as an ambassador of those ideas for nearly a century. But there were other measures beyond just beer that made Portland a natural fit for the craft beer revolution. The city's residents had a long streak of do-it-yourself culture, one in which Portlanders fiercely supported locally made foods and crafts.[29] For all of these reasons, Portland seemed primed to embrace the craft beer movement, even in the aftermath of Cartwright's failure. Yet, in a story that is now part of the industry lore, Portlanders almost watched their newest craft brewery follow a similar path.[30]

On Friday, November 27, 1984, the day after Thanksgiving, brand-new Columbia River Brewing held an opening reception. The owners, Willamette Valley winemakers Dick and Nancy Ponzi, believed that Portland was ripe for such an experiment. They hoped that they had learned something from the Courys' troubles at Cartwrights. Having secured a spot on Northwest Marshall Street in Portland, the Ponzi's put their trust in head brewer Karl Ockert, a winemaker, too, but one who had also graduated from the brewing program at the University of California at Davis. While a risk in many regards, his beers did not disappoint.[31] Excitement mounted. Yet in an unexpected turn of events that night, the new brewery almost folded. After starting with eighteen kegs there remained just two in the morning. Ockert recalled, "We . . . gave away almost all the beer."[32]

Cheering tongue-in-cheek for the crowds to quaff more than their share of suds that night was Bob Weisskirchen, a representative from the

FIGURE 22. Mike McMenamin, Brian McMenamin, and Ron Wolf (the McMenamins' first head brewer), preparing to imbibe a blackberry beer, Hillsdale, Oregon, 1985. Courtesy *Oregonian* archives.

Blitz-Weinhard Brewery. He recognized the potential for economic failure, and thus less competition, if the crowd consumed more beer than the brewery was prepared to offer. Ockert even remembers that Weisskirchen assisted in the drinking fete and called upon the crowd to "[d]rink up, we'll put them out of business."[33] But the drink-them-out-of-business strategy failed. The Columbia River Brewery staved off ruin that night, and in doing so set the stage for Portland's own craft beer revolution.[34] Ockert brewed roughly four thousand gallons of beer in the first year of business. The next year, the brewery—by then having changed its name to the now-familiar BridgePort Brewing Company—produced over thirty thousand gallons. Success continued thereon with the opening of a brewpub in 1986 and the incorporation of bottling for retail sales by the end of the decade.[35]

The Ponzis and Ockert were not alone. In the shadows of the Blitz-Weinhard facility in Northwest Portland over the next two years, three other craft breweries opened. Kurt and Rob Widmer joined the revolution later in 1984 when they opened their Widmer Brothers Brewery on Northwest

Lovejoy Street. The following year, pub-owning brothers Mike and Brian McMenamin began brewing and selling their own beer at their McMenamin establishments, including the Hillsdale Pub—the first brewpub in Oregon— just southwest of downtown Portland. (See figure 22.) Finally, in 1986, Art Larrance, Fred Bowman, and Clive Johnson opened the Portland Brewing Company on Northwest Flanders, initially making a name for themselves by crafting Bert Grant's recipes, along with their popular Portland Ale.[36]

During this boom in beer making, Portland's Fred Eckhardt (who began to write a regular beer column for the *Oregonian*) made a bold claim early in 1986, declaring Portland the "No. 1 beer town in the nation." He noted that, along with the Blitz-Weinhard Brewery, the small brewers came together to form "the largest brewing district in the Western hemisphere."[37] He was not exaggerating. There was nowhere else like it, with five different breweries operating within a small area of the city. What was more, the beer was by all accounts incredibly good. The new breweries offered a revelation for Portlanders, serving rich, malty, and hoppy beer varieties unknown to most consumers prior to that time. Although the brews could be found only on the taps of some of the region's alehouses and brewpubs, including the famous Horsebrass (operated by the late Don Younger) and Produce Row (a McMenamin establishment) pubs in the southeast section of the city, they inspired consumers and brewers alike to buy into the revolution.[38] Before the year was through, world beer expert Michael Jackson declared Portland the leader of the craft beer revolution in the country.[39]

The success of Portland's craft brewers unfolded along paths similar to those of their counterparts in Northern California and Washington State. Namely, all of the individuals involved knew that they were fighting an uphill battle in a world that favored big beer. To meet those challenges, they often ordered malt and hops together to get better deals; they talked through their plans for building their breweries and shared ideas about the trade.[40] The new brewers also received guidance from the Oregon Brew Crew, Oregon's home brewers' club, founded in 1979.[41] At no time were these collective efforts more evident than when nearly the whole group came together in Salem to lobby the passing of HR 1044 in 1985—a state law that allowed brewers to sell beer at their own breweries. Because they could not get shelf space at regular retail outlets, this was vital. One historian suggests that the Brewpub Bill, as it came to be known, was "arguably the single most important episode in the history of Oregon brewing."[42] While this might be a little overstated given the long history of brewing in the state, it was a huge victory for the brewing pioneers. The results were transformative. The early craft breweries quickly opened brewpubs and changed the culture of the city.[43] Further helping to create new devotees, the craft brewers launched the Oregon Brewers Festival beginning in 1988. Under the watchful eyes of Eckhardt and Portland Brewing's Art Larrance, the festival grew every year and continues as an important part of Portland's cultural scene to this day.[44]

In the wake of the Rose City's first successful craft breweries and brewpubs, others sprang up like mushrooms in the Pacific Northwest rain. Upstream on the Columbia River, Irene Firmat and Jerome Chicvara opened the Hood River Brewery (now Full Sail Brewing Company) in 1987. A year later on the other side of the Cascades, in Bend, Gary Fish began selling his beers at the Deschutes Brewery. Then in 1988 Jack Joyce, Rob Strasser, and Bob Woodell opened Rogue Ales to the south in Ashland, before moving to Newport on the coast a year later. There, master brewer John Maier became celebrated for his innovative beers, including Dead Guy Ale and Rogue Imperial Stout.[45] Several more breweries opened in Portland, too. Hair of the Dog Brewing Company (1993), Lucky Labrador Brewing Company (1994), New Old Lompoc Brewing Company (1996), Alameda Brewing Company (1996), and more McMenamin locations also joined to transform Portland into a craft beer and brewpub city. Even amid some setbacks that tempered craft beer growth in the mid- to late 1990s (which included the failure of some popular craft brewers due to the overextending of their products in the marketplace), Portland's craft beer and brewpub scene kept growing. As early as the 1990s, beer enthusiasts took a page from Eckhardt and called Portland the Craft Beer Capital of the World. Following a 1994 article in the *Willamette Week,* one of the city's most prominent cultural magazines, Portland residents lovingly labeled their city "Beervana."[46]

While diversity of beer styles and a thriving seasonal beer-making culture highlighted what it meant to live in the world's craft beer capital, the wolf of the willow carried on as the defining ingredient of the new brews. The Portland, Oregon, and Pacific Coast craft brewers followed in Ken Grossman's and Bert Grant's footsteps by using a heavy hand in hopping their vats and integrating the hybrid varieties released by Haunold and his colleagues and grown by regional farmers. The Cascade continued to shine as the most popular hop, with many of the most famous beers of the revolution—including Deschutes's Mirror Pond Pale Ale, Portland Brewing Company's MacTarnahan's Amber Ale, and even Widmer's Hefeweizen—depending on it.[47] Far from forgotten, Haunold's other aroma hops also drew substantial attention over time, contributing to a multitude of craft beers that tasted and smelled unlike any other in history. If one thing was clear about the craft beer revolution, it was that hops came to overshadow all other ingredients in the minds of brewers and beer drinkers alike. Fred Eckhardt captured the new meaning of hops when he noted, "The very nature of hops has changed since the advent of our craft beer revolution. Now we have brewers using hops just for the fun of it! We have drinkers searching for hoppy beers just for effect! We have hop growers creating new hop types just because they can!"[48] Beer lovers and critics simply called hoppy beer the West Coast style.

The trend in hopped-up recipes did not dissipate upon the growth of craft beer. In fact, it expanded. Perhaps at no time was this more evident than after BridgePort Brewing Company unleashed its first India pale ale in 1994, a beer more hoppy than anything on the market at the time; it featured Cascade, Chinook, Goldings, Crystal, and Ahtanum hops. *Oregonian* beer writer John Foyston noted at the time, "The new pale ale is even more extravagantly hopped than is normal in Northwest microbrews. Every barrel contains two pounds of five different varieties of hops. That contrasts to the more usual figure of two ounces per barrel."[49] The success of that India pale ale signaled a sign of things to come, with Portland brewers and others across the country engaged in what would be termed the "hop wars," or the unending quest to add more types of hops and more volume than ever before. In doing so, nearly all craft brewers turned toward the India pale ale as a means of satisfying their customers with as hoppy a beer as possible. For many brewers, the India pale ale became their signature beer.

Though hop flavor and aroma might not have been the center of every single beer crafted in the past thirty years, there is little question that the hop became the signature ingredient of the craft beer revolution. The list of beers with *hop* in their name grew by the month, and the hop's image became a marketing point for countless beer labels. Big beer even got in the game, adding specialty hoppy-beer lines to their standard light lagers and buying out local craft brewers, such as Seattle's Redhook. As if that were not enough attention to hops, astronaut Bill Readdy stowed nine ounces of Cascade hops aboard the spacecraft *Discovery* in 1992—hops that, upon his return, would contribute to the first beer crafted with ingredients that had reached beyond earthly limits. Back on the ground, Sierra Nevada and other brewers also took to using hops in new ways, including fresh-hopping recipes—or adding freshly picked hops to the kettle.[50]

Without a doubt, *Humulus lupulus L.*, the wolf of the willow, or the common hop, had achieved notoriety like never in the past because of the late-twentieth-century revolution in beer. Underpinning this all, of course, is the long history of hop agriculture in the Willamette Valley; it is critical to remember that the craft beer revolution would not have unfolded as it did were it not for that. A remaining question occurs: How did the craft beer revolution, in turn, transform the region's hop farmers?

A VALLEY OF AROMA HOPS

Although it might be hard to believe given the celebrated status of the hop, the late-twentieth-century revolution in beer did little early on to affect the American hop industry. Big beer, after all, still made up approximately 95 percent of the domestic market share in volume sold. This being the case, hop farmers and dealers, and the agricultural experiment stations of the Pacific Northwest, maintained strong relationships with the Anheuser-Busch, Miller, Pabst, and Coors breweries of the world. The hop industry unequivocally directed its research and business decisions with those companies in mind. This is not to say that the industry avoided relationships with smaller brewers. Ken Grossman of Sierra Nevada Brewing provided a good example of how some in the craft beer crowd sought out those connections, having visited the Pacific Northwest to sample various hop varieties before deciding on the Cascade for his famous pale ale. Bert Grant and others followed in those footsteps. But these interactions were few and far between. Overall, the early craft beer revolution had a minuscule impact on the region's hop industry, with hop growers and craft brewers seldom interacting.

The truth was that the hop industry had already undergone a revolution of its own that proved more influential than the emerging revolution in beer. Al Haunold and his colleagues' hybrids helped transform the Pacific Northwest commercial hop-growing landscape beginning in the early 1970s. Across the region, farmers turned to Cascade, then Willamette, and then Nugget. In the Willamette Valley that meant an upsurge in total acreage by the 1980s and into the 1990s, nearly doubling the amount of land dedicated to hops from the bleakest era two decades prior.[51] Other events had much more of an impact on the country's hop growers at the time. Hop-acreage controls imposed by several hop-marketing orders (dating to the Great Depression era) ended on December 31, 1986. This change allowed hop farmers to increase acreage if they desired; it also allowed new hop growers to enter the business. Although Congress resolved to continue a plan that dated back to 1966, the USDA made the decision to scrap it by the end of the year. Hop grower Ed Crosby, for one, spoke out against the decision when he noted, "We had 20 profitable years under the market order." He did not want it to end.[52] Other changes in disease management, technology, and branding also demanded more attention than they did from the emerging craft brewing crowd. Growers also embraced the arrival of the digital age by implementing digital irrigation and fertilizer systems and began using the Internet for marketing and sales.[53]

Over time, however, craft beer had an increasing impact on Pacific Northwest hop growers, and nowhere more so than in the Willamette Valley. Above all, early on in the movement craft brewers helped bring a new awareness of hops to their customers. From Anchor's Liberty Ale to BridgePort's India Pale Ale and all of those that followed, craft beers gave a new life and recognition to hops. The "hop wars" showed just how much the American public had pledged its allegiance to this new trend that put hops on a pedestal. In the same way that Americans became more aware of fine wine in the latter part of the twentieth century, they found an interest in craft beers and their hoppy profiles.

As the hop rose to stardom by the turn of the century, more noticeable changes emerged, ones that demonstrated a larger impact of the craft beer revolution on the American hop industry. While craft beer still made up less than 10 percent of beer sold by volume in the country at century's end, small brewers and hop farmers sought out one another to build relationships. One of the most important results of those relationships was that hop farmers became more agreeable to growing a diversity of hops on their acreage— particularly in the Willamette Valley. For example, while standard Oregon hop farmers might have grown predominantly Willamette and Cascade hops in the 1980s, they turned toward a wider selection of aroma hop varieties by the end of the 1990s.[54] They did so because craft brewers desired those novel, powerful aromas that were made available by the Willamette Valley's unique environment and climate. Regional hop-trading companies, including Yakima Chief and Hop Union, aided in the marketing and selling of those hops to local buyers and others across the world.[55]

By the turn of the century, a majority of domestic brewers specifically sought out aroma hops grown in the Willamette Valley as opposed to other regions. The statistics spoke loudly, with the Oregon Hop Commission estimating that brewers used 70 percent of Willamette Valley hops domestically; in comparison, around 70 percent of Yakima Valley hops were exported.[56]

These numbers emphasize that the Oregon industry had become a center of aroma hops. Whereas Washington, German, and Czech hop growers certainly produced more quantities of aroma hops, the Willamette Valley arguably produced more aroma varieties of a better quality than other hop-growing regions of the world, thus making itself the backbone of American craft beer flavor and aroma.[57]

Still, this is not quite the end of the story. In a complete about-face from previous centuries, the American craft beer movement stimulated changes to brewing across the globe. As one journalist notes, "It used to be that American craft brewers took cues from Belgium, Britain, and Germany. Now the industry has come full circle, and breweries worldwide are looking to the U.S. beer scene for inspiration." Another writer notes that American craft brewers were seen as "the most innovative, if not best, in the world."[58] Upon the onset of the new millennium, craft brewers in Italy, Scandinavia, England, and countless other countries found inspiration in the Pacific Coast of North America. They used generous quantities of hops, with a particular interest in the American hybrids. In doing so, the brewers followed their American counterparts in removing the long-standing stigma of extreme bitterness and pungency associated with the American product. This is not to say that all beer drinkers across the world welcomed the stronger beers (with many preferring their light lagers), but the new American influence in hoppy beers could not be ignored.

The new global craft beer culture had an effect on global hop farming, with many farmers of the world experimenting with Pacific Northwest hybrid hops. This has been particularly true in England. There, local craft brewers searched out American beer styles and American hops in an effort to create their own recipes, and in the Borough Market of Southeast London—a stone's throw from what was once one of the world's largest hop exchanges—craft beer vendors exalt their new wares by emphasizing that the hops were grown in Oregon and Washington. Perhaps, most significantly, Peter Darby, England's foremost hop breeder and expert, even suggests that the future of hop growing in his country rests on the ability to emulate the Willamette Valley's model of growing aroma hops for regional craft brewers. Evidence surrounding Darby's research facility near Canterbury offers proof. In those rocky hills of Kent, a commercial grower has planted a substantial acreage of Cascades. While this might have been inconceivable to generations prior, it marked just one more step in the global journey of the wolf of the willow.[59]

Epilogue

HOPTOPIA IN THE TWENTY-FIRST CENTURY

IN THE NEW MILLENNIUM, the craft brewing revolution continues apace. While big beer and its mass-produced lagers still control a significant majority of the market share per volume sold, craft beer's appeal has reached all types of consumers.[1] From coast to coast, small town to metropolis, newly established breweries serve local clientele. According to the Brewers Association, over four thousand breweries operate in the United States today, marking an increase of roughly twenty-five hundred in the past fifteen years alone. These numbers are even more astonishing given that from the late 1950s to the late 1980s there were fewer than two hundred breweries in the country.[2] The substantial upswing in craft brewing has brought about important social transformations within the industry. Perhaps most notably, women have reclaimed roles as professional beermakers. The Pink Boots Society, an organization created "to empower women beer professionals to advance their careers," has over two thousand members and is growing.[3] Similar trends in craft brewing stretch across the world.[4]

No city has embraced the craft beer revolution more than Portland, Oregon. As Christian Ettinger, founder of Hopworks Urban Brewery, on Southeast Powell Boulevard, notes, "The statistics speak for themselves."[5] He is right. As of the autumn of 2015, Portland holds tightly to the title Craft Beer Capital of the World with almost one hundred breweries in its metropolitan area. Given these numbers, it should be no surprise that Rose City residents drink local craft beer at a higher rate per capita than do the citizens of any other city on the planet.[6] Along with Ettinger's business, which opened in 2008, other award-winning Portland breweries of the new millennium include Breakside Brewery, Gigantic Brewing Company, Laurelwood Brewing Company, Migration Brewing Company, and Sasquatch Brewing Company—

but the list goes on. Portland's influence as a craft beer capital has also expanded up and down the Willamette Valley, with thirteen breweries in Eugene alone and over fifty total in the region. Oregon as a whole has well over two hundred breweries. In 2006, to celebrate this growth, Governor Ted Kulongoski designated July as Craft Beer Month. Simply put, craft beer has engrained itself into the cultural fabric of the region.[7]

Although new directions in beer making transcend the hop wars of years past by emphasizing new malts, yeasts, and brewing methods, the wolf of the willow still shines. Brewers continue to feature hop bines and cones on their labels and concoct creative beer names, such as Ninkasi Brewing Company's Tricerahops Double IPA, that demonstrate their adulation for the ingredient. There are many other signs of the hop's ongoing importance and celebrity as well. Rogue Ales and Spirits, of Newport, Oregon, recently purchased a working hopyard near Buena Vista on the Willamette River, vertically integrating the specialty agriculture directly into operations. A handful of growers have also obliged the craft beer industry's request for organic hops, long seen as a difficulty given the plant's susceptibility to pests and diseases. Gayle Goschie and the organic hops of Goschie Farms even became marketing tools when Hopworks Urban Brewery began brewing batches of Gayle's Pale Ale in 2010.[8] (See figure 23.) To emphasize the crop's ongoing stardom outside of just breweries, the towns of Hubbard and Independence have revived annual summer hop festivals. In 2009, the Portland suburb of Hillsboro chose to name its baseball team the Hops, anointing Barley the Hop as the mascot of the franchise. Of interest for scholars, Oregon State University, under the direction of Tiah Edmunson-Morton, has created the Oregon Hops and Brewing Archives, the first of its kind in the nation.[9]

In the twenty-first century, Oregon farmers cultivate an average of five thousand to six thousand acres of hops per year, producing between eight million and eleven million pounds of the crop. The work occurs exclusively in the Willamette Valley and under the direction of less than thirty families working an average of fewer than four hundred acres in hops. Although the size of hop farms has increased in recent decades and growers achieve higher yields per acre than at midcentury when the scourge of downy mildew raged, production sits at about a fifth of the output during the heyday a century prior. In terms of volume, Willamette Valley growers continue to take a backseat to their neighbors in Washington's Yakima Valley, who operate much

FIGURE 23. Gayle Goschie, third-generation hop grower, surrounded by her family's Silverton-area hop acreage, 2014. Photo by Eric Buist, hopstories.com. Courtesy Special Collections and Archives Research Center, Oregon State University.

larger hop farms under conditions of irrigated agriculture. Still, Oregon's "horn of plenty" produces about 5 percent of the world's crop.[10] The varieties grown emphasize the region's newer standing as an aroma hop capital, specializing in hops used for the craft beer industry. In the recent past this meant that, besides the high-alpha Nugget variety, the Willamette Valley hop harvest has centered on the Cascade, Willamette, Centennial, Mount Hood, and many other aroma types.[11]

Amid these developments, the way brewers and hop growers obtain new hop varieties has changed substantially. After 1995, when Alfred Haunold retired from his position with the USDA in Corvallis, the federal government initiated a monumental shift in plant breeding by allowing plant patents. The proprietary nature of hop breeding in the past two decades produced mixed results. The roles of Haunold's successor and other peers diminished, while the privatized nature of hop breeding created greater competition and a wider variety of hops available for brewers. Private companies continue to invest in breeding "super hop" varieties, or those that contain unforeseen amounts of bittering alpha acids and can be efficiently turned into extracts and pellets at cheaper rates. The names of these hops include

Apollo, Zeus, and Millennium, emphasizing strength and superiority (for marketing purposes), as opposed to Haunold's hops, which often had regional names.[12]

Other changes in the hop industry span the worlds of business and the environment—worlds that have always been present in the industry. Fluctuating rates of production can still be a problem, as was the case in 2006, when a "perfect storm" of events left many craft brewers without their favorite ingredient.[13] The hop shortage arose from a reduction of crops grown in the previous decade to adjust the market, a warehouse fire in Yakima that destroyed millions of pounds of hops, and bad weather in Europe that destroyed up to a third of the crops in the Czech Republic and Slovenia, as well as some in England and Germany.[14] Along with these developments, downy mildew continues to affect growers, despite the proliferation of resistant hop varieties. In the 1990s another debilitating disease, powdery mildew, appeared in Pacific Northwest hopyards. The pathogen has been more of a problem for Yakima Valley growers because it affects drier climates. But its presence still threatens Oregonians. Looking toward the future, hop growers will have to tackle the issue of climate change. A warmer region and less rainfall in the Willamette Valley may challenge how and if farmers continue to grow the crop.[15]

Overshadowing all of these issues for Willamette Valley hop growers for now, however, has been the influence of transnational beer conglomerates. In 2009, the Belgian-Brazilian corporation InBev purchased Anheuser-Busch, a major buyer of hops from the Pacific Northwest and the producer of nearly half of American beer by volume. Seeking cheaper supplies of hops, InBev joined a big-beer trend in pursuing crops grown in China and other places, as opposed to big beer's long-standing contracts with farmers in the Pacific Northwest and the Hallertau district of Germany, where quality hops mean higher prices. Those decisions have affected farmers whose families had relied on contracts with Anheuser-Busch for decades. The shift in big market purchases has left many growers questioning their future.[16]

Responses from Pacific Northwest growers have varied. In 2009, the Hop Growers of America (which includes members of the Oregon Hop Commission) met to discuss the rebranding of their organization. They hired a public-relations firm to design a new logo and changed their website in the hopes of bringing recognition to the high-valued hops under their cultivation. Growers have also continued to collaborate with the Pacific

Northwest agricultural experiment stations in an effort to find new uses for and buyers of hops outside of the beer industry. The most tantalizing and feasible use appears to be in animal feed. With rising health concerns surrounding industrialized meat and dairy production, scientists see the use of hops in animal feed as a means to provide organic, as opposed to synthetic, antibacterial agents.[17] Mostly, Willamette Valley growers have decided that their future rests on their relationship with craft brewers and the continued growth of that sector. There are compelling reasons to believe this gamble workable.

Together, the 2006 hop shortage and the Anheuser-Busch takeover by InBev inspired Willamette Valley growers to double down on their standing as a global center of aroma hops used by craft brewers. Doug Weathers, son of hop farmer Don Weathers and grandson of hop farmer Ray Kerr, explains that "the key is quality" for most Willamette Valley growers today; diversity is also important, with the family's Sodbuster Farms producing fourteen different hop varieties alone as of the 2013 season.[18] Many craft brewers up and down the West Coast agree. They continue to build strong relationships with growers, including the Weathers, Goschie, and Coleman families, who see their new role as serving almost exclusively the craft beer industry.[19] The formation of a Hopunion craft brewing division and the independent Indiehops, of Hubbard, Oregon, both of which cater to the craft beer and home-brewing crowds, suggests that the days of multiple-year contracts with big beer may be gone. To ensure quality, growers now mark each bale with the name of their farm so brewers know the provenance of their product.[20]

A PARTING WORD ON SPECIALTY CROPS AND SENSE OF PLACE

Hops never enveloped the Willamette Valley landscape in the way that tobacco and cotton dominated much of the American South in the past or that grains dominate the Great Plains today. Instead, the wolf of the willow has always been a specialty crop, produced with the nearly exclusive purpose of beer making in the past century and a half. But that does not mean the crop fails to tell larger stories about the Willamette Valley. It is precisely because of the scale and the fact that only a handful of regions in the world produce hops commercially that this story is unique. Perhaps in a subtle way at times, this book has tried to show that the Willamette Valley's multilayered hop and beer history helps to cast a "sense of place" or "sense of history" for its residents, offering an intimate way to understand landscapes and peoples, economy and science, and work and culture over time. Other specialty-crop histories might also lend themselves to finding meaning or "sense of place" in other parts of the world.

The Hoptopia story began with the Willamette Valley's local soils and climate and the introduction of a nonnative plant. In the nineteenth century, the crop offered small farmers a source of ready cash income and a steady supply of the brewer's gold for local beer makers. But this was not just a regional story. Cultivation required global exchanges of knowledge, plants, technology, and goods; successful marketing required reaching out to brewers across North America and overseas. In the twentieth century, the global identity of the Oregon industry expanded when hop growers and dealers rethought business practices involving scientific and marketing research. The efforts helped carry the Pacific Northwest industry through the difficulties of the Great War and Prohibition. Those earlier foundations in science and marketing, along with the resolve of the members of the agricultural and brewing communities, also allowed hop culture to grow and change upon the outbreak of downy mildew and then during the "bland" post–World War II period. All along, labor added another critical element, with peoples of different races, classes, genders, and ages finding varied meanings in the yearly harvest. The end of the story—a global craft beer revolution that embraced Oregon-produced hop hybrids—continues to connect Pacific Northwest agriculture and beer to people and places near and far.[21]

In the twenty-first century, many vestiges of hop agriculture in the Willamette Valley have vanished. The region no longer has fifteen hundred individual family hop growers, nor does it have the hundreds of hop dryers that dotted the landscape in the golden era of Willamette Valley hop production. The days of the Hop Fiesta are long gone. But hops on high wire trellises, a curiosity among the region's orchards and fields, remain. They are present along the country roads of Marion and Polk counties, near Silverton, Mount Angel, Aurora, St. Paul, Salem, and Independence. In the late summer, a driver on these back roads has a good chance of following behind a truck transporting hop bines cut from the trellis to the mechanical harvester somewhere down the road. Those plants and their harvest offer physical reminders that the Willamette Valley has been an agricultural utopia for the past century and a half and more recently became a hub for the craft beer revolution. So, too, do encounters with BridgePort IPA bottles or Budweiser cans that have images of hops imprinted on their labels. What might be most remarkable at the end of this story, however, is that Willamette Valley hops and beer history is vividly alive in the smells and tastes of craft beers across the world. Every sip offers an opportunity to connect with that rich and complex past.

NOTES

INTRODUCTION

1. A very small number of hop cones are used in soaps and the cosmetics industry, in teas, and in pharmaceuticals. One fairly recent report noted that only 2 percent of all commercial hops are purchased for the nonbrewing economy. But it also mentioned that that sector is growing. See Joh. Barth & Sohn, *Hops, 2006/07* (Nuremberg: Joh. Barth & Sohn, 2007).

2. Stan Hieronymus, *For the Love of Hops: The Practical Guide to Aroma, Bitterness, and the Culture of Hops* (Boulder, CO: Brewers Publications, 2012), 15–43.

3. Oregon Brewers Guild, "Oregon Beer: World Class, Made Local," *On Tap* (blog), accessed August 14, 2010, oregonbeer.org/beer. Oregon Brewers Guild, "Beer Facts," *On Tap* (blog), accessed September 21, 2009, oregonbeer.org; Pete Dunlop, *Portland Beer: Crafting the Road to Beervana* (Charleston, SC: History Press, 2013), 105. Also, see the popular blog *Beervana* (beervana.blogspot.com), which includes links to dozens of local and national websites on the same subject.

4. In the past decade, hop growing has reemerged in New York State and the Midwest, as well as in the Rocky Mountain states and California. The purpose of these crops is to serve local brewers, and they do not have the same impact on the global marketplace as do the Pacific Northwest states. Hop Growers of America, *2009 Statistical Report* (Moxee, WA: Hop Growers of America, 2010), 11. In the past twenty years, many craft breweries—such as Sierra Nevada (Chico, California), New Belgium (Fort Collins, Colorado), and Rogue Ales and Spirits (Newport, Oregon)—have established hopyards on their properties, but the crops are minimal and inadequate for their levels of brewing. These operations should not be confused with the Pacific Northwest states, which are the only three U.S. states that export hops as a commodity. Internationally, the leading national producers in general order in recent years are Germany, the United States, China, the Czech Republic, the United Kingdom, Slovenia, Poland, Australia, Spain, and Ukraine. For more statistics, see *The Barth Hop Reports* (dating from 1909 to the present),

which have been digitized and made available to the public for free: http://www
.barthhaasgroup.com/en/news-and-reports/the-barth-report-hops.

5. E. Meeker, "Hop Culture," *Oregonian* (Portland, OR), December 14, 1874; Oregon State Board of Agriculture, *The Resources of the State of Oregon: A Book of Statistical Information Treating upon Oregon as a Whole, and by Counties,* 3rd rev. ed. (Salem: Oregon State Board of Agriculture, 1898), 112–16.

6. Many other scholars have used commodity histories to explain unique regional histories. Some that shaped this book include Sterling Evans, *Bound in Twine: The History and Ecology of the Henequen-Wheat Complex for Mexico and the American and Canadian Plains, 1880–1950* (College Station: Texas A&M Press, 2007); John Soluri, *Banana Cultures: Agriculture, Consumption, and Environmental Change in Honduras and the United States* (Austin: University of Texas Press, 2005); Kristen Johannsen, *Ginseng Dreams: The Secret World of America's Most Valuable Plant* (Lexington: University of Kentucky Press, 2006); Michael Pollan, *The Botany of Desire: A Plant's Eye View of the World* (New York: Random House, 2001); Douglas Cazaux Sackman, *Orange Empire: California and the Fruits of Eden* (Berkeley: University of California Press, 2005); Jimmy M. Skaggs, *The Great Guano Rush: Entrepreneurs and American Overseas Expansion* (New York: St. Martin's Press, 1995); and Allen M. Young, *The Chocolate Tree: A Natural History of Cacao* (Washington, DC: Smithsonian Institution Press, 1994).

7. A tradition of referring of the Willamette Valley as Eden began with booster literature from Hall Jackson Kelley in the 1830s and carried into the resettlement period of the nineteenth and early twentieth centuries. Stewart Holbrook in his vast works, and then Ernest Callenbach in his novel *Ecotopia* (1975), continued the tradition during the mid–twentieth century. For more recent scholarship on the topic of Oregon and the Willamette Valley as Eden, see Robert Bunting, *The Pacific Raincoast: Environment and Culture in an American Eden, 1778–1900* (Lawrence: University of Kansas Press, 1997); and James J. Kopp, *Eden within Eden: Oregon's Utopian Heritage* (Corvallis: Oregon State University Press, 2009).

1. WOLF OF THE WILLOW

1. R. A. Neve, *Hops* (London: Chapman and Hall, 1991), 1–29; John Percival, "The Hop Plant," in *The Hop and Its Constituents: A Monograph on the Hop Plant,* ed. Alfred C. Chapman (London: Brewing Trade Review, 1905), 1–9; Heinrich Johann Barth, Christiane Klinke, and Claus Schmidt, *The Hop Atlas: The History and Geography of the Cultivated Plant* (Nuremberg: Joh. Barth & Sohn, 1994), 25–39; Stan Hieronymus, *For the Love of Hops: The Practical Guide to Aroma, Bitterness and the Culture of Hops* (Boulder, CO: Brewers Publications, 2013), 46.

2. Neve, *Hops,* 1–29; Edward L. Davis, "Morphological Complexes in Hops (*Humulus lupulus L.*) with Special Reference to the American Race," *Annals of the*

Missouri Botanical Garden 44, no. 4 (November 1957): 271–94; Ernest Small, "A Numerical and Nomenclatural Analysis of Morpho-Geographic Taxa of Humulus," *Systematic Botany* 3, no. 1 (Spring 1978): 37–76; Richard Hampton, Ernest Small, and Alfred Haunold, "Habitat and Variability of *Humulus lupulus* var. *lupuloides* in Upper Midwestern North America: A Critical Source of American Hop Germplasm," *Journal of the Torrey Botanical Society* 128, no. 1 (January–March 2001): 35–46; Josef Patzak, Vladimír Nesvadba, Alena Henychová, and Karel Krofta, "Assessment of the Genetic Diversity of Wild Hops (*Humulus lupulus* L.) in Europe Using Chemical and Molecular Analyses," *Biochemical Systematics and Ecology* 38, no. 2 (April 2010): 144–45. There are debates about whether *Humulus yunnanensis* is indeed its own species. Scientists last documented its existence in the wild in the 1970s, but subsequent researchers have failed to replicate those efforts. Alfred Haunold, interview by the author, Oregon State University, Corvallis, OR, June 29, 2009; Peter Darby, interview by the author, Wye Hops, China Farms, Canterbury, England, May 21, 2013.

3. For broad discussions on these topics, refer to E. C. Pielou, *After the Ice: The Return of Life to Glaciated North America* (Chicago: University of Chicago Press, 1991); and Hazel R. Delcourt and Paul A. Delcourt, *Quaternary Ecology: A Paeleocological Perspective* (London: Chapman and Hall, 1991).

4. Uwe Koetter and Martin Biendl, "Hops (*Humulus lupulus*): A Review of Its Historical and Medicinal Uses," *HerbalGram* 87 (2010): 44–57.

5. Graeme Barker, *The Agricultural Revolution in Prehistory: Why Did Foragers Become Farmers?* (Oxford: Oxford University Press, 2009), 42–103.

6. Brian Hayden, Neil Canuel, and Jennifer Shanse, "What Was Brewing in the Natufian?: An Archaelogical Assessment of Brewing Technology in the Epipaleolithic," *Journal of Archaeological Method and Theory* 20, no. 1 (March 2013): 102–50; Abigail Tucker, "Dig, Drink, and Be Merry," *Smithsonian,* July/August 2011, 38–48.

7. Ian Hornsey, *A History of Beer and Brewing* (Cambridge: Royal Society of Chemistry, 2003), 84–86.

8. For broad overviews of both beer and wine culture in Europe, see Hornsey, *History of Beer and Brewing;* and Patrick E. McGovern, Stuart J. Fleming, and Solomon H. Katz, *The Origins and Ancient History of Wine* (Amsterdam: Gordon and Breach, 1996).

9. Steven L. Sewell, "The Spatial Diffusion of Beer from Its Sumerian Origins to Today," in *The Geography of Beer: Regions, Environment, and Societies,* ed. Mark Patterson and Nancy Hoalst-Pullen (New York: Springer, 2014), 23–25.

10. Max Nelson, "The Geography of Beer in Europe from 1000 BC to AD 1000," in *The Geography of Beer: Regions, Environment, and Societies,* ed. Mark Patterson and Nancy Hoalst-Pullen (New York: Springer, 2014), 9–19.

11. Hornsey, *History of Beer and Brewing,* 269–76.

12. Hornsey, *History of Beer and Brewing,* 155–56; Koetter and Biendl, "Hops (*Humulus lupulus*)," 44–57. There is some debate over Pliny's naming of the hop, particularly "wolf of the willow" or "wolf of the willows." See Martyn Cornell, "So

What DID Pliny the Elder Say about Hops?," zythophile.co.uk, last modified March 14, 2010, accessed July 27, 2015, http://zythophile.co.uk/2010/03/14/so-what-did-pliny-the-elder-say-about-hops/.

13. Hornsey, *History of Beer and Brewing,* 303–14; Barth, Klinke, and Schmidt, *Hop Atlas,* 25–29.

14. Hornsey, *History of Beer and Brewing,* 303–9.

15. Neve, *Hops,* 25–26.

16. Mark Denny, *Froth!: The Science of Beer* (Baltimore: Johns Hopkins University Press, 2009), 17–19; Hornsey, *History of Beer and Brewing,* 303–22.

17. The creation of hills, of course, needed occur only during a hop plant's first spring. As a perennial, the plant stays alive for many years, with the below-ground portion, or the crown, surviving. For descriptions of preparing the hills, see Meeker, *Hop Culture in the United States,* 86; W. W. Stockberger, *Growing and Curing Hops,* U.S. Department of Agriculture, Farmer's Bulletin 304 (Washington, DC: Government Printing Office, 1907), 11.

18. Barth, Klinke, and Schmidt, *Hop Atlas,* 56–61.

19. Hornsey, *History of Beer and Brewing,* 303–22; Barth, Klinke, and Schmidt, *Hop Atlas,* 25–26; Patzak et al., "Genetic Diversity of Wild Hops," 136.

20. Hornsey, *History of Beer and Brewing,* 303–22; Richard W. Unger, *Beer in the Middle Ages and the Renaissance* (Philadelphia: University of Pennsylvania Press, 2004), 53–106.

21. Hornsey, *History of Beer and Brewing,* 326; Celia Cordle, *Out of the Hay and into the Hops: Hop Cultivation in Wealdon Kent and Hop Marketing in Southwark, 1744–2000* (Hatfield, UK: University of Hertfordshire Press, 2011), 1–13. For more on the general history of the English hop industry, see George Clinch, *English Hops: A History of Cultivation and Preparation for the Market from the Earliest Times* (London: McCorquodale, 1919).

22. Barth, Klinke, and Schmidt, *Hop Atlas,* 25–26; Peter Darby, "The History of Hop Breeding and Development," *Brewery History* 121 (2005): 94–112.

23. Hornsey, *History of Beer and Brewing,* 326–30; Unger, *Beer in the Middle Ages,* 37–52.

24. Barth, Klinke, and Schmidt, *Hop Atlas,* 25–38, 74–116.

25. Reprints of Scot's work appeared in 1576 and 1578. For more on Scot's significance, see Richard Filmer, *Hops and Hop Picking* (Oxford: Shire Publications, 2011), 9–14.

26. Margaret Lawrence, *The Encircling Hop: A History of Hops and Brewing* (Sittingbourne, UK: SAWD, 1990), 15–24; Cordle, *Out of the Hay,* 13, 45.

27. For good overviews on this subject, see Alfred Crosby, *Ecological Imperialism: The Biological Expansion of Europe, 900–1900* (Cambridge: Cambridge University Press, 1986); and Allan Kulikoff, *From British Peasants to Colonial American Farmers* (Chapel Hill: University of North Carolina Press, 2000). Although the history of beer is largely a European story focused on temperate climates, it should be noted that, over time, the transformation of American landscapes via Old World

plants, animals, and disease unfolded due to exchange with Asia and Africa, too. Perhaps the best example is rice cultivation in subtropical areas. See Judith A. Carney, *Black Rice: The African Origins of Rice Cultivation in the Americas* (Cambridge, MA: Harvard University Press, 2001); and Judith A. Carney and Richard Nicholas Rosomoff, *In the Shadow of Slavery: Africa's Botanical Legacy in the Atlantic World* (Berkeley: University of California Press, 2011).

28. Jake E. Haugland, "The Origins and Diaspora of the India Pale Ale," in *The Geography of Beer: Regions, Environment, and Societies,* ed. Mark Patterson and Nancy Hoalst-Pullen (New York: Springer, 2014), 119–21.

29. Amy Mittelman, *Brewing Battles: A History of American Beer* (New York: Algora, 2008) 5–6; Michael A. Tomlan, *Tinged with Gold: Hop Culture in the United States* (Athens: University of Georgia Press, 1992), 11–13.

30. Stanley Baron, *Brewed in America: The History of Beer and Ale in the United States* (Boston: Little, Brown, 1962), 10.

31. Tomlan, *Tinged with Gold,* 1–39. Also see Sarah Hand Meacham, *Every Home a Distillery: Alcohol, Gender and Technology in the Colonial Chesapeake* (Baltimore: Johns Hopkins University Press, 2009).

32. Tomlan, *Tinged with Gold,* 13–26.

33. Baron, *Brewed in America,* 122; Darby, interview.

34. Tomlan, *Tinged with Gold,* 13–26. For more on the Vermont hop industry, see Adam Krakowski, "A Bitter Past: Hop Farming in Nineteenth-Century Vermont," *Vermont History* 82, no. 2 (Summer/Fall 2014): 91–105.

35. Thomas G. Fessenden, *The New England Farmer: Containing Essays, Original and Selected, Relating to Agriculture and Domestic Economy. With Engravings, and the Prices of Country Produce,* vol. 2 (Boston: William Nichols, 1824), 52.

36. Great Britain, Parliament, House of Commons, *Report from the Select Committee on Hop Duties; Together with the Proceedings of the Committee, Minutes of Evidence and Appendix* (London: Great Britain. Parliament. House of Commons, 1857), 87.

37. For example, see Great Britain. Parliament. House of Commons. *Reports from the Select Committee on the Hop Industry. Together with the Proceedings of the Committee, Minutes of Evidence, and Appendix Home Work to Hop Industry* (London: Great Britain. Parliament. House of Commons: 1908., xiv); and "Hops," *Mark Lane Express Agricultural Journal,* December 5, 1898, p. 707.

38. Baron, *Brewed in America,* 68; Michael Pollan, *The Botany of Desire: A Plant's-Eye View of the World* (New York: Random House, 2001), 20–23.

39. W. J. Rorabaugh, *The Alcoholic Republic: An American Tradition* (New York: Oxford University Press, 1979), 5–21, 187–222.

40. Rorabaugh, *Alcoholic Republic,* 7.

41. Rorabaugh, *Alcoholic Republic,* 5–21, 187–222.

42. Tomlan, *Tinged with Gold,* 23–25.

43. Maureen Ogle, *Ambitious Brew: The Story of American Beer* (Orlando, FL: Harcourt, 2006), 15–16, 31–33, 74–79.

1. A. Branscomb, "Geology," in *Willamette River Basin Planning Atlas: Trajectories of Environmental and Ecological Change,* ed. David Hulse, Stan Gregory, and Joan Baker (Corvallis: Pacific Northwest Ecosystem Research Consortium and the Oregon State University Press, 2002), 8–9.

2. William G. Robbins, *Landscapes of Promise: The Oregon Story, 1800–1940* (Seattle: University of Washington Press, 1997), 26–27.

3. J. Baker, J. Van Sickle, and D. White, "Water Sources and Allocation," in *Willamette River Basin Planning Atlas: Trajectories of Environmental and Ecological Change,* ed. David Hulse, Stan Gregory, and Joan Baker (Corvallis: Pacific Northwest Ecosystem Research Consortium and the Oregon State University Press, 2002), 34; Robert Bunting, *The Pacific Raincoast: Environmental and Culture in an American Eden, 1778–1900* (Lawrence: University of Kansas Press, 1997), 5.

4. James R. Habeck, "The Original Vegetation of the Mid-Willamette Valley, Oregon," *Northwest Science* 35, no. 2 (May 1961): 65–77; Robbins, *Landscapes of Promise,* 23–31; Bunting, *Pacific Raincoast,* 5–8.

5. Eugene S. Hunn with James Selam and family, *Nch'i-wána, "The Big River": Mid-Columbia Indians and Their Land* (Seattle: University of Washington Press, 1990), 138–268.

6. For the latest science that attributes megafauna extinction to climate, see Alan Cooper, Chris Turney, Konrad A. Hughen, Barry W. Brook, H. Gregory McDonald, and Corey J. A. Bradshaw, "Abrupt Warming Events Drove Late Pleistocene Megafaunal Turnover," *Science* 349, no. 6248: 602–6; published online July 23, 2015, http://science.sciencemag.org/content/349/6248/602.long.

7. Robert Boyd, "Strategies of Indian Burning in the Willamette Valley," in *Indians, Fire, and the Land in the Pacific Northwest,* ed. Robert Boyd (Corvallis: Oregon State University Press, 1999), 94–128; Hunn, *Nch'i-wána,* 138–200; Bunting, *Pacific Raincoast,* 8–18; C. Melvin Akins, *Archeology of Oregon* (Portland: US Department of the Interior. Bureau of Land Management, 1993), 188. For a larger historical overview of the impact and purpose of Indian burning, see Thomas R. Vale, *Fire, Native Peoples, and the Natural Landscape* (Washington, DC: Island Press, 2002). For comparative discussion of American Indians and their landscapes, see Richard White, *The Roots of Dependency: Subsistence, Environment, and Social Change among the Choctaws, Pawnees, and Navajos* (Lincoln: University of Nebraska Press, 1983); and Richard White, *The Middle Ground: Indians, Empires, and Republics in the Great Lakes Region, 1650–1815* (Cambridge: Cambridge University Press, 1991).

8. Stan Hieronymus, *For the Love of Hops: The Practical Guide to Aroma, Bitterness and the Culture of Hops* (Boulder, CO: Brewers Publications, 2013), 46; Ernest Small, "A Numerical Nomenclatural Analysis of Morpho-Geographic Taxa of Humulus," *Systematic Botany* 3, no. 1 (Spring 1978): 43–49. Generally, there is some uncertainty surrounding the deep-time evolutionary history of *Humulus lupulus L.* Some experts, for example, are not sure if the Eurasia-to–North America migration is supported by evidence.

9. The author wishes to thank Jim Oliphant and Kim Hummer for their answers to my questions about wild North American hops. For some of their recent work on the topic, see J. Smith, J. Oliphant, and K.E. Hummer, "Plant Exploration for Native Hop in the Southwestern United States," *Plant Genetic Resources Newsletter* 147 (September 2006): 29–37; N.V. Bassil, B.S. Gilmore, J.M. Oliphant, K.E. Hummer, and J.A. Henning, "Genic SSRs for European and North American Hop (*Humulus lupulus L.*)," *Genetic Resources and Crop Evolution* 55, no. 7 (November 2008): 959–69; and K.E. Hummer, "Wild *Humulus* Genetic Resources at the U.S. National Clonal Germplasm Repository," in *Proceedings of the First International Humulus Symposium,* ed. K.E. Hummer and J.A. Henning (Leuven, Belgium: ACTA Horticulturae, 2004), 75. For the prehistoric migration of plants and the role of glaciation in determining plant distributions, see Paul A. Delcourt and Hazel R. Delcourt, "Paleoclimates, Paleovegetation, and Paleofloras of North American North of Mexico During the Late Quaternary," chap. 4 of *Flora of North America,* vol. 1, ed. Flora of North America Editorial Committee (New York and Oxford: Flora of North America Editorial Committee, 1993); and E.C. Pielou, *After the Ice Age: The Return of Life to Glaciated North America* (Chicago: University of Chicago Press, 1991).

10. John A. Hussey, "The Fort Vancouver Farm," unpublished manuscript (National Park Service, n.d.), 3–10, http://www.nps.gov/fova/learn/historyculture /historical-studies.htm. The author wishes to thank Doug Wilson and Greg Shine of Fort Vancouver National Historic Site for all of their research assistance on hop growing and brewing at Fort Vancouver.

11. Hussey, "Fort Vancouver Farm," 9–12; Dorothy Morrison, *Outpost: John McLoughlin and the Far Northwest* (Portland: Oregon Historical Society Press, 1999), 152–56. The National Park Service has a wealth of historical and archeological studies related to Fort Vancouver. See the "Online Publications" page of the National Park Service's Fort Vancouver website: http://www.nps.gov/fova/learn /historyculture/online-publications.htm.

12. William A. Bowen, *The Willamette Valley: Migration and Settlement of the Oregon Frontier* (Seattle: University of Washington Press, 1978), 8.

13. Hussey, "Fort Vancouver Farm," 12–27.

14. Hussey, "Fort Vancouver Farm," 89.

15. Hussey, "Fort Vancouver Farm," 35.

16. Lt. Charles Wilkes, [June?] 1841, quoted in "Diary of Wilkes in the Northwest (Continued)," *Washington Historical Quarterly* 17, no. 1 (January 1926): 62.

17. Gov. Etoline to McLoughlin, New Archangel, September 1, 1843, in *The Letters of John McLoughlin, from Fort Vancouver to the Governor and Committee,* 2nd ser., *1839–44,* ed. E.E. Rich (Toronto: Champlain Society, 1943), 330–31; Herbert Beaver to Benjamin Harrison, March 19, 1838, in *Reports and Letters, 1836–1838, of Herbert Beaver, Chaplain to the Hudson's Bay Company and Missionary to the Indians at Fort Vancouver,* ed. Thomas E. Jessett (Portland, OR: Champoeg Press, 1959), 80.

18. Norman H. Clark, *The Dry Years: Prohibition and Social Change in Washington,* rev. ed. (Seattle: University of Washington Press, 1988), 3–20.

19. Lester A. Ross, *Fort Vancouver, 1829–1860: A Historical Archeological Investigation of the Goods Imported and Manufactured by the Hudson's Bay Company,* part 1 (Vancouver, WA: Office of Archeology and Historic Preservation/Fort Vancouver National Historic Site, 1976), 780.

20. "H. Bingham to J. Everts, February 16, 1829," *Oregon Historical Quarterly* 30 (September 1929): 264. Also see a forthcoming article for the *Oregon Historical Quarterly* by Tim Hills on the origins of brewing in the Portland area.

21. See a forthcoming article for the *Oregon Historical Quarterly* by Tim Hills on the origins of brewing in the Portland area.

22. United States, Department of the Interior, Census Office, *Report on the Productions of Agriculture at the Tenth Census (June 1, 1880), Embracing General Statistics and Monographs on Cereal Production, Flour-Milling, Tobacco Culture, Manufacture and Movement of Tobacco, Meat Production* (Washington, DC: Government Printing Office, 1883), 23.

23. Bunting, *Pacific Raincoast,* 39–40.

24. Earl Pomeroy, *The Pacific Slope: A History of California, Oregon, Washington, Idaho, Utah, and Nevada,* rev. ed. (Reno: University of Nevada Press, 1991), 24–26.

25. Clark, *Dry Years,* 9–16.

26. For overviews of Pacific Northwest settlement, see Pomeroy, *Pacific Slope;* Bowen, *Willamette Valley;* D. W. Meinig, *The Great Columbia Plain: A Historical Geography, 1805–1910* (Seattle: University of Washington, Press, 1968); and Dorothy O. Johansen and Charles M. Gates, *Empire of the Columbia: A History of the Pacific Northwest* (New York: Harper & Row, 1967). For specific discussion of the Donation Land Act, see Robbins, *Landscapes of Promise,* 83–85; and Bunting, *Pacific Raincoast,* 93–97.

27. Bunting, *Pacific Raincoast,* 72–76; Dean May, *Three Frontiers: Family, Land, and Society in the American West, 1850–1900* (Cambridge: Cambridge University Press, 1994), 23–27, 148.

28. Robbins, *Landscapes of Promise,* 92; Bunting, *Pacific Raincoast,* 72–76.

29. Robert Bunting, "The Environment and Settler Society in Western Oregon," *Pacific Historical Review* 64, no. 3 (August 1995): 418.

30. For American perceptions of wildlife in this period, particularly the distinction between "beneficial" and "noxious" species, see Thomas R. Dunlap, *Saving America's Wildlife: Ecology and the American Mind, 1850–1990* (Princeton, NJ: Princeton University Press, 1988); Karl Jacoby, *Crimes against Nature: Squatters, Poachers, Thieves, and the Hidden History of American Conservation* (Berkeley: University of California Press, 2014); Richard Judd, *Common Lands, Common People: The Origins of Conservation in Northern New England* (Cambridge, MA: Harvard University Press, 1997); Andrew C. Isenberg, *The Destruction of the Bison* (Cambridge: Cambridge University Press, 2000); Louis S. Warren, *The Hunter's Game: Poachers and Conservationists in Twentieth-Century America* (New Haven, CT: Yale University Press, 1997); and Donald Worster, *Nature's Economy: A History of Ecological Ideas* (Cambridge: Cambridge University Press, 1994).

31. Robbins, *Landscapes of Promise*, 72–74, 77–78, 97–99, 106. The story of a rising agricultural empire in the Willamette Valley is also the story of American Indian dispossession. For recent overviews of how the arrival of Euro-American cultures affected regional tribes, see Laura Berg, ed., *The First Oregonians,* 2nd ed. (Portland: Oregon Council for the Humanities, 2007); and Stephen Dow Beckham, *Oregon Indians: Voices from Two Centuries* (Corvallis: Oregon State University Press, 2006).

32. H. O. Lang, ed., *History of the Willamette Valley, Being a Description of the Valley and Its Resources and Settlement by White Men, and Its Subsequent History; Together with Personal Reminiscences of Its Early Pioneers* (Portland, OR: Himes and Lang, 1885), 447.

33. Robbins, *Landscapes of Promise*, 99; Bowen, *Willamette Valley*, 88–90.

34. Robbins, *Landscapes of Promise*, 98.

35. Lang, *History of the Willamette Valley*, 547; Robbins, *Landscapes of Promise*, 103–7; Rodman Paul, "The Great California Grain War: The Grangers Challenge the Wheat King," *Pacific Historical Review* 27, no. 4 (November 1958): 331–48. For larger discussions of Willamette Valley agriculture and geography in the nineteenth century, see James R. Gibson, *Farming the Frontier: The Agricultural Opening of the Oregon Country, 1786–1846* (Seattle: University of Washington Press, 1985).

36. Robbins, *Landscapes of Promise*, 103.

37. There is a growing body of scholarship on the urban American West, particularly related to environmental history. First and foremost, see William Cronon, *Nature's Metropolis: Chicago and the Great West* (New York: Norton, 1991). For the larger West and Pacific Northwest, see Carl Abbott, *How Cities Won the West: Four Centuries of Urban Change in Western North America* (Albuquerque: University of New Mexico Press, 2008); and Char Miller, ed., *Cities and Nature in the American West* (Reno: University of Nevada Press, 2010). For Portland, see Carl Abbott, *Greater Portland: Urban Life and Landscape in the Pacific Northwest* (Philadelphia: University of Pennsylvania Press, 2001). For Seattle, see Matthew W. Klingle, *Emerald City: An Environmental History of Seattle* (New Haven, CT: Yale University Press, 2008); and Coll Thrush, *Native Seattle: Histories from the Crossing-Over Place* (Seattle: University of Washington, Press, 2007).

38. Richard Maxwell Brown, *Strain of Violence: Historical Studies of American Violence and Vigilantism* (New York: Oxford University Press, 1975), 101–2; Dean L. May, *Three Frontiers: Family, Land, and Society in the American West, 1850–1900* (Cambridge: Cambridge University Press, 1994), 148; Linda Nash, *Inescapable Ecologies: A History of Environment, Disease, and Knowledge* (Berkeley: University of California Press, 2006), 1–8.

39. Doctor E. White, *Ten Years in Oregon: Travels and Adventures of Doctor E. White and Lady West of the Rocky Mountains,* comp. A. J. Allen (Ithaca, NY: Mack, Andrus, 1848), 316–17.

40. For celebratory literary works from late-nineteenth- and early-twentieth-century Oregon writers, see Joaquin Miller, Sam Simpson, Eva Emery Dye, Frederic Homer Balch, and H. L. Davis.

41. James J. Kopp, *Eden within Eden: Oregon's Utopian Heritage* (Corvallis: Oregon State University Press, 2009), 39–87.

42. Clark, *Dry Years,* 17.

3. HOP FEVER

1. The information in this paragraph derives from a forthcoming article by Tim Hills in the *Oregon Historical Quarterly.* Also see Gary Meier and Gloria Meier, *Brewed in the Pacific Northwest: A History of Beer Making in Oregon and Washington* (Seattle: Fjord Press, 1991), 30–163; and Pete Dunlop, *Portland Beer: Crafting the Road to Beervana* (Charleston, SC: History Press, 2013), 20–23.

2. See the forthcoming article by Hills in the *Oregon Historical Quarterly.*

3. Aukjen T. Ingraham, "Henry Weinhard and Portland's City Brewery," *Oregon Historical Quarterly* 102, no. 2 (Summer 2001): 181–86; Dunlop, *Portland Beer,* 29–34.

4. See the forthcoming article by Hills in the *Oregon Historical Quarterly.*

5. Tom Stout, ed., *Montana, Its Story and Biography: A History of Aboriginal and Territorial Montana and Three Decades of Statehood* (Chicago: American Historical Society, 1921), 2: 400.

6. Meier and Meier, *Brewed in the Pacific Northwest,* 30–163. For brewing east of the Cascades, see Herman Ronnernberg, *Beer and Brewing in the Inland Northwest: 1850–1950* (Moscow: University of Idaho Press, 1993).

7. Stanley Baron, *Brewed in America: The History of Beer and Ale in the United States* (Boston: Little, Brown, 1962), 200.

8. Prior to the Isaac Wood and the Meeker family's hop experiment in Western Washington, a handful of farmers had planted desirable English Cluster hops in Oregon. A source from the Oregon Agricultural Extension Service suggested that "[h]ops have been grown in Oregon since 1857, when a few vines were planted near Silverton in Marion County. But an absolute date of the first attempts appears unclear." See Federal Cooperative Extension Service, Oregon State College, *Oregon's First Century of Farming: A Statistical Record of Achievements and Adjustments in Oregon Agriculture, 1859–1958* (Corvallis: Federal Cooperative Extension Service, Oregon State College, 1959), 19.

9. Ezra Meeker, *Ox-Team Days on the Oregon Trail,* ed. Howard R. Driggs (Yonkers-on-Hudson, NY: World Book Company, 1922), 155.

10. The author would like to thank Dennis Larsen of the Ezra Meeker Historical Society for his assistance in uncovering the history of the Meeker family's first hop plantings. In his various writings, Ezra Meeker continually misrepresented both his first year of planting as 1864 or 1865, and the brewer for whom he planted as Charles Wood, not Isaac Wood. Larsen has drawn from various publications, including those with remarks from Ezra Meeker's brother John Valentine Meeker, to paint a more accurate picture. Additionally, there is debate over who planted the first hopyard. Some newspapers from the late nineteenth century indicate that Messrs. Meade and

Thompson of Puyallup began their hopyard a year or two prior to Jacob Meeker. The debate will likely be left unsettled. Regardless, it was clear that Ezra Meeker took the charge of promoting the hop industry in the Pacific Northwest. For Isaac Wood's brewery, see the *Pioneer and Democrat* (Olympia, Washington Territory), July 22, 1859; and the *Pioneer and Democrat* (Olympia, Washington Territory), May 24, 1861. For the Meeker family's first hop plantings, see "Hops and Their Cultivation," *Tacoma Daily Ledger,* July 25, 1894, 2; "Hop Growing in the Pacific Northwest," *Pacific Rural Press* 24, no. 9 (August 26, 1882): 134; Ezra Meeker, *Hop Culture in the United States: A Practical Treatise on Hop Growing in Washington Territory from the Cutting to Bale* (Puyallup, WA: Ezra Meeker, 1883), 8; and Meeker, *Ox-Team Days,* 155. For the Thompson and Meade debate, see "The Hop Fields of the Puyallup," *Snohomish (WA) Evening Star,* September 29, 1877; and the *Pacific Rural Press,* August 26, 1882.

11. "Oregon's First Hopyards," *Oregon Native Son and Historical Magazine* 1, no. 4 (August 1899): 180.

12. "First Hop Yard," *Eugene City Herald,* September 8, 1899, 220. Alternate publications refer to George Leasure as "George Leisure." I use the spelling "Leasure" because the majority of sources use it.

13. H. O. Lang, ed., *History of the Willamette Valley, Being a Description of the Valley and Its Resources, with an Account of Its Discovery and Settlement by White Men, and Its Subsequent History; Together with Personal Reminiscences of Its Early Pioneers* (Portland, OR: Himes and Lang, 1885), 562; "Alexander Seavey, Lane County Pioneer, Ends Well-Filled Life," *Oregonian,* February 2, 1908; M. Bennett and G. Fitzsimons, "Historic American Engineering Record: James W. Seavey Hop Driers" (unpublished document, Historic American Engineering Record), 10–11.

14. "Oregon Hops," *Oregonian,* March 10, 1873.

15. United States, Department of the Interior, Census Office, *Report on the Productions of Agriculture at the Tenth Census (June 1, 1880), Embracing General Statistics and Monographs on Cereal Production, Flour-Milling, Tobacco Culture, Manufacture and Movement of Tobacco, Meat Production* (Washington, DC: Government Printing Office, 1883), 10–15.

16. Census Office, *Report on the Productions of Agriculture,* 10–15, xx.

17. Census Office, *Report on the Productions of Agriculture,* 10, 15; "The First in Hop Production," *Oregonian,* January 1, 1902.

18. "The Agricultural Prospect," *New York Times,* August 1, 1882, 2. Belgian wheat, for example, was at half output for the year. "European Crop Prospects," *New York Times,* August 10, 1882, 3; "Failure of the Wisconsin Hop Crop," *New York Times,* September 4, 1882, 2.

19. Meeker, *Hop Culture in the United States,* 3.

20. E. Meeker, "Hop Culture," *Oregonian,* December 14, 1874.

21. William Cronon, *Nature's Metropolis: Chicago and the Great West* (New York: Norton, 1991), 74–93; William G. Robbins, *Landscapes of Promise: The Oregon Story, 1800–1940* (Seattle: University of Washington Press, 1997), 102–16.

22. There is a long body of scholarship on the history of railroads in the American West. In particularly, see Cronon, *Nature's Metropolis;* James B. Hedges, *Henry*

Villard and the Railways of the Northwest (New Haven, CT: Yale University Press, 1930); Michael P. Malone, *James J. Hill: Empire Builder of the Northwest* (Norman: University of Oklahoma Press, 1996); D. C. Jesse Burkhardt, *Backwoods Railroads: Branchlines and Shortlines of Western Oregon* (Pullman: Washington State University Press, 1994); Richard J. Orsi, *Sunset Limited: The Southern Pacific Railroad and the Development of the American West, 1850–1930* (Berkeley: University of California Press, 2005); and Richard White, *Railroaded: The Transcontinentals and the Making of Modern America* (New York: Norton, 2011). For California specialty crops, see Kevin Starr, *Americans and the California Dream, 1850–1915* (New York: Oxford University Press, 1973); Kevin Starr, *Inventing the Dream: California through the Progressive Era* (New York: Oxford University Press, 1985); Donald J. Pisani, *From the Family Farm to Agribusiness: The Irrigation Crusade in California and the West, 1850–1931* (Berkeley: University of California Press, 1984); David Vaught, *Cultivating California: Growers, Specialty Crops, and Labor, 1875–1920* (Baltimore: Johns Hopkins University Press, 1999); and Carey McWilliams, *California: The Great Exception* (Berkeley: University of California Press, 1998), 108–26.

23. Starr, *Inventing the Dream*, 161–62; Vaught, *Cultivating California,* 95–114.

24. Northern Pacific Railroad, *The Northern Pacific Railroad: Sketch of Its History; Delineations of Its Transcontinental Line; Its Features as a Great Through Route from the Great Lakes to the Pacific Ocean; Its Relations to the Chief Water Ways of the Continent; and, a Description of the Soils and Climates of the Region's Traversed by It as to Their Adaptability to Agricultural Production; with Descriptive and Statistical Exhibits of the Counties on and Near Its Line in Minnesota and Dakota (for the Information of Those Seeking New Homes and Profitable Investments)* (Chicago: Rand, McNally, 1882), 15.

25. "How Oregon Prospers," *New York Times,* May 23, 1886, 10.

26. "A Gem of the Willamette," *West Shore* 13, 1887, 629.

27. For example, see C. W. Mott, General Emigration Agent, "All About Fruit and Hop Raising, Dairying and General Farming, Lumbering, Fishing and Mining in Western Washington" (St. Paul, MN: C. W. Mott, General Emigration Agent, c. 1907), 7, 9. There, he noted, "[t]he deep alluvial soil of the valleys is excellent for general farming and for hay crops, and is considered the best hop land in the world. Hops grow to great perfection and yield more heavily than in any other hop regions of the East or of Europe. The crop runs from 1,500 to 3,000 pounds to the acre. . . . Hops are one of the staple crops of western Washington. They were first grown in the Puyallup Valley, but they are now being cultivated also in the White River, Skagit, Snohomish, Chehalis, and other fertile valleys of western Washington. A larger average yield is realized in this part of the country than anywhere else in the United States, and the quality is the finest."

28. Meeker, *Hop Culture in the United States,* 83.

29. Meeker, *Hop Culture in the United States,* 10–33; "A Cold Bleach on Hops," *Pacific Rural Press* 16, no. 10 (September 1878): 156; "Washington Territory Hop Houses," *Pacific Rural Press* 28, no. 6 (August 1884): 109. According to a later source, "The use of sulphur not only gives the hops the desirable yellow color, but makes

them more uniform in appearance, thus increasing their salability. Many dealers are guided more by color than by other qualities, and such dealers have been known to rate unsulphured hops as inferior, while sulphured hops from the same field were classed as choice." See W. W. Stockberger, *Growing and Curing Hops,* United States Department of Agriculture, Farmers' Bulletin 304 (Washington, DC: Government Printing Office, 1907), 26.

30. "On a Model Hop Farm," *Oregonian,* October 6, 1895.

31. In the Pacific Northwest this resulted in the decrease of three-hundred- to five-hundred-acre farms and the increase of those between ten and one hundred acres. Elizabeth Louise Keeler, "The Landscape of Horticultural Crops in the Northern Willamette Valley from 1850–1920" (Ph.D. diss., University of Oregon, 1994), 58–78; Oregon State University, *Oregon's First Century of Farming,* 2–3. Some of the competition emerged from a back-to-the-land movement or country-life movement of the late nineteenth and early twentieth centuries. Fed up with industrial life, many urbanites dropped their increasingly modern lives and resettled on farms. All of this activity reflected the changing rural landscape in the Willamette Valley, as the 1890s marked the decade in which there was no more unsettled land in the region. For the country-life movement in Oregon, see Keeler, "Landscape of Horticultural Crops," 3–5, 58–78. For the movement more generally, see L. H. Bailey, *The Country-Life Movement in the United States* (New York: MacMillan, 1911); William L. Bowers, *The Country Life Movement in America, 1900–1920* (Port Washington, NY: Kennikat Press, 1974); and Gene Wunderlich, *American Country Life: A Legacy* (Lanham, MD: University Press of America, 2003). Also see the journal of the movement, *Country Life.*

32. Keeler, "Landscape of Horticultural Crops," 2–5; Oregon State University, *Oregon's First Century of Farming,* 2–3, 7, 8.

33. *Tacoma Herald,* October 16, 1879, 3.

34. "San Francisco's Trade with the South Pacific Colonies," *New York Times,* December 29, 1870; Colin Carmichael to J. W. Dole Esq. (Boston, Mass.), April 16, 1897, Colin Carmichael Collection, Washington State Historical Society, Tacoma; Colin Carmichael to Messrs. J. W. Simonds (New York), June 16, 1897, Colin Carmichael Collection, Washington State Historical Society, Tacoma; Colin Carmichael to Messrs. F. W. S & Son, October 18, 1897, Colin Carmichael Collection, Washington State Historical Society, Tacoma.

35. J. P. Meeker to E. Meeker & Co., October 14, 1894; J. P. Meeker to E. Meeker & Co., October 16, 1894; Mr. E. C. Horst to Ezra Meeker, August 21, 1895; all in box 3, Ezra Meeker Collection, Washington State Historical Society, Tacoma (hereafter Meeker Collection).

36. Dunlop, *Portland Beer,* 41.

37. Folder "Lindsay Bird Co. 1892," folder 1B "Financial Records, Jan-June, 1893," folder 2 "Financial Records, July 1893," folder 3A "Financial Records (Aug 1–18, 1893)," folder 11C "Hop Correspondence (November 20–30, 1894)," all in box 3, Meeker Collection. These folders and other ledgers in the Meeker collection show the firm's far reach, as the company not only contracted with merchants from New

York, England, and Germany, but also dealt directly with individual brewers in American cities such as New Orleans and with middlemen in Australia and around the Pacific.

38. Meeker, *Hop Culture in the United States*, 45.

39. Michael A. Tomlan, *Tinged with Gold: Hop Culture in the United States* (Athens: University of Georgia Press, 1992), 33; S. R. Dennison and Oliver MacDonagh, *Guinness, 1886–1939: From Incorporation to the Second World War* (Cork, Ireland: Cork University Press, 1998), 112–14; United States Bureau of Agriculture, Bureau of Agricultural Economics, *Outlook for Hops From the Pacific Coast* (Washington, DC: US Bureau of Agriculture, Bureau of Agricultural Economics, 1948), 28; William A. Schoenfeld, John Marshall, Jr., and Paul C. Newman, *A Compendium of Hop Statistics (of interest to the Pacific Coast States)* (Portland, OR, 1930), 2; "E. Clemens Horst Called by Death," *Pacific Hop Grower*, May 1940; *Tacoma Daily Ledger*, June 29, 1886.

40. Tomlan, *Tinged with Gold*, 50; Emanuel Gross, *Hops: In Their Botanical, Agricultural, and Technical Aspect and as an Article of Commerce*, trans. Charles Salter (London: Scott, Greenwood, 1900), 38–41.

41. George L. Sulerud, *An Economic Study of the Hop Industry in Oregon* (Corvallis: Agricultural Experiment Station, Oregon State Agricultural College, 1931), 28–31; G. R. Hoerner and Frank Rabak, *Production of Hops*, U.S. Department of Agriculture, Farmer's Bulletin no. 1842 (Washington, DC: Government Printing Office, 1940), 6–7.

42. H. T. French and C. D. Thompson, *The Hop Louse*, Oregon Agricultural Experiment Station Bulletin no. 24 (Corvallis, OR: Agricultural College Printing Office, 1893), 9–11.

43. Meeker, *Hop Culture in the United States*, 3; Howard Jaslyn to E. Meeker and Co., June 2, 1892, box 3, Meeker Collection; Gladys Shafer, "Success, Adversity Marked Meeker's Life," *Tacoma Sunday Ledger-News Tribune*, March 19, 1961, 3; Frank L. Green, *Ezra Meeker—Pioneer: A Guide to the Ezra Meeker Papers in the Library of the Washington State Historical Society* (Tacoma: Washington State Historical Society, 1969).

44. Patrick Joseph King, "Labor and Mechanization: The Hop Industry in Yakima Valley, 1866–1950" (master's thesis, Washington State University, 2008).

45. Meeker, *Ox-Team Days*, 155–59; Wm. Sanders (on behalf of the Central Experimental Farm, Department of Agriculture, Ottawa) to E. Meeker Esq., March 30, 1894, box 3, Meeker Collection. This letter is one in a series of correspondence from Canadian scientists seeking to exchange hops and information. It indicated that a Canadian representative from Ottawa would be visiting Meeker and the larger Pacific Northwest to learn about hop growing and to inform the "Canadian Governmental Experimental farms."

46. Meeker, *Hop Culture in the United States*, 4–17.

47. Meeker, *Hop Culture in the United States*, 10.

48. Gross, *Hops*, 96–97.

49. Meeker, *Hop Culture in the United States*, 83.

50. Howard Jaslyn to E. Meeker and Co., June 2, 1892, box 3, Meeker Collection; John A. Shafer to E. Meeker & Son, August 17, 1893, box 3, Meeker Collection.

51. *Oregonian,* December 14, 1874.

52. "Eugene Items," *Oregonian,* September 18, 1876.

53. During the 1870s, the *Pacific Rural Press* produced a series of seven articles called "Hints on Hop Growing," and there were literally hundreds more suggestions. For just a few examples, see "Hints on Hop Growing," *Pacific Rural Press* 8, no. 3 (July 1874); "Commercial Points for Hop Growers," *Pacific Rural Press* 9, no. 19 (May 1875), 309; "Hop Harvest and Outlook," *Pacific Rural Press* 20, no. 10 (September 1880); F. L. Washburn, "The Hop Louse in Oregon," *Pacific Rural Press* 42, no. 21 (November 1891), 431; and "Sweet Clover in a Hop Field," *Pacific Rural Press* 69, no. 17 (April 1905), 258. There are literally hundreds more examples.

54. Oregon State Board of Agriculture, *The Resources of the State of Oregon: A Book of Statistical Information Treating upon Oregon as a Whole, and by Counties,* 3rd ed. (Salem: Oregon State Board of Agriculture, 1898), 124.

55. Herbert Myrick, *The Hop: Its Culture and Cure, Marketing and Manufacture* (1899; reprint, New York: O. Judd, 1904), 11.

56. Gross, *Hops,* 9.

57. One *New York Times* article suggested: "The Hop Crop of Washington and Oregon is in a more promising condition than it was last year. The lice, which have caused such trouble in the yards of England and of New-York State, did no great damage here until 1890, and in some yards there were few lice until last Summer." "A Promising Hop Crop," *New York Times,* July 25, 1892.

58. George K. Holmes, *Hop Crop of the United States, 1790–1911* (Washington, DC: U.S. Department of Agriculture, Bureau of Statistics, 1912), 4–8.

59. Myrick, *The Hop,* 19. The number of court cases surrounding hop growers was staggering. For an overview of some of these cases surrounding Oregon farmers, see: "Hop Grower Wins: Dealer Fails to Recover Possession of Crop," *Oregonian,* August 4, 1903; "Get $6000 on Hop Suit: Lilienthal Bros. Awarded Verdict in United States Court," *Oregonian,* January 19, 1905; "Chinese Win Hop Case: Supreme Court Affirms Verdict against Lachmund & Pincus," *Oregonian,* June 24, 1909; "Krebs Win in Hop Suit Twice before Court Finally Disposed Of," *Oregonian,* October 6, 1909; "Long Damage Suit over Hops Ends: Defendant Company Least at Fault, Supreme Court Says on Rehearing," *Oregonian,* December 14, 1910.

60. Gross, *Hops,* 9, 310.

61. Hoerner and Rabak, *Production of Hops,* 8; J.S. Hough, D.E. Briggs, R. Stevens, and T. W. Young, *Malting and Brewing Science,* vol. 2, *Hopped Wort and Beer* (London: Chapman & Hall, 1982), 392; Peter Darby, interview with author, Canterbury, England, May 22, 2013.

62. Gross, *Hops,* 17.

63. Myrick, *The Hop,* 9; Meeker, *Hop Culture in the United States,* 4.

64. "The Hop Situation," *Oregonian,* September 12, 1894.

65. "Do Not Raise Hops," *Oregonian,* March 13 1896.

66. "Hop Growers' Association," *Oregonian,* June 18, 1877; "Hop Men Organize," *Oregonian,* October 26, 1899.

67. Mark Wyman, *Hoboes: Bindlestiffs, Fruit Tramps, and the Harvesting of the West* (New York: Hill and Wang, 2010), 3–7. Also see Kathleen E. Hudson Cooler, "Hop Agriculture in Oregon: The First Century" (master's thesis, Portland State University, 1986), 36–37.

68. Paul H. Landis, "The Hop Industry, a Social and Economic Problem," *Economic Geography* 15, no. 1 (January 1939): 88; Carl F. Reuss, Paul H. Landis, and Richard Wakefield, *Migratory Farm Labor and the Hop Industry on the Pacific Coast with Special Application to Problems of the Yakima Valley, Washington,* Rural Sociology Series in Farm Labor, no. 3 (Pullman: State College of Washington Agricultural Experiment Station, 1938), 5–6.

69. Sulerud, *Hop Industry in Oregon,* 28–30.

4. HOP-PICKING TIME

1. The Chicago connection here is important because sociologists trained in this tradition focused on empirical observations and ground-up studies. See Martin Bulmer, *The Chicago School of Sociology: Institutionalization, Diversity, and the Rise of Sociological Research* (Chicago: University of Chicago Press, 1984), 1–8.

2. Annie Marion MacLean, "Two Weeks in Department Stores," *American Journal of Sociology* 4, no. 6 (May 1899): 721–41; Annie Marion MacLean, "The Sweat-Shop in Summer," *American Journal of Sociology* 9, no. 3 (November 1903): 289–309.

3. Annie Marion MacLean, "With Oregon Hop Pickers," *American Journal of Sociology* 15, no. 1 (July 1909): 83. For the previous studies mentioned, see her two books: Annie Marion MacLean *Wage Earning Women,* introduction by Grace H. Dodge (New York: MacMillan, 1910); Annie Marion MacLean, *Women Workers and Society* (Chicago: A. C. McClurg, 1916).

4. MacLean, "With Oregon Hop Pickers," 84.

5. Many well-known larger growers provided similar transportation options, both by railroad and by river steamer. Some unscrupulous growers, however, would not provide efficient transportation home, if they provided it at all.

6. MacLean, "With Oregon Hop Pickers": 84–86.

7. For more on the complexities of urban life at the turn of the twentieth century, see T. J. Jackson Lears, *No Place of Grace: Antimodernism and the Transformation of American Culture* (New York: Pantheon, 1981).

8. For newspaper accounts of the period, see "Picking Hops," *Oregonian,* August 18, 1909; and "Hop-Picking Time Is Holiday in Oregon," *Oregonian,* September 24, 1911. For reminiscences from the end of the twentieth century, see Nancie Fadeley, "Hopping into History," *Oregon Business Journal,* December 1990, 65; and Freda W. Jenkins, *You Picked What?* (Freda W. Jenkins, 1993), 25. For a critical perspective, see David Vaught, *Cultivating California: Growers, Specialty Crops, and Labor, 1875–1920* (Baltimore: Johns Hopkins University Press, 1999), 90.

9. MacLean, "With Oregon Hop Pickers," 95.

10. The reason for the great expansion in the first half of the nineteenth century related to changes in the brewing industry. Beer consumption expanded with the introduction of lagers in England, and this, in turn, led to the need for more hops. See Ian Hornsey, *A History of Beer and Brewing* (Cambridge: Royal Society of Chemistry, 2003), 508–22. Additionally, it is important to note that Kent was not England's only hop-growing area. Sussex, Surrey, Hampshire, Worcestershire, and Herefordshire provided competition and added to the hop economy, but with much lower production than Kent. See Celia Cordle, *Out of the Hay and into the Hops: Hop Cultivation in Wealdon Kent and Hop Marketing in Southwark, 1744–2000* (Hatfield, UK: University of Hertfordshire Press, 2011), 9–10.

11. As early as 1849, Londoners commonly referred to the hop-picking time as an "excursion." See Margaret Lawrence, *The Encircling Hop: A History of Hops and Brewing* (Sittingbourne, Kent, UK: SAWD, 1990), 15–24; Cordle, *Out of the Hay*, 13, 45.

12. Hop rash, or the clinically named "hop dermatitis," plagued some growers more than others. On many occasions medical scientists sought to better understand the skin condition. English growers continued to look for solutions into the 1950s. See J. S. Cookson and Ann Lawton, "Hop Dermatitis in Herefordshire," *British Medical Journal* 2 (August 15, 1953): 376–79.

13. Lawrence, *Encircling Hop*, 25–34, 47–60. Also see Cordle, *Out of the Hay*, 14–15. It should be noted that the engineers designed the first railways to Kent from London for marketing purposes, not just the transportation of pickers. For overviews of the British railways, see Jack Simmons, *The Railways of Britain: An Historical Introduction* (London: Macmillan, 1968); John R. Kellett, *The Impact of Railways on Victorian Cities* (Toronto: University of Toronto Press, 1969); and Christian Wolmar, *Fire and Steam: A New History of the Railways in Britain* (London: Atlantic Books, 2008).

14. Lawrence, *Encircling Hop*, 19–50; Cordle, *Out of the Hay*, 5–80.

15. Lawrence, *Encircling Hop*, 35–38.

16. "Three Weeks with the Hop-Pickers," *Littell's Living Age*, 5th ser., vol. 20, no. 1750 (December 29, 1877): 791.

17. Frank T. Maezials, John Foster, Mamie Dickens, and Adolphus W. Ward, *The Life of Charles Dickens* (New York: University Society, 1908), 248.

18. "Three Weeks with the Hop-Pickers," 790.

19. One of the leading histories of the importance of outdoor health and activity is Clifford Putney, *Muscular Christianity: Manhood and Sports in Protestant America, 1880–1920* (Cambridge, MA: Harvard University Press, 2001). For urban health and responses, including rural retreat, also see F. G. Gosling, *Before Freud: Neurasthenia and the American Medical Community, 1870–1910* (Urbana: University of Illinois Press, 1987); Cindy S. Aron, *Working at Play: A History of Vacations in the United States* (New York: Oxford University Press, 1999); Henry J. Tobias and Charles E. Woodhouse, *Santa Fe: A Modern History, 1880–1990* (Albuquerque: University of New Mexico Press, 2001); and Frank Uekötter, *The Age of Smoke:*

Environmental Policy in Germany and the United States, 1880–1970 (Pittsburgh: University of Pittsburgh Press, 2009).

20. "How to Spend a Summer Holiday: Records of Actual Experience: Hop-Picking in Kent," *Leisure Hour,* Summer 1902, 687.

21. For a photographic overview of twentieth-century hop picking in England, see Hilary Heffernan, *The Annual Hop: London to Kent* (Stroud, Gloucestershire, England: Chalford, 1996). For a collection of oral history quotes, see Hilary Heffernan, *Voices of Kent's Hop Gardens* (Stroud, Gloucestershire, England: Tempus, 2000).

22. Joan M. Jensen, "Women in the Hop Harvest from New York to Washington," in *Promise to the Land: Essays on Rural Women* (Albuquerque: University of New Mexico Press, 1991), 97–109. Also see Vincent DiGirolamo, "The Women of Wheatland: Female Consciousness and the 1913 Wheatland Hop Strike," *Labor History* 34, nos. 2–3 (1993): 236–55.

23. C.G. Brainard, "Hop-Picking in Central New York," *Country Life in America* 6, no. 4 (August 1904): 342.

24. Hop growing arose so quickly in New York that growers and railroad companies cooperated to establish railroads for the "express purpose of bringing the hop pickers." See Michael A. Tomlan, *Tinged with Gold: Hop Culture in the United States* (Athens: University of Georgia Press, 1992), 121.

25. Frances M. Smith, "A Hop Field Illustrated," *Frank Leslie's Popular Monthly,* July–December 1893, 507.

26. Some have speculated that the term *hop* meaning "dance" originated in the hop fields. But it is unclear, at least according to the *Oxford English Dictionary.* The Lindy Hop and other jazz dances had origins in African-American slave culture, and the term *hop* more likely referred to the physical movement of hopping.

27. For example, see Maud Ellen Baggarley, "Joy: A Hop-Yard Story," *Young Woman's Journal* 17, no. 11 (November 1906): 496–99; and Smith, "Hop Field Illustrated," 507.

28. Henry H. Johnson, "Hop-Picking Time," in *Ballads of the Farm and Home* (Elkhart, IN: Mennonite Publishing Company, 1902), 86.

29. Ninetta Eames, "In Hop-Picking Time," *Cosmopolitan,* November 1893, 27.

30. For western American rail travel, particularly by women, see Amy G. Richter, *Home on the Rails: Women, the Railroad and the Rise of Public Domesticity* (Chapel Hill: University of North Carolina Press, 2005); Anne Farrar Hyde, *An American Vision: Far Western Landscape and National Culture, 1820–1920* (New York: New York University Press, 1990).

31. "Among the Hop Pickers," *Oregonian,* September 26, 1888. For another example, see Susan Lord Currier, "Some Aspects of Washington Hop Fields," *Overland Monthly* 32 (December 1898): 541–44.

32. Vaught, *Cultivating California,* 90.

33. Iris Tarbell (Yankton, Oregon) to Anna (Tarbell?), October 13, 1895, Tarbell-Brown Family Papers, Oregon Historical Society, Portland.

34. Hamill recalled that the main purpose for the family hop-picking trip in those years was to buy supplies for his father's fishing business. Robert

M. Hamill, "Hop Picking—Hamill Family 1904 and 1906," manuscript—Hop Picking—Hamill Family 1904 and 1906, Benton County Historical Society, Philomath, OR.

35. Amanda Grim, oral history by Benton County Historical Society, Philomath, OR, April 6, 1982.

36. In many instances, pole pullers of certain races or ethnicities worked in segregated yards according to their cultural background. See Tomlan, *Tinged with Gold*, 142–43.

37. Ezra Meeker, *Hop Culture in the United States: Being a Practical Treatise on Hop Growing in Washington Territory from the Cutting to Bale* (Puyallup, WA: Ezra Meeker, 1883), 24.

38. Tomlan, *Tinged with Gold*, 143.

39. Tomlan, *Tinged with Gold*, 143–44; Mollie Schneider Willman, "Wire Down: Memories of the Hop Harvest," unpublished reminiscence compiled and edited by Vickie William Burns (St. Paul, Oregon, c. 1995); Fadeley, "Hopping into History," 65.

40. Kathleen E. Hudson Cooler, "Hop Agriculture in Oregon: The First Century" (master's thesis, Portland State University, 1986), 59.

41. "Yamhill Woman Sets New Hop Gathering Record," 1912, Scrapbook Collection, #44 138, Oregon Historical Society, Portland.

42. For changes to American society and culture in the late=nineteenth- and early-twentieth-century consumer culture, see Lawrence B. Glickman, *A Living Wage: American Workers and the Making of Consumer Society* (Ithaca, NY: Cornell University Press, 1997); Susan Strasser, Charles McGovern, and Matthias Judt, *Getting and Spending: European and American Consumer Societies in the Twentieth Century* (Cambridge: Cambridge University Press, 1998); and Susan J. Matt, *Keeping Up with the Joneses: Envy in American Consumer Society, 1890–1930* (Philadelphia: University of Pennsylvania Press, 2003).

43. Sidney Newton, oral history by Kathleen Hudson Cooler, Benton County Historical Society, Philomath, OR, March 24, 1982.

44. Again, by no means was providing transportation unique to the Willamette Valley or Pacific Coast hop industry; growers in Kent and New York had done the same thing. Nor were the practices unique to the hop industry. As one scholar noted in the mid–twentieth century about the turn-of-the-century Midwest, "Wherever there were berries, fruit, or vegetables to be gathered, women and children were employed. By 1900 flat cars and steamboats carried thousands of women and children from Chicago and Detroit to the berry fields." See LaWanda F. Cox, "The American Agricultural Wage Earner, 1865–1900: The Emergence of a Modern Labor Problem," *Agricultural History* 22, no. 2 (April 1948): 107.

45. Tomlan, *Tinged with Gold*, 140–41.

46. Mark Wyman, *Hoboes: Bindlestiffs, Fruit Tramps, and the Harvesting of the American West* (New York: Hill and Wang, 2010), 3–7; Patricia Nelson Limerick, *The Legacy of Conquest: The Unbroken Past of the American West* (New York: Norton, 1987), 18–32, 259–92.

47. Vaught, *Cultivating California*, 68.

48. Meeker, *Hop Culture in the United States*, 18.

49. Meeker, *Hop Culture in the United States*, 18.

50. "Eugene Items," *Oregonian*, September 18, 1876.

51. "Indian Hop-Pickers," *Overland Monthly*, 2nd ser., vol. 17, no. 98 (February 1891): 164; William Bauer, "Working for Identity: Race, Ethnicity, and the Market Economy in Northern California, 1875–1936," in *Native Pathways: American Indian Culture and Economic Development in the Twentieth Century*, ed. Brian Hosmer and Colleen O'Neil (Boulder: University Press of Colorado, 2004): 242–45.

52. "News of the Northwest," *Oregonian*, July 21, 1885, 2.

53. "Puyallup Valley Notes," *Oregonian*, September 25, 1882. Also see the discussion in Coll Thrush, *Native Seattle: Histories from the Crossing-Over Place* (Seattle: University of Washington Press, 2007), 109–11.

54. "Indian Hop-Pickers," 162.

55. Clyde Ellis, "Five Dollars a Week to Be a 'Regular Indian': Shows, Exhibitions, and the Economics of Indian Dancing, 1880–1930," in *American Indian Culture and Economic Development in the Twentieth Century*, ed. Brian Hosmer and Colleen O'Neil (Boulder: University Press of Colorado, 2004): 184–208; Louis S. Warren, *Buffalo Bill's America: William Cody and the Wild West Show* (New York: Alfred A. Knopf, 2005).

56. Paige Raibmon, *Authentic Indians: Episodes of Encounter from the Late Nineteenth-Century Northwest Coast* (Durham, NC: Duke University Press, 2005), 88–93, 126, 141, 217. For more of this topic, see Philip J. Deloria, *Playing Indian* (New Haven, CT: Yale University Press, 1998).

57. Wallace Nash, *The Settler's Handbook to Oregon* (Portland: J. K. Gill, 1904), 127.

58. Gale Evans, oral history by Daniel C. Robertson, April 7, 1982, Benton County Historical Society, Philomath, OR.

59. Bauer, "Working for Identity," 242–45; Eric V. Meeks, "The Tohono O'odham, Wage Labor, and the Resistant Adaptation, 1900–1930," *Western Historical Quarterly* 34, no. 4 (Winter 2003): 468–89; Robert B. Campbell, "Newlands, Old Lands: Native American Labor, Agrarian Ideology, and the Progressive-Era State in the Making of the Newlands Reclamation Project, 1902–1926," *Pacific Historical Review* 71, no. 2 (May 2002): 203–38. A panel at the 2010 annual meeting of the Western Historical Association indicates that many scholars are in the midst of broadening historical understanding of Indian wage labor in the American West. Arguably, because of the associated tourist industry discussed at length by Paige Raibmon and Cole Thrush, the hop fields provided the most visible example of the indigenous transition in the cash economy via wage labor, but the panel provides other important insights that have been both published and unpublished at the time of this writing. The presenters and their papers included Michael Magliari, "Indian Slavery in California's Sacramento Valley, 1850–1867"; William J. Bauer, Jr., "California Indian Work and the Making of Anthropology"; and Robert Walls, "Coastal Salish Loggers and the Intercultural Landscape of Woods Work."

60. George Ehret, *Twenty-Five Years of Brewing with an Illustrated History of American Beer* (New York: George Ehret, 1891), 111. Ehret was one of New York's most prominent brewers.

61. "Indian Hop-Pickers," 164; Alexandra Harmon, *Indians in the Making: Ethnic Relations and Indian Identities around the Puget Sound* (Berkeley: University of California Press, 1998), 106; Raibmon, *Authentic Indians,* 109–10; Richard Steven Street, *Beasts of the Field: A Narrative History of California Farmworkers 1769–1913* (Stanford, CA: Stanford University Press, 2004), 157. Street's interpretation of California Indian labor differed greatly from my study of the Willamette Valley. He suggests that after genocide and indentured servitude following the discovery of gold in California, Indian farmworkers faced debt, disease, and alcohol abuse. He notes that by the early twentieth century, when California agriculture greatly expanded, "Indians would not be around in large enough numbers to furnish the required labor." For a comparison with southwestern American Indian economies and cultures, see Erika Marie Bsumek, *Indian-Made: Navajo Culture in the Marketplace, 1868–1940* (Lawrence: University Press of Kansas, 2008).

62. "The Scheme to Introduce Child Labor in the Hop Fields," *Oregonian,* September 14, 1888; "Solving a Problem," *Oregonian,* September 14, 1888.

63. "Child Labor in the Hop Fields"; "Solving a Problem."

64. D. P. Durst, "White Labor at Hop-Picking," *Pacific Rural Press* 36, no. 8 (August 1888): 149. In this article, Durst noted, "In other words, when the whites commenced picking they picked a small one-third of the crop, but now they pick over one-half. They do the work well and appear to be well satisfied doing it. In a few more years the whites will pick all of the hops in Yuba country."

65. "Siwash," *Overland Monthly,* 2nd ser., vol. 20, no. 119 (November 1892): 502.

66. "All among the Hops," *Oregonian,* September 20, 1888.

67. *West Shore* 9, no. 6 (June 1883): 139.

68. Sucheng Chan, *This Bittersweet Soil: The Chinese in California Agriculture: 1860–1910* (Berkeley: University of California Press, 1986), 7, 20–23, 30–31.

69. Carey McWilliams, *Factories in the Field: The Story of Migratory Farm Labor in California* (Boston: Little, Brown, 1939), 68.

70. Vaught, *Cultivating California,* 71.

71. Street, *Beasts of the Field,* 242.

72. Chan, *This Bittersweet Soil,* 201–2, 332–33. Also see McWilliams, *Factories in the Field,* 59. In 1880, Oregon was the home to 9,510 Chinese immigrants. In 1910 that number dropped to 6,468 and again in 1920 to 2,151. There were also smaller numbers of Japanese immigrants in Oregon. In 1880 the census recorded only 2 residents. By 1900 that number jumped to 2,522, and in 1920 it jumped again to 12,971. Many of the Japanese agriculturalists had moved to the Hood River area in the Columbia Gorge, but Asian immigrants also worked in canneries, logging, and railroad building. See Linda Tamura, *The Hood River Issei: An Oral History of Japanese Settlers in Oregon's Hood River Valley* (Urbana: University of Illinois Press, 1993); and Wyman, *Hoboes,* 91–96.

73. "Hops and Hop Picking," *Oregonian*, August 6, 1885. For similar sentiments, see "Lebanon Notes," *Oregonian*, July 1, 1885.

74. Wyman, *Hoboes*, 91. Also see Ming Kee, oral history by Daniel C. Robertson, April 19, 1982, Benton County Historical Society, Philomath, OR.

75. "Why Chinese Were Hired," *Oregonian*, September 7, 1880.

76. "Harvesting of the Hop Crop," *Oregonian*, August 28, 1889.

77. "Notes of the Northwest," *West Shore*, 10, no. 10 (October 1884): 336.

78. "Girls Best Hop Pickers," *Chico Enterprise*, August 9, 1911.

79. "All among the Hops."

80. For a larger discussion, see Chan, *This Bittersweet Soil;* and Erika Lee, *At America's Gates: Chinese Immigration during the Exclusion Era, 1882–1943* (Chapel Hill: University of North Carolina Press, 2003).

81. "Chinese Hop-Pickers," *Oregonian*, September 6, 1893, 8; "All Quiet in Butteville," *Oregonian*, September 8, 1893, 8; "Hop-Pickers Are Disappointed: Preference Given Oriental Laborers," *Oregonian*, September 9, 1904, 4; Hudson Cooler, "Hop Agriculture in Oregon," 49–50; Ruth Kirk and Carmela Alexander, *Exploring Washington's Past: A Road Guide to History*, rev. ed. (Seattle: University of Washington Press, 1995), 346; Tomlan, *Tinged with Gold*, 138; Raibmon, *Authentic Indians*, 79; Carlos Schwantes, "Protest in a Promised Land: Unemployment, Disinheritance, and the Origin of Labor Militancy in the Pacific Northwest, 1885–1886," *Western Historical Quarterly* 13, no. 4 (October 1982): 373–90.

82. Cox, "American Agricultural Wage Earner," 101–3.

83. Vaught, *Cultivating California*, 119–28.

84. Tomlan, *Tinged with Gold*, 128.

85. Vaught, *Cultivating California*, 90.

86. Ottilia Willi, "Hop-Picking in the Pleasanton Valley," *Out West* 19, no. 2 (August 1903): 158.

87. Willi, "Hop-Picking in the Pleasanton Valley," 158.

88. Kee, oral history; Harvey Kaser, oral history by Kathleen Hudson Cooler, April 16, 1982, Benton County Historical Society, Philomath, OR; Fadeley, "Hopping into History," 65; Jenkins, *You Picked What?*, 25.

5. HOP CENTER OF THE WORLD

1. For a broad overview of the decade's depression and its social and political realities, see H. W. Brands, *The Reckless Decade: American in the 1890s* (Chicago: University of Chicago, Press, 2002).

2. Frank L. Green, *Ezra Meeker—Pioneer: A Guide to the Ezra Meeker Papers in the Library of the Washington State Historical Society* (Tacoma: Washington State Historical Society, 1969), 14–18. For a full account of Meeker's experiences in the Yukon, see Dennis M. Larsen, *Slick as a Mitten: Ezra Meeker's Klondike Enterprise* (Pullman: Washington State University Press, 2009). Also, for a history

of the Klondike gold rush, see Katherine Morse, *The Nature of Gold: An Environmental History of the Klondike Gold Rush* (Seattle: University of Washington Press, 2010).

3. D. B. DeLoach, *Outlook for Hops from the Pacific Coast* (Washington, DC: US Department of Agriculture, Bureau of Agricultural Economics, 1948), 2–29. By the first decade of the twentieth century, Josephine, Jackson, and Douglas counties stood as the only others in Oregon to claim commercial hop agriculture. There, hop farmers continued to produce small quantities of hops throughout the following decades. For more, see "The First in Hop Production," *Oregonian,* January 1, 1902; and George L. Sulerud, *An Economic Study of the Hop Industry in Oregon,* Station Bulletin 288 (Corvallis: Agricultural Experiment Station, Oregon State Agricultural College, 1931), 7, 25–26.

4. St. Benedict's Abbey, *Benedictine Hop Growers of the Willamette Valley* (Mount Angel, OR: Benedictine Press, 1935).

5. United States Bureau of Foreign Commerce, *Report upon the Commercial Relations of the United States with Foreign Countries for the Year 1877* (Washington, DC: Government Printing Office, 1878), 336.

6. For the numbers, see Department of Commerce and Labor, Bureau of the Census, *Thirteenth Census of the United States, Taken in the Year 1910, Abstract of the Census* (Washington, DC: Government Printing Office, 1913), 408; and DeLoach, *Outlook for Hops,* 2, 28. For various examples of references to the "Hop Center of the World" (which, at times was also the "Hop Capital of the World"), see "Monmouth Beginning Wonderful Era of Industrial Development," *Oregonian,* July 9, 1911; "Polk Country to Exhibit," *Oregonian,* August 23, 1912; "Hop Picking to Start," *Oregonian,* August 31, 1922; "Independence Busy on Eve of Hop Harvest," *Oregonian,* September 4, 1927; Sidney P. Newton, *Early History of Independence, Oregon* (Independence, OR: Sidney P. Newton, 1971), 64–69; and O. P. Hoff, *Sixth Biennial Report of the Bureau of Labor Statistics and Inspector of Factories and Workshops of the State of Oregon, 1915* (Salem, OR: State Printing Department, 1914), 149, 176.

7. For California's influence, see Kevin Starr, *Inventing the Dream: California through the Progressive Era* (New York: Oxford University Press, 1985), 45–47, 134, 158–64; and David Vaught, *Cultivating California: Growers, Specialty Crops, and Labor, 1875–1920* (Baltimore: Johns Hopkins University Press, 1999), 95–114. For the most extensive account specific to California agricultural cooperatives, see Clarke A. Chambers, *California Farm Organizations: A Historical Study of the Grange, the Farm Bureau, and the Associated Farmers, 1929–1941* (Berkeley: University of California Press, 1952).

8. Historians of the Pacific Northwest have long realized that its natural resources have been essential to its identity. Two edited collections offer good historical and historiographic insight into what is an extensive topic. See Dale D. Goble and Paul W. Hirt, eds., *Northwest Lands, Northwest Peoples: Readings in Environmental History* (Seattle: University of Washington Press, 1999); and William G. Robbins and Katrine Barber, eds., *Nature's Northwest: The North Pacific Slope in the Twentieth Century* (Tucson: University of Arizona Press, 2011).

9. Lewis and Clark Centennial Exposition Commission, *Report of the Lewis and Clark Centennial Exposition Commission for the State* (Portland, OR: Lewis and Clark Centennial Exposition Commission, 1906), 12.

10. Carl Abbott, *The Great Extravaganza: Portland and the Lewis and Clark Exposition,* 3rd ed. (Portland: Oregon Historical Society, 2004), xv. World's fairs have been a recent topic of interest for historians to frame nationalism and empire in the late nineteenth and early twentieth centuries. Many specific studies of individual fairs have examined their economic, social, cultural, and nationalistic reach. Capturing the breadth of the fairs themselves is John E. Findling and Kimberley D. Pelle, eds., *Historical Dictionary of World's Fairs and Expositions, 1851–1988* (New York: Greenwood Press, 1990).

11. For explanations of the impact of the Spanish-American War and American empire, see Robert E. Hannigan, *The New World Power: American Foreign Policy, 1898–1917* (Philadelphia: University of Pennsylvania Press, 2002); Ivan Musicant, *Empire by Default: The Spanish-American War and the Dawn of the American Century* (New York: H. Holt, 1998); and Paul T. McCartney, *Power and Progress: American National Identity, the War of 1898, and the Rise of American Imperialism* (Baton Rouge: Louisiana State University Press, 2006).

12. The literature on immigration during this period is extensive. For examples, see Dorothy B. Fujita-Rony, *American Workers, Colonial Power: Philippine Seattle and the Transpacific West, 1919–1941* (Berkeley: University of California Press, 2003); Mae M. Ngai, *Impossible Subjects: Illegal Aliens and the Making of Modern America* (Princeton, NJ: Princeton University Press, 2004); and Christina A. Ziegler-McPherson, *Americanization in the States: Immigrant Social Welfare Policy, Citizenship, and National Identity in the United States, 1908–1929* (Gainesville: University of Florida Press: 2009).

13. Lewis and Clark Centennial Exposition Commission, *Report,* 48–57.

14. Paul Landis, "The Hop Industry: A Social and Economic Problem," *Economic Geography* 22, no. 2 (January 1939): 85; Kathleen E. Hudson Cooler, "Hop Agriculture in Oregon: The First Century" (master's thesis, Portland State University, 1986), 40.

15. Michael A. Tomlan, *Tinged with Gold: Hop Culture in the United States* (Athens: University of Georgia Press, 1992), 157–214. Tomlan's book has an entire chapter on hop dryers. It is one of the reasons this project does not go into great detail about them.

16. Sidney W. Newton, *Early History of Independence, Oregon* (Salem, OR: Sidney W. Newton, 1971), 64–65; "Old Photograph Shows Era of Hops," *Capital Journal* (Salem, OR), November 14, 1958.

17. Wallace Nash, *The Settler's Handbook to Oregon* (Portland: J. K. Gill, 1904), 125.

18. E. Meeker, *Hop Culture in the United States: A Practical Treatise on Hop Growing in Washington Territory from the Cutting to Bale* (Puyallup, WA: Ezra Meeker, 1883), 10–33; United States Bureau of the Census, *Agriculture: Hops: With Data on Acreage and Production in Selected States and Counties; and Trade Census*

11th, 1890 Bull. no. 143 (Washington, DC: US Bureau of the Census, 1891); Oregon State Board of Agriculture, *The Resources of the State of Oregon: A Book of Statistical Information Treating upon Oregon as a Whole, and by Counties,* 3rd rev. ed. (Salem: Oregon State Board of Agriculture, 1898), 112. Regarding the ongoing debates of the best area to grow hops, one 1930's report documented that 64.4 percent of the valley's hop farmers "grew on bottomlands," 32.1 percent grew "on the main valley floor," and 3.5 percent "on residual soils or hill lands." See Otis W. Freeman, "Hop Industry of the Pacific Coast States," *Economic Geography* 12, no. 2 (April 1936): 162.

19. Lorenzo D. Snook, "Improvement Hop-Trellis," Letters Patent no. 77,775, United States Patent Office, May 12, 1868; "Hop-Growing on Wires," *Pacific Rural Press* 8, no. 25 (December 1874): 387; "Trellising for Hopyards," *Pacific Rural Press* 55, no. 5 (February 1893), 98; "Spraying Hops and Training Vines," *Pacific Rural Press* 55, no. 17 (April 1893), 378. For one of the earliest descriptions of the trellis system in Oregon (as adapted from English growers), see H. T. French and C. D. Thompson, *Agriculture: Potatoes, Roots* (Corvallis: Oregon Agricultural Experiment Station, 1893), 12–13. Also see W. W. Stockberger, *Growing and Curing Hops,* United States Department of Agriculture, Farmers' Bulletin 304 (Washington, DC: Government Printing Office, 1907), 13–18. The trellis system became ubiquitous by the 1930s and 1940s.

20. Otis W. Freeman, "Hop Industry of the Pacific Coast States." *Economic Geography* 12, no. 2 (April 1936): 156.

21. Freeman, "Hop Industry," 156.

22. G. R. Hoerner and Frank Rabak, *Production of Hops,* U.S. Department of Agriculture, Farmer's Bulletin 1842 (Washington, DC: Government Printing Office, 1940), 12.

23. Hoerner and Rabak, *Production of Hops,* 15.

24. "Machine-Picked vs. Hand-Picked Hops: A Revolution in the Hop Industry," *Scientific American,* supplement, no. 1770 (December 4, 1909): 361.

25. Vaught, *Cultivating California,* 133–34.

26. The U.S. patent office approved dozens of patents related to the hop industry throughout the late nineteenth and early twentieth centuries. Most, however, dealt with how to improve poling (and removing them upon harvest) and drying and curing hops. For some examples, see Jonathan Whitney, "Improvement in Hop-Driers," Letters Patent no. 75,503, March 10, 1868; W. F. Waterhouse, "Improvement in Hop-Driers," Letters Patent no. 80,578, August 4, 1868; Garret J. Olendorf, "Improved Mode of Poling Hops," Letters Patent no. 84,299, November 24, 1868; Stephen V. Barns, "Improved Hop-Pole Sharpener," Letters Patent no. 93,269, August 3, 1869; and Jacob Seeger and John Boyd, "Improvement in Preserving and Using Hops in Brewing," Letters Patent no. 121,902, December 12, 1871.

27. American staple-crop farmers benefitted from motorized farm instruments much earlier than farmers of specialty crops. By the 1920s, three million tractors filled the nation. See Bruce L. Gardner, *American Agriculture in the Twentieth Century: How It Flourished and What It Cost* (Cambridge, MA: Harvard University Press, 2002), 11–12.

28. Ming Kee, oral history with Daniel C. Robertson, April 19, 1982, Benton County Historical Society, Philomath, OR.

29. Harvey Kaser, oral history with Kathleen Hudson Cooler, April 16, 1982, Benton County Historical Society, Philomath, OR.

30. L. D. Wood, "The Future of the Hop Industry," *Pacific Hop Grower,* January 1935, 4.

31. Herbert Myrick, *The Hop: Its Culture and Cure, Market and Manufacture* (1899; reprint, New York: O. Judd, 1904), 9.

32. Starr, *Inventing the Dream,* 161–62; Vaught, *Cultivating California,* 95–114; Paul F. Starrs, "The Navel of California and Other Oranges: Images of California and the Orange Crate," *California Geographer* 28 (1988): 1–41; Gordon T. McClelland and Jay T. Last, *California Orange Box Labels: An Illustrated History* (Santa Ana, CA: Hillcrest Press, 1986).

33. Myrick, *The Hop,* 9; Meeker, *Hop Culture,* 4; Hudson Cooler, "Hop Agriculture in Oregon," 36–37; "The Hop Situation," *Oregonian,* September 12, 1894; "Do Not Raise Hops," *Oregonian,* March 13, 1896; "Hop Growers' Association," *Oregonian,* June 18, 1877; "Hop Men Organize," *Oregonian,* October 26, 1899.

34. Vaught, *Cultivating California,* 13. Along with works on the country-life movement, the historiography on populism is rich with examples of this type of thinking from the late nineteenth and early twentieth centuries. In particular, see John D. Hicks, *The Populist Revolt: A History of the Farmer's Alliance and the People's Party* (Minneapolis: University of Minnesota Press, 1931); Richard Hofstadter, *The Age of Reform: From Bryan to F.D.R.* (New York: Vintage Books, 1955); C. Vann Woodward, *The Burden of Southern History* (Baton Rouge: Louisiana State University Press, 1960); Walter T. K. Nugent, *The Tolerant Populists: Kansas Populism and Nativism* (Chicago: University of Chicago Press, 1963); Lawrence Goodwyn, *The Populist Moment: A Short History of the Agrarian Revolt in America* (Oxford: Oxford University Press, 1978); and Michael Kazin, *The Populist Persuasion: An American History* (Ithaca, NY: Cornell University Press, 1998).

35. Oregon State Board of Horticulture, *Third Biennial Report of the Oregon State Board of Horticulture to the Legislative Assembly, Eighteenth Regular Session* (Salem: Oregon State Board of Horticulture, 1894), 244.

36. Oregon State Board of Horticulture, *Fourth Biennial Report of the Oregon State Board of Horticulture to the Legislative Assembly, Nineteenth Regular Session* (Salem: Oregon State Board of Horticulture, 1897), 66.

37. Maureen Ogle, *Ambitious Brew: The Story of American Beer* (Orlando, FL: Harcourt, 2006), 90.

38. This ad ran for several years and in many newspapers. For another example, see *Washington Times,* August 9, 1908.

39. Similarly, Pabst ran this ad across several years and in many newspapers. For another example, see *Washington Sentinel,* March 30, 1901.

40. Nancie Fadeley, "Hopping into History: Oregon's Hop Industry," *Oregon Business Journal* 13, no 12 (December 1990): 65. In this overview of the Willamette Valley hop industry, the author explained that "[t]he J. W. Seavey Hop Company

calendars carried a convincing anti-Prohibition message: 'HOPS are so wholesome!'"

41. Martin Heidegger Stack, "Liquid Bread: An Examination of the American Brewing Industry, 1865–1940" (Ph.D. diss., Department of Economics, University of Notre Dame, 1998), 113–78.

42. Heinrich Johann Barth, assisted by Christiane Klinke, *The History of the Family Enterprise: Joh. Barth & Sohn, Nuremberg* (Nuremberg: Joh. Barth & Sohn, 1994), 5–17; S. S. Steiner, Inc., *Steiner,* rev. ed. (S. S. Steiner, 2004), 7–11.

43. Ogle, *Ambitious Brew,* 90; Agricultural Experiment Station, Oregon State Agricultural College, *An Economic Study of the Hop Industry in Oregon* (Corvallis: Agricultural Experiment Station, Oregon State Agricultural College, 1931), 45; Heinrich Johann Barth, Christiane Klinke, and Claus Schmidt, *The Hop Atlas: The History and Geography of the Cultivated Plant* (Nuremberg: Joh. Barth & Sohn, 1994), 9–13.

44. "Testimony of E. Clemens Horst to the Commission of Industrial Relations," in *The Seasonal Labor Problem in Agriculture, Industrial Relations: Final Report and Testimony Submitted to Congress by the Commission on Industrial Relations,* vol. 5 (Washington, DC: Government Printing Office, 1916), 4923, 4926, 4931. The claim may be dubious, since Barth and Steiner's holdings in Europe certainly competed with Horst's in the United States. Furthermore, the German companies also seriously moved into the American market during the 1930s. For an example of the E. Clemens Horst Co.'s advertisement, see *Dun's International Review* 36, no. 3 (November 1920): 151.

45. Wheatland Historical Society, *Wheatland* (Chicago: Arcadia, 2009), 91. The book notes also that "Horstville had company housing, a dining hall, a tent city for seasonal workers, and a company store where workers could purchase food and dry goods."

46. Myrick, *The Hop,* 288; "Testimony of E. Clemens Horst," 4923–26; "E. Clemens Horst Called by Death," *Pacific Hop Grower,* May 1940, 3.

47. Tomlan, *Tinged with Gold,* 103; S. R. Dennison and Oliver MacDonagh, *Guinness, 1886–1939: From Incorporation to the Second World War* (Cork, Ireland: Cork University Press, 1998), 112–14. Although Horst did not pen a manifesto as did Meeker to disseminate knowledge, he did publish often in agricultural and regional journals. For example, see E. Clemons Horst, "Hop Growing in California," *California's Magazine* 1, no. 1 (July 1915): 565–68.

48. "Hop Cultivation in America," *Journal of the Royal Society of Arts* 56, no. 2,886 (March 13, 1908): 449.

49. Dennison and MacDonagh, *Guinness,* 112–14. While Oregon growers rejoiced at the news, English growers opposed it in full force. The inability of English growers to compete with Americans by this time inspired a movement for high tariffs on imported hops. An article originally printed in the *Times* of London and made available for the *Oregonian* demonstrated that Parliament took the threat very seriously by 1908. Even in years prior they had been collecting reports on the topic. "Keep Out Oregon Hops," *Oregonian* (reprinted from the *Times* [London]), April

26, 1908, 11. Also see "English Hopmen Protest," *Oregonian,* March 29, 1908, 2; Great Britain, The Tariff Commission, *The Tariff Commission,* vol. 3, *Report of the Agricultural Committee* (London: Tariff Commission, 1906), 1085–90; and United States, Sixtieth Congress, "Tariff Hearings before the Committee on Ways and Means of the House of Representatives, Sixtieth Congress" (Washington, DC: Government Printing Office, 1908), 604–9.

50. J. F. Brown, *Guinness and Hops* (London: Arthur Guinness Son, 1980), 149.

51. The decline of English production was a common topic among those in the Pacific Coast hop industry during the early twentieth century. See, for example, "Decline of Hop Growing in England," *Pacific Rural Press* 78, no. 4 (July 1909): 63; and Sulerud, *Hop Industry in Oregon,* 36–37.

52. "Revolution in the Hop Trade: Big London Firm of Wiggins, Richardson & Co. Will Come to Oregon," *Oregonian,* August 3, 1906.

53. "The English Hop Market," *Pacific Rural Press* 52, no. 10 (September 1896): 147.

54. Ruth Kirk and Carmela Alexander, *Exploring Washington's Past: A Road Guide to History,* rev. ed. (Seattle: University of Washington Press, 1995), 437. In a fascinating public argument, Klaber and California's Durst debated the merits of unionization in print in the *Oregonian* from 1905 to 1908. While Durst diligently fought for the cause, Klaber could see no benefit. For the debate, see Herman Klaber, "Says It Will Fail: Herman Klaber's Opinion of Hopgrowers' Union," *Oregonian,* December 8, 1907, 8; "Hopmen's Union Meets with Favor: First of a Series of Oregon Hopgrowers' Meetings Is Held in Woodburn," *Oregonian,* December 24, 1907, 6; M. H. Durst, "Plans of the Union: Durst Further Explains Hopgrowers Association, Klaber Is Contradicted," *Oregonian,* December 15, 1907, 10; M. H. Durst, "Left in the Hands of Oregon Growers: Fate of Hop Association Rests with the Producers of This State," *Oregonian,* January 3, 1908, 16.

55. Arch Sloper, "Hop-Trellis-Wire Dropper," Patent no. 934,023, United States Patent Office, September 14, 1909; "An Automated Baler," *Pacific Hop Grower,* July 1934; "Oregon Hop Men View Robot Picker," *Pacific Hop Grower,* November 1938, 3; *Oregon Hop Grower,* July 1934, 7. See also "An Automated Baler," *Pacific Hop Grower,* July 1934, 7; this article noted, "An electric hop baler has been developed by Mr. Arch Sloper, well known hop grower of Independence, Oregon, that is said to possess exceptionally good operating characteristics and bids to be a popular product for hop farms having a need for a mechanical press."

56. Joseph Gaston, *The Centennial History of Oregon, 1811–1912* (Chicago: S. J. Clarke, 1912), 4:1091–92; Lynda Sekora, "Historic American Engineering Record: James W. Seavey Hop Driers," Hop Agriculture Collection, Benton County Historical Society, Philomath, OR.

57. Gaston, *Centennial History of Oregon,* 4:1092. Even though he remained solely in the Willamette Valley, Seavey was as interested as Horst in expansion. Charles Staley, a farmer from near Salem, recalled that Seavey's company bought out his entire diversified farm and planted it in hops. After the purchase, Staley ended up working a short time for the company. See Charles Staley, oral history by Mickey Peterson, April 2, 1982, Benton County Historical Society, Philomath, OR.

58. For an overview of positivism, see Michael Singer, *The Legacy of Positivism* (New York: Palgrave MacMillan), 2005. For its relationship to education, including the development of hard and soft sciences, see Lawrence A. Cremin, *American Education: The Metropolitan Experience, 1876–1980* (New York: Harper and Row, 1988).

59. Pete Dunlop, *Portland Beer: Crafting the Road to Beervana* (Charleston, SC: History Press, 2013), 31; Ogle, *Ambitious Brew*, 63–64.

60. Ogle, *Ambitious Brew,* 55; See also Stack, "Liquid Bread."

61. Charles E. Bevington, "Average Year for the Brewing Trade," *New York Times,* January 8, 1905.

62. Ernst Hantke, "Pacific Hops and Fertilizers," *Transactions of the American Brewing Institute* 2 (September 1902-September 1904): 226–33. For more of Horst's collaborations, see William B. Parker, *The Red Spider on Hops in the Sacramento Valley of California* (Washington, DC: US Department of Agriculture, Bureau of Entomology, 1913), 9.

63. For more information generally on these developments and the impact of these acts, see Coy F. Cross, *Justin Smith Morrill: Father of the Land-Grant Colleges* (East Lansing: Michigan State University Press, 1999); and Ralph D. Christy and Lionel Williamson, eds., *A Century of Service: Land-Grant Colleges and Universities, 1890–1990* (New Brunswick, NJ: Transaction, 1992). For the impact more specifically in Oregon, see Oregon Agricultural Experiment Station, *100 Years of Progress: The Agricultural Experiment Station, Oregon State University, 1888–1988* (Corvallis: Oregon State University, Agricultural Experiment Station, 1990).

64. Wayne D. Rasmussen, *Taking the University to the People: Seventy-Five Years of Cooperative Extension* (Ames: Iowa State University Press, 1989), 4.

65. R. H. McDowell, *Hops,* Nevada Agricultural Bulletin 35 (Reno: Nevada Agricultural Experiment Station, 1896), 4.

66. Later publications from the Government Printing Office, including *Growing and Curing Hops* (1907), *Hops in Principal Countries: Their Supply, Foreign Trade, and Consumption, with Statistics of Beer Brewing* (1907), and *Necessity for New Standards of Hop Valuation* (1909), further demonstrated a new attention to the expansion of American hop growing and increased market shares for the nation, including Latin America and East Asia. For examples of Oregon scientists and growers examining the global scientific and agricultural literature that existed around the turn of the century, see H. T. French and C. D. Thompson, *The Hop Louse,* Oregon Agricultural Experiment Station Bulletin no. 24 (Corvallis, OR: Agricultural College Printing Office, 1893), 10–12; and H. V. Tartar and B. Pilkington, *Hop Investigations,* Bulletin no. 114 (Corvallis: Oregon Agricultural College Press, 1913), 9–10.

67. French and Thompson, *Hop Louse,* 10–12; Oregon State Agricultural College, *Annual Report of the President of the Board of Regents of the State Agricultural College to the Governor of Oregon, Legislative Assembly, Eighteenth Regular Session, 1895* (Salem: Oregon State Agricultural College, 1894), 25, 29; Oregon State Agricultural College, *Annual Report of the Oregon Agricultural College and Experiment Station for the Year Ending June 30, 1989* (Salem: Oregon State Agricultural College, 1899), 16.

68. Tartar and Pilkington, *Hop Investigations,* 22.

69. Oregon Agricultural Experiment Station, *100 Years of Progress,* 29–31; James Withycombe, "Report of Office of Experiment Stations: Oregon," *Annual Report of the Office of Experiment Stations for the Year Ended June 30, 1911* (Washington, DC: US Department of Agriculture, 1911), 182.

70. Tartar and Pilkington, *Hop Investigations,* 6.

71. Hoerner and Rabak, *Production of Hops,* 3–4.

72. Wye College Department of Hop Research, *Annual Report, 1953* (Wye, England: Wye College Department of Hop Research, 1953), 5; Peter Darby, "The History of Hop Breeding and Development," *Brewery History* 121 (Winter 2005): 94–112; Peter Darby, interview by the author, Wye Hops, China Farm, Canterbury, England, May 22, 2013; Alfred Haunold, interview by the author, Oregon State University, Corvallis, June 29, 2009.

73. E. S. Salmon, *Four Seedlings of the Canterbury Golding* (Canterbury, England: Wye College, 1944), 245–46; J. S. Hough, D. E. Briggs, R. Stevens, and T. W. Young, *Malting and Brewing Science,* vol. 2, *Hopped Wort and Beer* (London: Chapman & Hall, 1982), 406; J. F. Brown, *Guinness and Hops* (London: Arthur Guinness Son, 1980), 56–63.

74. The author wishes to thank Peter Darby of Wye Hops for showing me around the archive and making sense of it all. Also, see Peter Darby, "The History of Hop Breeding and Development," *Brewing History* 121 (2005): 94–112.

75. "Fifth Anniversary Dinner of the American Brewing Institute," *Transactions of the American Brewing Institute* 3 (September 1904–September 1907): 124.

76. "Fifth Anniversary Dinner," 124.

77. For the financial impact of the Smith-Lever Act, see Gardner, *American Agriculture,* 182.

78. Emma Seckle Marshall, "Hop Culture in Oregon," *Sunset,* November 1903–April 1904, 243.

79. C. E. Woodson, *Benton County, Oregon: Illustrated* (Corvallis: Benton County Citizens' League, 1905), 11–13.

80. Woodson, *Benton County,* 35.

81. Joseph Gaston, *The Centennial History of Oregon, 1811–1912* (Chicago: S. J. Clarke, 1912), 1:544.

82. "Initiative, Referendum and Recall: 1912–1914," *Oregon Blue Book,* accessed May 10, 2010, http://bluebook.state.or.us/state/elections/elections12.htm.

6. THE SURPRISE OF PROHIBITION

1. For a further discussion of the relationship between women's suffrage and anti-alcohol campaigns, see Catherine Gilbert Murdock, *Domesticating Drink: Women, Men, and Alcohol in America, 1870–1940* (Baltimore: Johns Hopkins University Press, 1998), 9–69.

2. The ballot measure, along with women's suffrage, was one of the first major victories of the "Oregon System" approved by voters in 1902 to allow citizen-supported initiatives, referenda, and recall. "Initiative, Referendum and Recall: 1912–1914," *Oregon Blue Book,* accessed May 10, 2010, http://bluebook.state.or.us /state/elections/elections12.htm. It is also important to note, however, that counties and then cities within counties in Oregon also had dry options following 1904. For the success of women's suffrage on the local and national levels, see Rebecca J. Mead, *How the Vote Was Won: Woman Suffrage in the Western United States, 1868–1914* (New York: New York University Press, 2004); and Allison L. Sneider, *Suffragists in an Imperial Age: U.S. Expansion and the Woman Question, 1870–1929* (New York: Oxford University Press, 2008). Also see Steven L. Piott, *Giving Voters a Voice: The Origins of the Initiative and Referendum in America* (Columbia: University of Missouri Press, 2003); Robert D. Johnston, *The Radical Middle Class: Populist Democracy and the Question of Capitalism in Progressive Era Portland, Oregon* (Princeton, NJ: Princeton University Press, 2003); and Richard J. Ellis, *Democratic Delusions: The Initiative Process in America* (Lawrence: University of Kansas Press, 2002).

3. John J. Rumbarger, *Profits, Power, and Prohibition: Alcohol Reform and the Industrializing of American, 1800–1930* (Albany: State University of New York Press, 1989), xxiv.

4. This testimony can be found in United States Senate, Committee on the Judiciary, *Proposing an Amendment to the Constitution Prohibiting the Sale, Manufacture, and Importation of Intoxicating Liquors: Hearings before a Subcommittee of the Committee on the Judiciary of the United States on S.J. Res. 88 and S.J. Res. 50, 63rd Cong., 2nd sess., 1914* (Washington, DC: Government Printing Office, 1914).

5. W. J. Rorabaugh, *The Alcoholic Republic: An American Tradition* (New York: Oxford University Press, 1979), 109–10.

6. Charles Stelzle, *Why Prohibition!* (New York: George H. Doran, 1918), 22.

7. U.S. Constitution, amendment 18, sec. 1.

8. "Intercede for Brewers: Oregon Hopmen Petition Senator Fulton to Oppose Legislation," *Oregonian,* January 22, 1908, 7.

9. "Their Trade Drying Up Hopgrowers Call for Fight on Prohibition," *Oregonian,* March 8, 1908, 3.

10. "Brewer Issues Circular: Gustave Pabst Seeks to Counteract Spread of Prohibition Movement," *Oregonian,* February 9, 1908, 9.

11. National Wholesale Liquor Dealers Association of America, *The Anti-Prohibition Manual: A Summary of Facts and Figures Dealing with Prohibition* (Cincinnati: National Wholesale Liquor Dealers Association of America, 1917), 13; United States Brewers' Association, *The 1918 Yearbook of the United States Brewers' Association* (New York: United States Brewers' Association, 1919), 110.

12. "Declare Saloons Must Be Orderly: Oregon Brewers Will Aid in Regulation of Drink-Selling Resorts," *Oregonian,* December 19, 1907, 12.

13. Norman H. Clark, *The Dry Years: Prohibition and Social Change in Washington,* rev. ed. (Seattle: University of Washington Press, 1988), 56.

14. Pete Dunlop, *Portland Beer: Crafting the Road to Beervana* (Charleston, SC: History Press, 2013), 35.

15. See advertisement in the *Oregonian,* May 31, 1908, 11.

16. "A New Beverage Placed on the Market by Henry Weinhard," *Oregonian,* August 29, 1908, 10.

17. "Initiative, Referendum and Recall: 1912–1914," *Oregon Blue Book,* accessed May 10, 2010, http://bluebook.state.or.us/state/elections/elections12.htm.

18. "Notes Relating to the Brewing Industry," *Scientific Station for Pure Products* 12, no. 1 (January 1916): 450.

19. "Notes Relating to the Brewing Industry," 450.

20. United States Department of Agriculture, Statistical Reporting Service, *Hops by States, 1915–69,* Statistical Bulletin no. 469 (Washington, DC: United States Department of Agriculture, Statistical Reporting Service, 1971), 4–6. For more context, see George L. Sulerud, *An Economic Study of the Hop Industry in Oregon* (Corvallis: Agricultural Experiment Station, Oregon State University, 1931); and Michael A. Tomlan, *Tinged with Gold: Hop Culture in the United States* (Athens: University of Georgia Press, 1992), 30–39.

21. "Some Breweries Still to Operate; Weinhard Plant, Largest in Portland, Will Manufacture Non-Intoxicant; Many Out of Work," *Oregonian,* November 14, 1915.

22. "H. V. Tarter to A. B. Cordley," June 12, 1914, and "Report of Work Engaged in by Chemical Department of Experiment Station during Fiscal Year July 1, 1917 to June 30, 1918," folder Hops (Adams #3), 1907–1919, RG 25 Agricultural Experiment Station, subgroup 3, ser. 5, Research Project Records, University Archive, Oregon State University, Corvallis.

23. Patents for multiple-use fruit, vegetable, and hop dryers existed earlier; farmers had just chosen not to invest in them because of cost or uncertainly about the technologies. See, for example, "Butts' Fruit, Vegetable, and Hop Drier," *Pacific Rural Press* 17, nos. 26, 28 (June 1879): 428.

24. E. Clemens Horst, "The New Dried Vegetable Industry," *Statistical Report of the California State Board of Agriculture for the Year 1918* (Sacramento: California State Printing Office, 1919), 154.

25. United States Department of Agriculture, *Hops by States,* 3.

26. Barth, Heinrich Johann, Christiane Klinke, and Claus Schmidt, *The Hop Atlas: The History and Geography of the Cultivated Plant* (Nuremberg: Joh. Barth & Sohn: 1994), 15. As the text notes, "Beer consumption worldwide was more than halved. Parallel to this development, hop acreage in Germany came down from 68,500 acres in 1914 to 22,000 acres in 1918, and prices were hovering at or below cost of production level with the exception of 1918, the year of a complete crop failure."

27. Sulerud, *Hop Industry in Oregon,* 7, 12–16.

28. William G. Robbins and Katrine Barber, *Nature's Northwest: The North Pacific Slope in the Twentieth Century* (Tucson: University of Arizona Press, 2011), 59–64.

29. Frederick Converse Beach, ed., *The Encyclopedia Americana,* vol. 8 (New York: Encyclopedia Americana Corporation, 1919), 377.

30. "Pacific States," *Dun's Review: A Journal of Finance and Trade—Domestic and Foreign* 27, no. 1335 (March 8, 1919): 6. For the decline in English hop cultivation, also see Herbert Myrick, *The Hop: Its Culture and Cure, Market and Manufacture* (1899; reprint, New York: O. Judd, 1904), 4.

31. Sulerud, *Hop Industry in Oregon,* 7, 12–16.

32. United States Department of Agriculture, *Hops by States,* 4–5.

33. Hop Industry Productivity Team, *The Hop Industry: Report of a Visit to the U.S.A. and Canada in 1950 of a Productivity Team Representing the Hop Industry* (London and New York: Anglo-American Council on Productivity, 1951), 22; "Coir Yarn," *Hopper* 6, no. 5 (August 1949): 9.

34. For an overview of the downy mildew disease, see A. Lebeda, P. T. N. Spencer-Phillips, and B. M. Cooke, *The Downy Mildews: Genetics, Molecular Biology and Control* (Dordrecht, Germany: Springer, 2008). For the first agricultural bulletin on the subject in the Pacific Northwest, see G. R. Hoerner, *Downy Mildew of Hops* (Corvallis: Oregon State Agricultural College Extension Service, 1933), 1–8.

35. Barth, Klinke, and Schmidt, *Hop Atlas,* 18.

36. For the economic crisis of American agriculture in the 1920s and responses to it, see John D. Black, *Agricultural Reform in the United States* (New York: McGraw-Hill, 1929); William D. Rowley, *M. L. Wilson and the Campaign for the Domestic Allotment* (Lincoln: University of Nebraska Press, 1970); David B. Danbom, *The Resisted Revolution: Urban America and the Industrialization of Agriculture, 1900–1930* (Ames: Iowa State University Press, 1979); Deborah Fitzgerald, *Every Farm a Factory: The Industrial Ideal in American Agriculture* (New Haven, CT: Yale University Press, 2003); and Sarah T. Philips, *This Land, This Nation: Conservation, Rural America and the New Deal* (Cambridge: Cambridge University Press, 2007).

37. For an economic study of global agriculture in the 1920s, see Giovanni Federico, "Not Guilty? Agriculture in the 1920s and the Great Depression," *Journal of Economic History* 65, no. 4 (December 2005): 949–76.

38. Herman Feldman, *Prohibition: Its Economic and Industrial Aspects* (New York: Appleton, 1927), 277.

39. For a story of the impact of Prohibition on family winemakers, see Vivienne Sosnowski, *When the Rivers Ran Red: An Amazing Story of Courage and Triumph in America's Wine Country* (New York: Palgrave McMillan, 2009).

40. Clark, *Dry Years,* 158; Amy Mittelman, *Brewing Battles: A History of American Beer* (New York: Algora, 2008), 95–96; Jewel Lansing, *Portland: People, Politics, and Power, 1851–2001* (Corvallis: Oregon State University Press, 2003), 309.

41. "Erect New Building: Gambrinus Brewery Plans Seven-Story Structure," *Oregonian,* October 1, 1909. "2-Story Brick Planned; Gambrinus Brewery Will Build at Oregon City," *Oregonian,* May 29, 1910.

42. Gary Meier and Gloria Meier, *Brewed in the Pacific Northwest: A History of Beer Making in Oregon and Washington* (Seattle: Fjord Press, 1991), 78; Clark, *Dry Years,* 146.

43. Meier and Meier, *Brewed in the Pacific Northwest* 79–80, 160–61.

44. Advertisement, *Oregonian,* May 28, 1924, 15; advertisement, *Oregonian,* November 25, 1925, 9.

45. "Fall of Brewery Makes 'Em Moan; Days of Old Recalled When Beer Wasn't Near," *Oregonian,* March 4, 1928.

46. Advertisement, *Oregonian,* June 9, 1919, 7; Dunlop, *Portland Beer,* 48.

47. Susan M. Gauss and Edward Beatty, "The World's Beer: The Historical Geography of Brewing in Mexico," in *The Geography of Beer: Regions, Environment, and Societies,* ed. Mark Patterson and Nancy Hoalst-Pullen (New York: Springer, 2014), 61–63.

48. Clark, *Dry Years,* 153.

49. Clark, *Dry Years,* 153.

50. Lansing, *Portland,* 309.

51. Earl Pomeroy, *The Pacific Slope: A History of California, Oregon, Washington, Idaho, Utah, and Nevada,* rev. ed. (Reno: University of Nevada, Press, 1991), 215–52.

52. Vincent DiGirolamo, "The Women of Wheatland: Female Consciousness and the 1913 Hop Strike," *Labor History* 34, no. 2 (1993): 236–55; Wheatland Historical Society, *Wheatland* (Chicago: Arcadia Publishing, 2009), 91. Also see discussion in David Vaught, *Cultivating California: Growers, Specialty Crops, and Labor, 1875–1920* (Baltimore: Johns Hopkins University Press, 1999), 134–45.

53. Kathleen E. Hudson Cooler, "Hop Agriculture in Oregon: The First Century" (master's thesis, Portland State University, 1986), 50–51.

54. Department of Commerce and Labor, Bureau of the Census, *Thirteenth Census of the United States, 1910* (Washington, DC: Government Printing Office, 1912), 441; Pomeroy, *Pacific Slope,* 138–40, 148–51, 161–62.

55. Vaught, *Cultivating California,* 90.

56. "A Benton County Hopyard," *Corvallis (OR) Gazette Times,* September 21, 1916.

57. "A Benton County Hopyard."

58. "NOTICE: Hop Pickers Wanted," *Corvallis (OR) Gazette Times,* April 19, 1918.

59. "A Benton County Hopyard."

60. Tomlan, *Tinged with Gold,* 154–55.

61. "Recreation in the Oregon Hop Fields," *Recreation* 7 (1923): 646–47.

62. Tomlan, *Tinged with Gold,* 124–26.

63. G. R. Hoerner and Frank Rabak, *Production of Hops,* U.S. Department of Agriculture, Farmer's Bulletin 1842 (Washington, DC: Government Printing Office, 1940), 26–27.

64. Hoerner, *Downy Mildew of Hops,* 1–8.

7. FIESTA AND FAMINE

1. H. A. Cornoyer, "Growers Back Recovery Plan," *Oregon Hop Grower* 1, no. 6 (August 1933): 2.

2. *New York Times,* November 9, 1932, 3.

3. For more on Roosevelt and the election, see Donald A. Richie, *Electing FDR: The New Deal Campaign of 1932* (Lawrence: University of Kansas Press, 2007).

4. Maureen Ogle, *Ambitious Brew: The Story of American Beer* (Orlando: Harcourt, 2006), 199–201; William Knoedelseder, *Bitter Brew: The Rise and Fall of Anheuser-Busch and America's Kings of Beer* (New York: Harper Collins, 2012), 10–12.

5. The experiment in Prohibition was underscored by such things, as one opposition group suggested, including "corruption," "failure of enforcement agents," "prevalence of moonshine," general "lawlessness," and, indeed economic consequence; see Association against the Prohibition Amendment, *A Criticism of National Prohibition* (Washington, DC: Association against the Prohibition Amendment, 1926), 21–30, 52–54.

6. George H. Copeland, "One Month of Beer Has Stimulated Industry," *New York Times,* May 14, 1933.

7. Jean Edward Smith, *FDR* (New York: Random House, 2007), 316.

8. Eleanor Roosevelt, My Day (newspaper column), United Features Syndicate, July 15, 1939.

9. Ogle, *Ambitious Brew,* 98–124.

10. Ogle, *Ambitious Brew,* 185–88.

11. Ogle, *Ambitious Brew,* 190–205.

12. Gary Meier and Gloria Meier, *Brewed in the Pacific Northwest: A History of Beer Making in Oregon and Washington* (Seattle: Fjord Press, 1991), 138–39; Pete Dunlop, *Portland Beer: Crafting the Road to Beervana* (Charleston, SC: History Press, 2013), 54–56.

13. Meier and Meier, *Brewed in the Pacific Northwest,* 101–2, 146–47.

14. Dunlop, *Portland Beer,* 64. Some new outfits, such as Portland's Rose City Brewery (1934–40), tried to get into the post-Prohibition market, but never found success.

15. Amy Mittelman, *Brewing Battles: A History of American Beer* (New York: Algora, 2008), 98–99; Martin Heidegger Stack, "Liquid Bread: An Examination of the American Brewing Industry, 1865–1940" (Ph.D. diss., Department of Economics, University of Notre Dame, 1998), 79.

16. "What! No Beer? Here's Sad News; There Won't Be Enough, Brewers Admit," *Oregonian* April 5, 1933.

17. Surprisingly, very little has been written on the Oregon Liquor Control Commission (OLCC), though records exist in the Oregon State Archives in Salem. For early enforcement, see Warren Niete, "Enforcing Oregon's State Alcohol Monopoly: Recollections from the 1950s," *Oregon Historical Quarterly* 115, no. 1 (Spring 2014): 90–105. For the Washington State Liquor Control Board, see Norman H. Clark, *The Dry Years: Prohibition and Social Change in Washington,* rev. ed. (Seattle: University of Washington Press, 1988), 243–45.

18. Leonard V. Harrison and Elizabeth Laine, *After Repeal: A Study of Liquor Control Administration* (New York: Harper and Brothers, 1936); Ogle, *Ambitious Brew,* 183–98.

19. Nathan Michael Corzine, "Right at Home: Freedom and Domesticity in the Language and Imagery of Beer Advertising, 1933–1960," *Journal of Social History* 43, no. 3 (Summer 2010): 846–47.

20. Ogle, *Ambitious Brew,* 208–17. The author notes that in 1935 Americans consumed approximately one-third of their beer in cans and bottles. By 1940, this number increased to approximately 50 percent, and twenty-years later the number reached 80 percent.

21. Corzine, "Right at Home," 843–47.

22. Mittelman, *Brewing Battles,* 99.

23. Joh. Barth & Sohn, *Hops 1935/36* (Nuremberg: Joh. Barth & Sohn, 1936), 3.

24. Ogle, *Ambitious Brew,* 219.

25. *Hopper,* April 1944, 3; Mittelman, *Brewing Battles,* 128.

26. D. B. DeLoach, "Outlook for Hops from the Pacific Coast" (Washington, DC: United States Department of Agriculture, Bureau of Agricultural Economics, 1948), 15–18. By the 1930s, the Pacific Coast hop growers had replaced the German connection to these areas. Demonstrating the intensity with which they looked after world markets, the following articles appeared in 1946–1947 alone: "Mexican Hops and Beer," *Hopper,* May 1946, 8; "Brazilian Hops," *Hopper,* June 1946, 8; "Poland Rebuilding the Hop Industries," *Hopper,* September 1946, 4; "English Hop Prospects," *Hopper,* September 1946, 5–6; "Czech Hop Crop Increasing," *Hopper,* October 1946, 6; "Good Hop Crop in Canada," *Hopper,* October 1946, 8; "Hop Situation in Yugoslavia," *Hopper,* November 1946, 10; "Hop Situation in Belgium," *Hopper,* February 1947, 10.

27. George L. Sulerud, *An Economic Study of the Hop Industry in Oregon* (Corvallis: Agricultural Experiment Station, Oregon State Agricultural College, 1931), 11–16.

28. United States Department of Agriculture, *Hops: By States, 1915–69,* Statistical Bulletin no. 469 (Washington, DC: US Department of Agriculture, Statistical Reporting Service, 1971), 5.

29. Barth & Sohn, "Hop Report for 1935/36," 3. Speaking to the incredible resurgence of American brewing after repeal, the Barth reports, the most important global source of hop data published annually, transitioned in the mid-1930s to using American measurements (hop acreage and American barrels), as opposed to European and British standards.

30. Ezra Meeker, *Hop Culture in the United States: Being a Practical Treatise on Hop Growing in Washington Territory from the Cutting to Bale* (Puyallup, WA: Ezra Meeker, 1883), 3.

31. Sulerud, *Hop Industry in Oregon,* 11.

32. S. S. Steiner, Inc., *Steiner,* rev. ed. (Atlantic Highlands, NJ: S. S. Steiner, 2004), 14–15; "Big Business in Hops," *Hopper,* May 1953, 6.

33. "Hop Growers' Association," *Oregonian,* June 18, 1877; "Hop Men Organize," *Oregonian,* October 26, 1899; "Hopgrowers Form State Association," *Oregonian,* March 21, 1915.

34. Sidney Jackson, *Oregon Hop Grower,* May 1933, 2.

35. Sulerud, *Hop Industry in Oregon,* 57.

36. Paul H. Landis, "The Hop Industry, a Social and Economic Problem," *Economic Geography* 15, no. 1 (January 1939): 86.

37. Landis, "Hop Industry," 87.

38. C. F. Noakes, "Quality Production of Hops: What It Means to Oregon Growers," *Oregon Hop Grower,* May 1933, 10.

39. Noakes, "Quality Production of Hops," 3–4.

40. For overviews of the federal government's approach to agricultural reform during the New Deal, see William D. Rowley, *M. L. Wilson and the Campaign for Domestic Allotment* (Lincoln: University of Nebraska Press, 1970); and Sarah T. Phillips, *This Land, This Nation: Conservation, Rural America, and the New Deal* (Cambridge: Cambridge University Press, 2007).

41. "Hop Root Control Vital," *Pacific Hop Grower,* March 1934, 2.

42. For more discussion on the continued speculative aspect of the hop market in the 1930s, see Otis W. Freeman, "Hop Industry of the Pacific Coast States," *Economic Geography* 12, no. 2 (April 1936): 162.

43. George Clinch, *English Hops: A History of Cultivation and Preparation for the Market from the Earliest Times* (London: McCorquodale, 1919), 55–56; United States, Congress, Senate, Committee on Agriculture and Forestry, "Amend the Agricultural Adjustment Act so as to Include Hops as a Basic Agricultural Commodity: A Report (to accompany S. 626)" (Washington, DC: Government Printing Office, 1935).

44. 75h Congress, 1ˢᵗ Session, House of Representatives, Report no. 1298, "Permitting Producers of Hops to Enter into Marketing Agreements under Agricultural Adjustment Act," July 23, 1937; "Marketing Agreement Completed," *Pacific Hop Grower,* December 1939.

45. "Marketing Agreement Completed," *Pacific Hop Grower,* December 1939.

46. *Pacific Hop Grower,* September 1939, 2.

47. "Dust Hops to Control Mildew," *Benton County [OR] Herald,* April 27, 1933.

48. See the description of USDA accession no. 64107, a cross by Salmon: "USDA Named Hop Variety Descriptions," freshhops.com, accessed August 15, 2011, www .freshops.com/hops/usda-named-hop-variety-descriptions#usda_id_64107.

49. "Independence Receives English Hops," *Pacific Hop Grower,* January 1936, 7. "New English Hop Doing Well Here," *Pacific Hop Grower,* March 1937, 5. Heinrich Johann Barth, Christiane Klinke, and Claus Schmidt, *The Hop Atlas: The History and Geography of the Cultivated Plant* (Nuremberg: Joh. Barth & Sohn, 1994), 19.

50. Quoted in "Mildew Cuts Hop Crop to Half Normal," *Oregon Hop Grower,* June 1934, 2.

51. "Hopland Fiesta Colorful Event," *Oregonian,* August 28, 1936; "Parade Feature of Hop Festival," *Oregonian,* August 27, 1938; Sidney P. Newton, *Early History of Independence, Oregon* (Independence, OR: Sidney P. Newton, 1971), 64–69.

52. "Many Employed in Yards," *Oregon Hop Grower,* September, 1933, 2; Landis, "Hop Industry," 88.

53. Earl Pomeroy, *The Pacific Slope: A History of California, Oregon, Washington, Idaho, Utah, and Nevada,* rev. ed. (Reno: University of Nevada, Press, 1991), 347–50.

54. Edwin B. Self, *Limbo City* (New York: Herald, 1946).

55. "Official Program of the Sixth Annual Hop Fiesta, August 24–25–26–27, 1939," Independence—Entertainment—Fairs and Pageants, Independence Heritage Museum, Independence, OR.

56. Michael A. Tomlan, *Tinged with Gold: Hop Culture in the United States* (Athens: University of Georgia Press, 1992), 124.

57. "Marjorie Plant Queen of First Hop Fiesta in Independence," *Enterprise (OR) File,* August 31, 1934; "Second Annual—Official Program—Hop Fiesta, August 29, 30, and 31, 1935, Independence, Oregon," Independence—Entertainment—Fairs and Pageants, Independence Heritage Museum, Independence, OR.

58. "Day and Night Air Circus Gives Promise of Thrill at Hop Fiesta," *Oregonian,* August 25, 1935.

59. "Hop Fiesta Plans Assure Big Time," *Oregonian,* August 15, 1935.

60. "Second Annual—Official Program—Hop Fiesta."

61. "Official Program of the Sixth Annual Hop Fiesta."

62. It is important to note that the culture surrounding other seasonal harvests could be fun and a time for social interaction, music, and games, too. The overwhelming difference in the hopyards was the effort of growers to provide entertainment for their workers, whereas other agricultural workers had to fend for themselves. For an excellent comparison, see Jose M. Alamillo, *Making Lemonade out of Lemons: Mexican American Labor and Leisure in a California Town, 1880–1960* (Urbana: University of Illinois Press, 2006).

63. "Hop Center of the World," *Community Builder* (Independence, OR), August 14, 1941; "Program for Independence Hop Fiesta, August 1941," Independence, Oregon, Independence—Entertainment—Fairs and Pageants, Independence Heritage Museum, Independence, OR.

64. "Lack of Pickers May Ruin Many Hops," *Oregon Hop Grower,* July 1935, 6.

65. See, in particular, two works focused on the social makeup of the hopyards: Carl F. Reuss, Paul H. Landis, and Richard Wakefield, *Migratory Farm Labor and the Hop Industry on the Pacific Coast with Special Application to Problems of the Yakima Valley, Washington,* Rural Sociology Series in Farm Labor, no. 3 (Pullman: State College of Washington Agricultural Experiment Station, 1938.); and Freeman, "Hop Industry," 157–58.

66. Tomlan, *Tinged with Gold,* 154–55.

67. For a New Dealer's overview of these perceived crises of Pacific Coast migratory labor, whose numbers the author estimated to be two hundred thousand or more, see Paul S. Taylor, "Migratory Agricultural Workers on the Pacific Coast," *American Sociological Review* 3, no. 2 (April 1938): 225–32.

68. "Child Labor Permitted," *Oregon Hop Grower,* September 1933, 2. The concerns of child labor existed prior to the New Deal. But efforts at inspection and reform materialized rapidly following stipulations in the National Recovery Act of

1933. For an earlier governmental inquiry, see Alice Channing, Ellen Nathalie Matthews, and Mary E. Skinner, *Child Labor in the Fruit and Hop Growing Districts of Oregon and Washington* (Washington, DC: US Department of Labor, Children's Bureau, 1926).

69. Marion Hathway, *The Migratory Worker and Family Life: The Mode of Living and Public Provision for the Needs of the Family of the Migratory Worker in Selected Industries of the State of Washington* (Chicago: University of Chicago Press, 1934), 2–4.

70. Anne Whiston Spirn, *Daring to Look: Dorothea Lange's Photographs and Reports from the Field* (Chicago: University of Chicago Press, 2008), 143–89. In 1939, the famed New Deal photographer Dorothea Lange surveyed migrant agricultural life in the Pacific Northwest. As an employee of the Farm Security Administration (originally the Resettlement Administration), she was tasked with documenting working conditions. In the Willamette Valley she found dirty fieldworkers, but not horrible conditions, as others had. The conditions in the hopyards that she photographed appeared to be sanitary, and though the faces of children are dirty and their appearances disheveled, there were no signs of mistreatment or overwork. This was a good sign for growers, workers, and the government, given some of the decade's previous events and problems.

71. Stuart Jamieson, *Labor Unionism in American Agriculture* (1946; reprint, New York: Arno Press, 1976), 204.

72. Jamieson, *Labor Unionism,* 204–5.

73. Jamieson, *Labor Unionism,* 207.

74. Historian Cletus Daniel argued that the story of agricultural labor in the West was about the "erosion of agrarian idealism," meaning that large-scale farming both minimized the individual family farm and turned laborers into mostly powerless cogs in a capitalist system. The efforts of farmworkers to organize in the I.W.W. and other unions represented their new role in industrial capitalism. See Cletus Daniel, *Bitter Harvest: A History of California Farmworkers, 1870–1941* (Ithaca, NY: Cornell University Press, 1981), 15, 104–41. For more on the I.W.W. in Western agriculture, see Greg Hall, *Harvest Wobblies: The Industrial Workers of the World and Agricultural Laborers in the American West, 1905–1930* (Corvallis: Oregon State University Press, 2001).

75. For more on Ginther's life growing up a socialist and labor leader in the West, see "R. D. Ginther, Workingman Artist and Historian of Skid Row," *California Historical Quarterly* 54, no. 3 (Fall 1975): 263–71. Ginther's paintings now reside permanently in the collections of the Washington State Historical Society. For online viewing, visit https://depts.washington.edu/depress/Ronald_Ginther_watercolors.shtml.

76. Freeman, "Hop Industry," 157–58.

77. Landis, "Hop Industry," 88–89.

78. Frank T. Maezials, John Foster, Mamie Dickens, and Adolphus W. Ward, *The Life of Charles Dickens* (New York: University Society, 1908), 248.

79. Landis, "Hop Industry," 89–92.

80. "E. Clemens Horst Called by Death," *Pacific Hop Grower,* May 1940, 3.
81. Tomlan, *Tinged with Gold,* 32.

8. AFTER THE HOP RUSH

1. United States Department of Agriculture, *Hops: By States, 1915–69,* Statistical Bulletin no. 469 (Washington, DC: US Department of Agriculture, Statistical Reporting Service, 1971), 5. In the postwar period, the Willamette Valley became a world leader in several other areas of agriculture, particularly orchard fruits, grass seed, and row crops like green beans. Plums and prunes earned farmers $181 million in 1939, the most of any other crop by nearly five times. See United States Department of Commerce, Bureau of the Census, *Sixteenth Census of the United States: 1940: Agriculture, Volume 1, First and Second Series, State Reports, Part 6: Statistics for Counties, Farms and Farm Property, With Related Information for Farms and Farm Operators, Livestock and Livestock Products, and Crops* (Washington, DC: Government Printing Office, 1942), 618–19; William G. Robbins, *Landscapes of Conflict: The Oregon Story 1940–2000* (Seattle: University of Washington Press, 2004), 79–113; and Floyd J. McKay, "Green Beans, Green Cash: Alderman's Post World War II Teenage Workforce," *Oregon Historical Quarterly* 11, no. 3 (Fall 2010): 372–86.

2. 75h Congress, 1st Session, House of Representatives, Report no. 1298, "Permitting Producers of Hops to Enter into Marketing Agreements under Agricultural Adjustment Act," July 23, 1937; "Marketing Agreement Completed," *Pacific Hop Grower,* December 1939; "All Hops Must Be Inspected," *Hopper,* August, 1944, 2; Fred E. Carver, "How Hop Inspection Service Will Operate," *Hopper,* August, 1944, 6; G. L. Becker, "Pacific Coast Hops to the Fore," *Hopper,* June 1945, 2.

3. "Independence Receives English Hops," *Pacific Hop Grower,* January 1936, 7; "New English Hop Doing Well Here, *Pacific Hop Grower,* March 1937, 5.

4. Herman Goschie and Vernice Goschie, interview by the author, Goschie Farms, Silverton, OR, August 18, 2008; John Annen, interview by the author, Annen Brothers Farm, Mt. Angel, OR, March 11, 2008; Harvey Kasar, oral history by Kathleen Hudson Cooler, Silverton, OR, April 16, 1982.

5. The concept of productivity, or the more efficient use of natural resources through engineering and science, became a central component of containment policy for the postwar American government. In Great Britain and Western Europe, the Marshall Plan entailed collaboration with agricultural practices centered on technology and efficiency. See David Ekbladh, *The Great American Mission: Modernization and the Construction of an American World Order* (Princeton, NJ: Princeton University Press, 2010).

6. Hop Industry Productivity Team, *The Hop Industry: Report of a Visit to the U.S.A. and Canada in 1950 of a Productivity Team Representing the Hop Industry* (London: Anglo-American Council on Productivity, 1951), 3–11. For the English hop industry, see Herbert Myrick, *The Hop: Its Culture and Cure, Marketing and*

Manufacture (1899; reprint, New York: O. Judd, 1904), 1–12; and Margaret Lawrence, *The Encircling Hop: A History of Hops and Brewing* (Sittingbourne, Kent, UK: SAWD, 1990), 1–24.

7. Christine E. Pfaff, *Harvests of Plenty: A History of the Yakima Irrigation Project, Washington* (Denver: United States Department of the Interior, Bureau of Reclamation, Technical Service Center, 2002), 37–125; Elbert E. Miller and Richard M. Highsmith, Jr., "The Hop Industry of the Pacific Coast," *Journal of Geography,* 49, no. 2 (February 1950): 63–77. For more on Yakima's geology, environment, and agriculture from a mid-twentieth-century perspective, see D. W. Meinig, *The Great Columbia Plain: A Historical Geography, 1805–1910* (Seattle: University of Washington Press, 1968), 6–7, 340–44, 441–42; and Richard M. Highsmith, Jr., and Elbert E. Miller, "Open Field Farming in the Yakima Valley, Washington," *Economic Geography* 28, no. 1 (January 1952): 74–87.

8. G. R. Hoerner and Frank Rabak, *Production of Hops*, Farmers' Bulletin no. 1842 (Washington, DC: US Department of Agriculture, 1940), 8. The question of seeded hops had been long answered in Europe, as Continental growers outlawed all male plants; in England cross-fertilization and therefore seeded hops were a fact of life because of the constant presence of wild hops growing in the hedgerows that divided farmsteads.

9. D. B. DeLoach, *Outlook for Hops from the Pacific Coast* (Washington, DC: US Department of Agriculture, Bureau of Agricultural Economics, 1948), 2–5; US Department of Agriculture, *Hops: By States, 1915–69*, Statistical Bulletin no. 469 (Washington, DC: US Department of Agriculture, Statistical Reporting Service, 1971), 2–8. More specifically, two social scientists reported in 1950, "The importance of the changes is seen in the fact that during the five-year period from 1940 to 1945, the acreage of hops in Washington increased 205 per cent while the acreage in Oregon decreased by 4 per cent." They continued, "The acreage of hops in Oregon is 62 per cent more than that of Washington but the production is 26 per cent less, due to the lower per yield acres." See Miller and Highsmith, "Hop Industry of the Pacific Coast," 63–71.

10. Although it is difficult to find the exact number of Willamette Valley hop growers in certain years, the June 1944 issue of the *Hopper* listed all Oregon members of the United States Hop Growers' Association: "Albany—J. R. & S. P. Linn, H. A. Luther, J. W. Morgan, Weather & Weather, Glen H. Wilfert, Lewis Winn, M. E. Winn, W. M. & J. A. Winn; Aurora—Anderson Bros. R. J. & A. F., Robert A. Armstrong, G. B. Bailey, John Bisanz, E. H. Crisell, Glenn Crisell, Fred Dentel, F. Ralph DuRette, H. J. Eldridge, James Feller, Howard Freeman, Theo. Freeman, E. M. Grim, Jing Guy, Delbert Haener, Thomas Jette, Jette Bros., John Kister, Peter Krupicka, Ella & James Leavy, Tong Lee, W. J. Miley, James E. Myers, J. J. Mortensen, Sun Onn & Dy Foon, Sam Pahl, Kilian Smith, Ray Yergen; Banks—Fred Hartwick, Wm. L. Moore; Brooks—Romeo Gourley; Canby—Fred Ganske, John Gelbrich, Gribble Bros., Frank Hein, Johnson Bros., Leon- [sic] F. J. & F. O. Kraxberger, H. C. Lemons, C. H. Lorenz, John Nordhausen, L. Rinehart & Son; Cornelius—John J. Buchanan; Dallas—Carver & Linn, Hugh Smith; Dayton—J. S. Gilkey; Donald—

Russell N. Lee, Lee Quan; Eugene—Anderson Bros. W. H. & C. I., J. O. Burgess, L. S. Christofferson, Andrew Erickson, James Erickson, James Hayes, Vance How, Fred M. Ireland, Joe Jacob; Forest Grove—O. N. Love, A. R. Nichols; Gales Creek— Emil C. Dober; Gaston—Charles G. Stiller; Grants Pass—S. J. & F. A. Christie, Ray W. Claudson, Charles B. Cook, DeArmond & Son, Howard Eisman, R. R. Every, Carl Geurin, Arthur King, Raymond A. Lathrop, Roy E. Lathrop, J. T. Middleton, D. G. Robertson, Fred N. Robertson, Paul Schroeder, R. E. Stephenson, Waite Hop Yard, Ada Weston; Gervais—Otton Berning, Donald Coleman, Robert Coleman, L. P. Forcier, A. W. Nusom, Fred Viesko; Halsey—Babb and Moody, C. S. Elliott, J. A. McLaren; Harrisburg—Grace Bristow, Tom Lowell, Norma E. Murphy, Stroda Bros.; Hillsboro—A. J. Ray & Son, John Sinclair; Hubbard—John H. Blosser, Ed Budreau, Glen C. Carothers, Ben Eppers, F. C. Eymon, A. E. Feller, Frank Gilles, Raymond Gilles, T. J. Hunt & Son, Geo. Leffler & Lawrence Pulley, A. G. Letten-maier, John & Vernon Miller, Pardey Bros., Eldon Pugh, L. W. Pulley, J. H. & Louie J. Schwabauer, Elmer Stauffer, Elmer Stauffer & Lester Wills, Lester Wills, Ralph Yergen; Independence—R. H. Ackerman, W. N. Acocks, J. P. Barker, Collins & Collins, Inc., S. J. Hoover & Son, A. M. McLaughlin, G. E. Newton, Russell Catlen Hop Co., Sloper Bros., Geo. Stapleton, F. E. Turner, Dean Walker, Walker & Walker, Homer H. Wood; Jefferson—Wm. & Walburga Krebs; Junction City— Frank Hayes, H. E. Hill, W. H. Hunt, Carl & Tom Kirk, W. G. Reetz-Marquam, E. J. Dahl, Arthur Olson. Molalla—Otto Lucht, W. G. Lucht, W. P. Nicholson, E. J. Seaman & Son; Mount Angel—Amon Bros., Henry Asboe, Butsch Bros. & Ertelt, Butte Creek Orchards, Carlson Bros. Oscar & Theo., Clement J. Duda, Carl Ertelt, Lawrence Gehrman, L. E. Gerlits, Harold Gregerson, Henry Humpert, Emil H. Jacobson, Andrew J. Lelek, Fred W. Lucht, P. J. May, Otto J. Schlottmann, Paul Schroeder, Elmer Thompson, J. P. Vandecoevering, A. Vanderbeck, Joe Walker, Ernest Westendorf, Martin Westendorf, G. G. White, C. L. Willig, R. F. Zollner; Murphey—B. M. Clute; Newberg—Walter & A. F. Everest, Peter B. Kirk, C. S. Mullen; Orting—Herman Bohn; Portland—J. W. Seavey Hop Corp., Willamette Hop Co., Williams & Hart, Ralph Williams, Jr., Williams & Thacker; Saint Paul— L. H. Bunning, A. B. Connor, E. C. Davidson, Edw. F. Davidson, Fred & John Davidson, R. S. Davidson, Ray Kerr, Lester Kirk, M. H. Merten, Steve J. Mertens, Henry Raymond, Marguerite Simon, J. E. Smith Hop Co., S. J. Smith, P. M. Spyrup, G. B. Wolf, Phillip Wolf; Salem—R. G. DeSart, Homer Gouley, Sr. & Jr., W. W. Graham, Hartley & Craig, C. E. Ireland, Mrs. A. M. Jerman, Jerman & Chittenden, Louis Lachmund, C. W. Paulus, Kola McClellan, D. P. MacCarthy, E. A. Miller, D. C. & John D. Minto, Oregon Hopq [sic] Company, John J. Roberts Co., Almon Winn, Wood & Nelson, Chas. A. Yergen. Sherwood—Ralph Baker, Guy Chapman, Fred Elmert, H. C. Stein; Silverton—George Elton & Loren Henjum, Elmer J. Gehring, Otto Gehring, C. H. Goschie, Herman [G]oschie [sic], A. Hari, Walter I. Hari, Hans Jacob, F. D. Kaser, Harvey Kaser, Ben Kaufman, George Kaufman, Fred Krug, H. Kuenzi, John Moe, Albert Overlund, Ed. Overlund, John Overlund, Oscar Overlund, Reidar Poverud, Oscar Satern, Peter Scymanski, Ernest Stadeli, Fred Stadeli, Paul Stadeli; Woodburn—John Beck, Albert B. Bing, Wayne Brantner, Cooke Bros., E. W. Crosby, John Drescher, Frank DuBois, Leonard Ferschweiler, Ray J. Glatt, J. N. & C. J. Gooding, Oscar Gregerson, H. H. Jacobson, Chas. R. Johnson, Kaufman Bros., Nick Krebs, Mrs. Grace McKay, Lena Neumann, Albert Pederson, R. H. Pomeroy, Mrs. Dora Prather, Nick & Joseph Serres, E. T. Tweed, L. H. West, Drexel White; West Woodburn—Wm. Wengenroth." From "United States Hop Growers' Membership in Oregon," *Hopper,* June 1944, 6–7.

11. Maureen Ogle, *Ambitious Brew: The Story of American Beer* (Orlando, FL: Harcourt, 2006), 225; DeLoach, *Outlook for Hops,* v.

12. For the rise of modern marketing and the rise of brand recognition as a means to meet the needs of coast-to-coast market shares, see Pamela Walker Laird, *Advertising Progress: American Business and the Rise of Consumer Marketing* (Baltimore: Johns Hopkins University Press, 1998), 156–57, 184–85, 201–3. Also see Richard Ohmann, *Selling Culture: Magazines, Markets, and Class at the Turn of the Century* (London: Verso, 1996); and Elspeth H. Brown, Catherine Gudis, and Marina Moskowitz, eds., *Cultures of Commerce: Representations and American Business Culture, 1877–1960* (New York: Palgrave Macmillan, 2006).

13. A. M. McGahan, "The Emergence of the National Brewing Oligopoly: Competition in the American Market, 1933–1958," *Business History Review* 65, no. 2 (Summer 1991): 229–84.

14. Ogle, *Ambitious Brew,* 217–26, 246. For a broader discussion, see historian Lizabeth Cohen's work on postwar consumerism and the notion of "market segmentation," as she notes how corporations expanded their influence across the nation by specifications on consumer types. See Lizabeth Cohen, *A Consumers' Republic: The Politics of Mass Consumption in Postwar America* (New York: Vintage, 2003). For a comparative perspective, see Susan Strasser, Charles McGovern, and Matthias Judt, eds., *Getting and Spending: European and American Consumer Societies in the Twentieth Century* (Cambridge: Cambridge University Press, 1998).

15. Amy Mittelman, *Brewing Battles: A History of American Beer* (New York: Algora, 2008), 158; Kathleen E. Hudson Cooler, "Hop Agriculture in Oregon: The First Century" (master's thesis, Portland State University, 1986), 76; Carl Spielvogel, "Beer Taste Trend Cuts Crop of Hops: Bitter Fruit of Vine Is Used Less by Brewers Today," *New York Times,* November 7, 1955, 40.

16. Miller and Highsmith, "Hop Industry of the Pacific Coast," 67–70.

17. DeLoach, *Hops from the Pacific Coast,* v.

18. DeLoach, *Hops from the Pacific Coast,* 11.

19. For contemporary accounts of these issues in the mid–twentieth century, see George Wythe, *The United States and Inter-American Relations: A Contemporary Appraisal* (Gainesville: University of Florida Press, 1964); and David Riesman, *The Lonely Crowd: A Study of the Changing American Character* (New Haven, CT: Yale University Press, 1950).

20. Ogle, *Ambitious Brew,* 228–31. The idea of the "blanding" of the American palette is Ogle's.

21. For postwar consumer choices, see Cohen, *Consumer's Republic;* and Eric Schlosser, *Fast Food Nation: The Dark Side of the All-American Meal* (Boston:

Houghton Mifflin, 2001). For discussion of American economic growth and beliefs in economic superiority, see Robert M. Collins, *More: The Politics of Economic Growth in Postwar American* (Oxford: Oxford University Press, 2000); and Ekbladh, *Great American Mission.*

22. D. D. Hill and D. E. Bullis, *Summary of Hop Grade Investigations and Hop Quality as the Brewer Sees It,* Brewers' Hop Research Institute, Circular no. 5 (Chicago: Brewers' Hop Research Institute, 1942), 11–16. The specifics were as follows: representatives of the institute judged samples for color, aroma, amount of lupulin, color and condition of lupulin, and general appearance. The writers noted, "Quality of hops is much more difficult to define and describe than is quality in most agricultural commodities." And they called many of the qualities "intangible." But they did determine that the American industry still needed to take measures for improvement. After analyzing 120 responses from brewers on hop quality, they insinuated that changes were needed across the Pacific Coast hop industry, particularly in the Willamette Valley. Most notably, brewers no longer wanted seeded hops because of their belief that they added extra weight and deterred from the flavor of beer. Five out of six brewers also wanted greater lupulin content overall in their hops, fewer broken cones in their bales, and, in general, better rules for grading hop quality. Another key point was that because of inconsistencies in domestic hops and a perception that European hops were still superior, over 50 percent of brewers continued to look to Europe before the Pacific Coast for their hops.

23. G. R. Hoerner, *Brewers Hop Research Institute: Circular no. 1* (Chicago: Brewers' Hop Research Institute, 1940), 1–7.

24. D. D. Hill and D. E. Bullis, *Second Annual Report of Investigations to Develop Hop Grades* (Corvallis, OR: Brewer's Hop Research Institute, 1941–42), 1–3.

25. According to a 1930's study, only about a third or two-fifths of growers entered into multiple-year contracts. See George L. Sulerud, *An Economic Study of the Hop Industry in Oregon* (Corvallis: Agricultural Experiment Station, Oregon State Agricultural College, 1931), 55.

26. Goschie and Goschie, interview; Alfred Haunold, interviews by the author, Oregon State University, Corvallis, March 18, 2008, and June 29, 2009.

27. Goschie and Goschie, interview; John Annen, interview by the author, Annen Brothers Farm, Mt. Angel, OR, March 11, 2008.

28. "The Seventh Hop Growers Convention," *Hopper,* February 1953, 2–10.

29. "Seventh Hop Growers Convention," 10.

30. For reclamation histories that cover this period, see William D. Rowley, *The Bureau of Reclamation: Origins and Growth to 1945,* vol. 1 (Denver: US Department of the Interior, Bureau of Reclamation, 2006); Donald Worster, *Rivers of Empire: Water, Aridity, and Growth of the American West* (New York: Oxford University Press, 1985); Marc Reisner, *Cadillac Desert: The American West and Its Disappearing Water* (New York: Viking, 1986); and Donald J. Pisani, *Water and American Government: The Reclamation Bureau, National Water Policy, and the West, 1902–1935* (Berkeley: University of California Press, 2002).

31. Deborah Fitzgerald, *Every Farm a Factory: The Industrial Ideal in American Agriculture* (London: Yale University Press, 2003). The notion of "every farm a factory" or "factories in the field" roots to the New Deal–era writer Carey McWilliams, who brought the nation's attention to the challenging working conditions of California farmworkers. See Carey McWilliams, *Factories in the Field: The Story of Migratory Farm Labor in California* (Boston: Little, Brown, 1939). Additionally, there is a growing body of scholarship that looks at these modernizing processes of American agriculture, including intense reclamation and industrialized farming, as instrumental in global policy and application. See, for example, Sterling Evans, *Bound in Twine: The History and Ecology of Henequen-Wheat Complex for Mexico and the American and Canadian Plains, 1880–1950* (College Station: Texas A&M University Press, 2007), 197–240; John Soluri, *Banana Cultures: Agriculture, Consumption, and Environmental Change in Honduras and the Unites States* (Austin: University of Texas Press, 2005), 75–127; and Arturo Warman, *Corn and Capitalism: How a Botanical Bastard Grew to Global Dominance,* trans. Nancy L. Westrate (Chapel Hill: University of North Carolina Press, 2003), 174–231.

32. R. Douglas Hurt, *Problems of Plenty: The American Farmer in the Twentieth Century* (Chicago: Ivan R. Dee, 2003), 122.

33. Hurt, *Problems of Plenty,* 122. For general works on industrial changes in post–World War II American agriculture, see Paul K. Conkin, *A Revolution down on the Farm: The Transformation of American Agriculture Since 1929* (Lexington: University Press of Kentucky, 2008); Bruce L. Gardner, *American Agriculture in the Twentieth Century: How It Flourished and What It Cost* (Cambridge, MA: Harvard University Press, 2002); Milton C. Hallberg, *Economic Trends in U.S. Agriculture and Food Systems since World War II* (Ames: Iowa State University Press, 2001); John T. Schlebecker, *Whereby We Thrive: A History of American Farming, 1607–1972* (Ames: Iowa State University Press, 1975); and Andrew P. Duffin, *Plowed Under: Agriculture and Environment in the Palouse* (Seattle: University of Washington Press, 2007).

34. "Big Business in Hops," *Hopper,* May 1953, 6. Also see S. S. Steiner, Inc., *Steiner's Guide to American Hops* (New York: S. S. Steiner, 1973); and Heinrich J. Barth and Christiane Klinke, *The History of a Family Enterprise: Joh. Barth & Sohn, Nuremberg,* trans. Ernest Sinauer, Chris Krason, and Alan Ross (Nuremberg: Joh. Barth & Son, 1994).

35. "Local Women Relieve Labor Shortages," *Hopper,* July 1944, 2.

36. "The Emergency Farm Labor Situation," *Hopper,* March 1945, 2–3; Patrick Joseph King, "Labor and Mechanization: The Hop Industry in Yakima Valley, 1866–1950" (master's thesis, Washington State University, 2008), 71–72. To understand the broader implications of the bracero movement in the region, see Erasmo Gamboa, *Mexican Labor and World War II: Braceros in the Pacific Northwest, 1942–1947* (Austin: University of Texas Press, 1990).

37. "Picking Hops by Machines," *Hopper,* October 1944, 8.

38. Goschie and Goschie, interview; Hoerner and Rabak, *Production of Hops,* 19.

39. European growers did not adopt mechanical harvesters until later in the 1950s and into the 1960s. See Hop Industry Productivity Team, *Hop Industry,* 29–46; and R. A. Neve, *Hops* (London: Chapman and Hall, 1991), 79.

40. Goschie and Goschie, interview; Miller and Highsmith, "Hop Industry of the Pacific Coast," 66–67; "Questionnaire Highlights from 1982–83," Charles K. Martin and Eugene C. Cole, box 1, Kathleen Hudson Cooler Papers, MSS 1979–735.0001, Benton County Historical Society, Philomath, OR.

41. Hoerner and Rabak, *Production of Hops,* 3–4; "Tobacco Dust Has Also Gone to War," *Hopper,* February 1945, 10.; Miller and Highsmith, "Hop Industry of the Pacific Coast," 64. In addition to legumes, growers also utilized winter grains, turnips, or wild mustard as nitrogen-producing cover crops.

42. "Fifty Million Lady Bugs Can't Be Wrong," *Hopper,* August 1945, 6.

43. Wye College Department of Hop Research, *Annual Report, 1953* (Wye, England: Wye College Department of Hop Research, 1953), 5; "Independence Receives English Hops," *Pacific Hop Grower,* January 1936, 7; "New English Hop Doing Well Here," *Pacific Hop Grower,* March 1937.

Plant breeding, of course, is an activity that dates to the agricultural revolution ten thousand years ago. The notion of hybridizing useful species proliferated with corn, rice, and other grains over the centuries; and more recently, specialty crops have emerged through cross-pollination and grafting programs. For his plant-breeding work in New York and California, Luther Burbank has been rightfully christened as the father of many important crops we have today: the Russet Burbank potato, elephant garlic, California poppies, and scores of other botanical brethren. As much as the legendary Burbank championed new species of plants, though, he never generated a new variety of hop—a crucial Pacific Coast cash crop that emerged at the same time that he expanded his work in Santa Rosa, California. See Jane S. Smith, *The Garden of Invention: Luther Burbank and the Business of Breeding Plants* (New York: Penguin, 2009). On the expansive influence of horticulture in the United States, see U. P. Hedrick, *A History of Horticulture in America to 1860* (New York: Oxford University Press, 1950); and Philip J. Pauly, *Fruits and Plains: The Horticultural Transformation of America* (Cambridge: Harvard University Press, 2007).

44. Barth and Klinke, *History of the Family Enterprise,* 21–22; Peter Darby, interview by the author, Wye Hops, China Farm, Canterbury, England, May 22, 2013; Haunold, interviews.

45. Hurt, *Problems of Plenty,* 116.

46. For overviews of DDT in America and Rachel Carson's work, see Rachel Carson, *Silent Spring* (Boston: Houghton Mifflin, 1962); Thomas Dunlap, *DDT: Scientists, Citizens, and Public Policy* (Princeton, NJ: Princeton University Press, 1981); Samuel P. Hays, *Beauty, Health, and Permanence: Environmental Politics in the United States, 1955–1985* (Cambridge: Cambridge University Press, 1987), 171–206; Mark H. Lytle, *The Gentle Subversive: Rachel Carson, Silent Spring, and the Rise of the Environmental Movement* (Oxford: Oxford University Press, 2007); and Naomi Oreskes and Eric M. Conway, *Merchants of Doubt: How a Handful of Scien-*

tists *Obscured the Truth on Issues from Tobacco Smoke to Global Warming* (New York: Bloomsbury Press, 2010), 216–33.

47. Hoerner and Rabak, *Production of Hops,* 27–38.

48. "D—D—T," *Hopper,* October 1945, 9–10.

49. "A Substitute for Nicotine," *Hopper,* October 1946, 4.

50. "Substitute for Nicotine," 4.

51. "Experience with New Insecticide," *Hopper,* February 1947, 8.

52. *Hopper,* December 1947, 10.

53. *Hopper,* February 1950, 9.

54. *Hopper,* August 1951, 5.

55. Charles R. Joshston, "Fertilization and Irrigation of Willamette Valley Upland Hops," *Hopper,* May 1952, 7–8.

56. Robbins, *Landscapes of Conflict,* 79–113.

57. A year after she published *Silent Spring,* Carson also spoke to Congress on the ills of DDT and pesticides. Of her many biographies, see Lytle, *Gentle Subversive;* and Kathleen Dean Moore, *Rachel Carson: Legacy and Challenge* (Albany: State of New York University Press, 2008). For other histories on toxins, see Sandra Steingraber, *Living Downstream: An Ecologist's Look at Cancer and the Environment* (Reading, MA: Addison-Wesley, 1997); and Sandra Steingraber, *Having Faith: An Ecologist's Journey to Motherhood* (New York: Berkley Books, 2003). For how DDT and toxins fit into the larger modern environmental movement, see Robert Gottlieb, *Forcing the Spring: The Transformation of the American Environmental Movement* (Washington DC: Island Press, 1993), 81–116; and Hays, *Beauty, Health, and Permanence,* 171–206.

58. "Oregon Hop Growers Association: May 11, 1955, Articles of Corporation, By-Laws," box 3, Oregon Hop Growers Association Records, Oregon State University Archives, Corvallis, OR; Oregon Hop Commission, Minutes, June 3, 1964.

59. "Exit the Hop Growers Friend," *Hopper,* June 1953, 9–10.

60. Oregon Hop Commission, Minutes, June 3, 1964.

61. Maureen Coleman, interview by the author, St. Paul, OR, March, 12, 2008. This vignette also stems from informal conversations with other members of the family.

62. Oregon Hop Commission, Minutes, June 25, 1964, 1; "Elaine Annen Obituary," *Salem [OR] Statesmen Journal,* March 11, 2014.

63. Oregon Hop Commission, Minutes, November 17, 1964, 2–3.

64. Alfred Haunold, interviews.

9. CASCADE

1. Oregon State Agricultural College, *Annual Report of the President of the Board of Regents of the State Agricultural College to the Governor of Oregon, Legislative Assembly, Eighteenth Regular Session, 1895* (Salem: Oregon State Agricultural College, 1894), 29. Also see: H. V. Tartar and B. Pilkington, *Hop Investigations,* Bulletin no. 114 (Corvallis: Oregon Agricultural College Press, 1913).

2. G. R. Hoerner and Frank Rabak, *Production of Hops*, U.S. Department of Agriculture, Farmers' Bulletin no. 1842 (Washington, DC: Government Printing Office, 1940), 6; R. A. Neve, "Notes on a Visit to the U.S.A.," in Wye College Department of Hop Research, *Annual Report, 1963* (Wye, England: Wye College Department of Hop Research, 1963), 46.

3. "News and Notes," *Journal of the American Society of Agronomy* 14, no. 8 (November 1922): 326. Bressman's papers reside in the archive of Iowa State University.

4. For several decades after its publication, *Corn and Corn Growing* was a staple in agriculture science classrooms. Subsequent editions were published in 1925, 1928, 1937, and 1939. Henry A. Wallace published the first edition, however. See Henry A. Wallace and E. N. Bressman, *Corn and Corn Growing* (Des Moines, IA: Wallace, 1923).

5. For histories of Mendel and genetics, see Diane B. Paul and Barbara A. Kimmelman, "Mendel in America: Theory and Practice, 1900–1919," in *The American Development of Biology*, ed. Ronald Rainger, Keith Benson, and Jane Maienschein (Philadelphia: University of Pennsylvania Press, 1988), 281–310; Peter J. Bowler, *The Mendelian Revolution: The Emergence of Hereditarian Concepts in Modern Science and Society* (Baltimore: Johns Hopkins University Press, 1989); and Robin Marantz, *The Monk in the Garden: The Lost and Found Genius of Gregor Mendel, the Father of Genetics* (Boston: Houghton Mifflin, 2000).

6. E. N. Bressman, "Report of Hop Breeding Project, Sept. 3, 1930, to Dec. 31, 1931," unpublished report, Corvallis: Oregon State Agricultural College, 1931, 65, box 13, record group 25, Agricultural Experiment Station, University Archive, Oregon State University, Corvallis.

7. E. N. Bressman, "Report of Hop Breeding Project, Jan. 1, 1931, to Dec. 31, 1931," unpublished report, Corvallis: Oregon State Agricultural College, 1931, 6, box 21, record group 25, Agricultural Experiment Station, University Archive, Oregon State University, Corvallis. Other countries were also active in the national and global exchanges of information. For example, see Helen R. Pearce, *The Hop Industry in Australia* (Carlton, Australia: Melbourne University Press, 1976).

8. E. N. Bressman, "Report of Hop Breeding Project, Sept. 3, 1930, to Dec. 31, 1931," unpublished report, Corvallis: Oregon State Agricultural College, 1931, 10, box 21, record group 25, Agricultural Experiment Station, University Archive, Oregon State University, Corvallis.

9. E. N. Bressman, "Report of Hop Breeding Project, Jan. 1, 1932 to Dec. 31, 1932," unpublished report, Corvallis, OR, Oregon State Agricultural College, 1931, 16, box, 21, record group 25, Agricultural Experiment Station, University Archive, Oregon State University, Corvallis.

10. E. N. Bressman, "Report of Hop Breeding Project, Jan. 1, 1932, to Dec. 31, 1932," unpublished report, Corvallis: Oregon State Agricultural College, 1931, 6, box 21, record group 25, Agricultural Experiment Station, University Archive, Oregon State University, Corvallis. For an official program description, see E. N. Bressman, "Developing New Varieties of Hops," *Science,* n.s., vol. 74, no. 1912 (August 21, 1931): 202–3.

11. E.N. Bressman, "Report of Hop Breeding Project, Sept. 3, 1930, to Dec. 31, 1931," unpublished report, Corvallis: Oregon State Agricultural College, 1931, 65; Bressman, "Developing New Varieties of Hops," 203. On general breeding and plant hybridity, see Noel Kingsbury, *Hybrid: The History and Science of Plant Breeding* (Chicago: University of Chicago Press, 2009); and Timothy J. Motley, Myree Zerega, and Hugh Cross, eds., *Darwin's Harvest: New Approaches to the Origins, Evolution, and Conservation of Crops* (New York: Columbia University Press, 2006).

12. E.N. Bressman, "Report of Hop Breeding Project, Sept. 3, 1930, to Dec. 31, 1931," 8, box 21, record group 25, Agricultural Experiment Station, University Archive, Oregon State University, Corvallis.

13. E.N. Bressman, "Report of Hop Breeding Project, Sept. 3, 1930, to Dec. 31, 1931," unpublished report, Corvallis: Oregon State Agricultural College, 1931, 24–29, 44–46.

14. E.N. Bressman, "Report of Hop Breeding Project, Jan. 1, 1933, to Dec. 31, 1933," unpublished report, Corvallis: Oregon State Agricultural College, 1934, 136, box, 21, record group 25, Agricultural Experiment Station, University Archive, Oregon State University, Corvallis.

15. E.N. Bressman, "Report of Hop Breeding Project, Jan. 1, 1933, to Dec. 31, 1933," unpublished report, Corvallis: Oregon State Agricultural College, 1934, 157–58.

16. Alfred Haunold, S.T. Likens, C.E. Horner, S.N. Brooks, and C.E. Zimmermann, "One-Half Century of Hop Research by the US Department of Agriculture," *Journal of the American Society of Brewing Chemists* 24, no. 3 (1985): 123–25.

17. R.E. Fore, "Report of Cooperative Hop Breeding Project: Division of Drug and Related Plants, Bureau of Plant Industry, United States Department of Agriculture and Oregon Experiment Station, Corvallis Oregon, Jan. 1, 1937, to Dec. 31, 1937," unpublished report, Corvallis: Oregon State Agricultural College, 1937, 3–4, box 21, record group 25, Agricultural Experiment Station, University Archive, Oregon State University, Corvallis. Also see D.D. Hill, "Report: Hop Analyses, 1938–1940," Corvallis: US Department of Agriculture and Oregon Experiment Station, 1940," unpublished report; and J.D. Sather, D.E. Bullis, D.D. Hill, "Hop Grade Studies: 1949 Crop," Corvallis: Oregon Agricultural Station in Cooperation with Grain Division, Production and Marketing Administration, U.S.D.A., 1950, unpublished report.

18. R.E. Fore, "Report of Cooperative Hop Breeding Project: Division of Drug and Related Plants, Bureau of Plant Industry, United States Department of Agriculture and Oregon Experiment Station, Corvallis, Oregon, Jan. 1, 1937, to Dec. 31, 1937," unpublished report, Corvallis: Oregon State Agricultural College, 1937, 3–4, box 21, record group 25, Agricultural Experiment Station, University Archive, Oregon State University, Corvallis.

19. G.R. Hoerner, "The Hop Press: A Memorandum of What's Brewin' (Compiled for the Convenience of County Agents)," Corvallis: Oregon State College and US Department of Agriculture Cooperating, October 18, 1948.

20. Hoerner, "Hop Press," 2; G.R. Hoerner, "The Hop Press: A Memorandum of What's Brewin' (Compiled for the Convenience of County Agents), Corvallis:

Oregon State College and US Department of Agriculture Cooperating, May 6, 1949, 2–3; G. R. Hoerner, "The Hop Press: A Memorandum of What's Brewin'" (Compiled for the Convenience of County Agents), Corvallis: Oregon State College and US Department of Agriculture Cooperating, July 18, 1950, 2–8.

21. Stanley N. Brooks, Chester E. Horner, and S. T. Likens, "1958 Annual Report of Hop Investigations," unpublished report, Corvallis: Oregon Agricultural Experiment Station, 1959, 5–6, box 13, record group 25, Agricultural Experiment Station, University Archive, Oregon State University, Corvallis.

22. Stanley N. Brooks, Chester E. Horner, Sam. T. Likens, and Charles E. Zimmermann "1961 Annual Report of Hop Investigations," unpublished report, Corvallis: Oregon Agricultural Experiment Station, 1962, 5, box 13, record group 25, Agricultural Experiment Station, University Archive, Oregon State University, Corvallis.

23. Stanley N. Brooks, Chester E. Horner, Sam. T. Likens, and Charles E. Zimmermann, "1963 Annual Report of Hop Investigations," unpublished report, Corvallis: Oregon Agricultural Experiment Station, 1964, 2, box 13, record group 25, Agricultural Experiment Station, University Archive, Oregon State University, Corvallis; Neve, "Notes on a Visit," 46–49.

24. Alfred Haunold, interview by the author, Oregon State University, Corvallis, June 29, 2009; Alfred Haunold, "Personal Reflections on 30 Years of Hop Breeding," unpublished document, Corvallis, OR: Alfred Haunold, 2007.

25. Haunold, interview.

26. Haunold, interview; Haunold, "Personal Reflections."

27. Stanley N. Brooks, Chester E. Horner, Sam. T. Likens, Alfred Haunold, and Charles E. Zimmermann, "1966 Annual Report of Hop Investigations," unpublished report, Corvallis, Oregon: Oilseed and Industrial Crops Research Branch, Crops Research Division, Agricultural Research Service, United States Department of Agriculture, in cooperation with Oregon Agricultural Experiment Station, 1967, 24. For more on the Mt. Angel hop growers, see St. Benedict's Abbey, *Benedictine Hop Growers of the Willamette Valley* (Mount Angel, OR: Benedictine Press, 1935).

28. Haunold, interview.

29. Haunold et al., "One-Half Century of Hop Research," 123–25.

30. Haunold, interview; Minutes, Oregon Hop Commission, November 13, 1969, 3; Minutes, Oregon Hop Commission, February 2, 1970, 2; Minutes, Oregon Hop Commission, July 14, 1970, 3; Minutes, Oregon Hop Commission, April 18, 1972.

31. C. E. Horner, Alfred Haunold, and S. T. Likens, "1969 Annual Report of Hop and Mint Investigations: Breeding, Genetics, Chemistry, Pathology of Hops and Mint," unpublished report, Corvallis: Oilseed and Industrial Crops Research Branch, Plant Science Research Division, Agricultural Research Service, United States Department of Agriculture, in cooperation with Oregon Agricultural Experiment Station, 1970, 8.

32. C. E. Horner, Alfred Haunold, and S. T. Likens, "1969 Annual Report of Hop and Mint Investigations: Breeding, Genetics, Chemistry, Pathology of Hops

and Mint," unpublished report, Corvallis: Oilseed and Industrial Crops Research Branch, Plant Science Research Division, Agricultural Research Service, United States Department of Agricultural, in cooperation with Oregon Agricultural Experiment Station, 1970, 8.

33. S. N. Brooks, C. E. Horner, S. T. Likens and C. E. Zimmermann, "Registration of Cascade Hop (Reg. no. 1)," *Crop Science* 12, no. 3 (1972): 394.

34. Haunold, interview.

35. Haunold, interview; "Hops Variety Excites Producers," *Idaho State Journal,* September 17, 1974, B1.

36. Joh. Barth & Sohn, *Hops 1974/75* (Nuremberg: Joh. Barth & Sohn, 1975), 17.

37. Haunold, interview; Haunold, "Personal Reflections"; Alfred Haunold, "Travel Report of a Visit to Anheuser-Busch, Inc., St. Louis, Missouri, March 26–28, 1974," in Alfred Haunold, "1974 Annual Report: Hop Breeding and Genetics," Corvallis: US Department of Agriculture, 1974, unpublished report, 53–54.

38. Haunold, interview.

39. "Hops Variety Excites Producers"; Rodger Williams, "Cascade: How Adolph Coors Helped Launch the Most Popular US Aroma Hop and the Craft Beer Revolution," *Hop Pursuit* (blog), January 24, 2010, indiehops.com, accessed May 10, 2010, http://inhoppursuit.blogspot.com/2010/01/cascade-how-adolph-coors-helped-launch.html.

40. Alfred Haunold, "Biographical Sketch," unpublished document, 2007, 1–3.

41. Haunold et al., "One-Half Century of Hop Research," 123–25.

42. Gail Nickerson, oral history with author and Tiah Edmunson-Morton, Corvallis, OR, August 8, 2014.

43. Haunold, interview.

44. For more on hop-breeding programs across the world, see Peter Darby, "The History of Hop Breeding and Development," *Brewery History* 121 (Winter 2005): 94–112. Other hop-breeding programs at the time included those in Germany, the Czech Republic, Denmark, Japan, Slovenia, the Ukraine, Poland, China, India, and New Zealand.

45. Alfred Haunold, interview. Also, see all of Haunold's yearly reports for the USDA throughout the 1970s: Alfred Haunold, Donald D. Roberts, S. T. Likens, and G. B. Nickerson, "1972 Report: Breeding, Genetics, and Chemistry" Corvallis: U.S. Department of Agriculture, 1972; Alfred Haunold and Donald D. Roberts, "1973 Annual Report: Hop Breeding and Genetics," Corvallis: U.S. Department of Agriculture, 1973; Alfred Haunold, "1974 Annual Report: Hop Breeding and Genetics," Corvallis: U.S. Department of Agriculture, 1974, (unpublished report); Alfred Haunold, S. T. Likens, and C. E. Horner, "1975 Annual Report: Hop Breeding, Genetics, Chemistry, and Pathology" Corvallis: U.S. Department of Agriculture, 1975 (unpublished report); Alfred Haunold, S. T. Likens, and C. E. Horner, "1976 Annual Report: Hop Breeding, Genetics, Chemistry, and Pathology" Corvallis: U.S. Department of Agriculture, 1976 (unpublished report); Alfred Haunold, S. T. Likens, and C. E. Horner, "1977 Annual Report: Hop Breeding, Genetics, Chemistry, and Pathology," Corvallis: U.S. Department of Agriculture, 1977

(unpublished report); Alfred Haunold, S. T. Likens, and C. E. Horner, "USDA-SEA, Hop Research: 1978 Annual Report," Corvallis: U.S. Department of Agriculture, 1978 (unpublished report).

46. C. E. Zimmermann, S. T. Likens, A. Haunold, C. E. Horner, and D. D. Roberts, "Registration of Comet Hop," *Crop Science* 15 (January–February 1975): 98.

47. Oregon Hop Commission, Minutes, January 16, 1974, 2.

48. Oregon Hop Commission, Minutes, April 30, 1975, 1.

49. Hop Research Council, box 3 Hop Research Council, "1983 Reports and Proposals to the Hop Research Council" (unpublished report), 1983; Alfred Haunold, "1983 Hop Research," unpublished report, Corvallis: United States Department of Agriculture and Oregon State University, 1983, 2–5.

50. George L. Sulerud, *An Economic Study of the Hop Industry in Oregon* (Corvallis: Agricultural Experiment Station, Oregon State Agricultural College, 1931), 288.

51. Alfred Haunold, "Polyploid Breeding with Hop *Humulus lupulus L.*," *Technical Quarterly* (Master Brewers Association of America) 9, no. 1 (1972): 36. According to Haunold, in 1963 Stan Brooks and Ray Neve had already started doubling the chromosome numbers of Fuggle in order to create a tetraploid. Yet after Neve returned to England the work did not continue, and Haunold had to start the project anew.

52. It is now accepted that polyploidy has occurred to 70 percent of angiosperms. The key, where naturally or artificially induced, was that plant species allow for extra sets of chromosomes to allow for future evolutionary diversity of traits. These are the reasons Haunold began polyploidy hop-breeding tests in the late 1960s—to seek out new varieties that met the goals of high yield, disease resistance, and storability, while all the while catering to the desires of brewers. It was a process that other plant breeders had turned to in large part by the 1940s. See Jane Masterson, "Stomatal Size in Fossil Plants: Evidence for Polyploidy in Majority of Angiosperms," *Science* 264, no. 5157 (April 1994): 421–24; and Joshua A. Udall and Jonathan F. Wendel, "Polyploidy and Crop Improvement," *Plant Genome (A Supplement to Crop Science)* 1 (November 2006): S3–S4, S6.

53. Alfred Haunold, S. T. Likens, C. E. Horner, C. E. Zimmermann, and D. D. Roberts, "Registration of the Columbia Hop," *Crop Science* 16 (September–October 1976): 738–39; Alfred Haunold, C. E. Horner, S. T. Likens, D. D. Roberts, and C. E. Zimmermann, "Registration of the Willamette Hop," *Crop Science* 16 (September–October 1976): 739.

54. Alfred Haunold, C. E. Horner, S. T. Likens, D. D. Roberts, and C. E. Zimmermann, "Registration of Willamette Hop (Reg. No. 6)," *Crop Science* 16 (September–October 1976): 739.

55. Alfred Haunold, Ulrich Gampert, and Gail B. Nickerson, "Annual Research Progress Report to Busch Agricultural Resources, Inc., April 1, 1991, to March 31, 1992" (unpublished report) (Corvallis: US Department of Agriculture and Oregon State University, 1992), Hop Research Council, OSU, box 2, Oregon State University.

56. See Alfred Haunold et al., yearly unpublished reports, 1972–1996, US Department of Agriculture, Corvallis, Agricultural Experiment State, record group 25, University Archive, Oregon State University, Corvallis.

57. Joh. Barth & Sohn, *Hops 1986/87* (Nuremberg: Joh. Barth & Sohn, 1987), 17. For expanded hop-growing regions around the world in the 1970s and 1980s, see Heinrich Johann Barth, Christiane Klinke, and Claus Schmidt, *The Hop Atlas: The History and Geography of the Cultivated Plant* (Nuremberg: Joh. Barth & Sohn, 1994); and Pearce, *Hop Industry in Australia*.

58. Barth & Sohn, *Hops 1986/87*, 14–16.

59. Haunold, interview; Alfred Haunold, S. T. Likens, G. B. Nickerson, and R. O. Hampton, "Registration of Nugget Hop," *Crop Science* 24 (May–June 1984): 618; Alfred Haunold and G. B. Nickerson, "Registration of Mt. Hood Hop," *Crop Science* 30, no. 2 (1990): 423; Alfred Haunold, G. B. Nickerson, U. Gampert, and P. A. Whitney, "Registration of 'Liberty' Hop," *Crop Science* 32, no. 4 (1992): 1071.

60. John Gever, "Tops in Hops: World-Famous Hop Breeder Alfred Haunold Retires," *Eugene Weekly,* July 27, 1995.

61. Haunold, interview.

62. J. Postman, K. Hummer, E. Stover, R. Krueger, P. Forsline, L. J. Grauke, F. Zee, B. Irish, and T. Ayala-Silva, "Fruit and Nut Genebanks in the USDA National Plant Germplasm System," *HortScience* 41, no. 5 (2006): 1188–94.

10. HOP WARS

1. Karl Ockert, interview by the author, BridgePort Brewing Co., Portland, OR, August 21, 2009; Art Larrance, interview by the author, Raccoon Lodge Brewery, Portland, OR, August 2009; John Foyston, "Tales of the Pioneers," *Oregonian,* June 29, 2010, advertising supplement, "Oregon Craft Beer Month."

2. See particularly Tom Acitelli, *The Audacity of Hops: The History of America's Craft Beer Revolution* (Chicago: Chicago Review Press, 2013); and Steve Hindy, *The Craft Beer Revolution: How a Band of Microbrewers Is Transforming the World's Favorite Drink* (New York: Palgrave Macmillan, 2014).

3. See, for example, Ken Grossman, *Beyond the Pale: The Story of Sierra Nevada Brewing Co.* (Hoboken, NJ: Wiley, 2013); Sam Calagione, *Brewing Up a Business: Adventures in Beer from the Founder of Dogfish Head Craft Brewery* (Hoboken, NJ: Wiley, 2011); and Greg Koch, Steve Wagner, and Randy Clemens, *The Craft of Stone Brewing Co.: Liquid Lore, Epic Recipes, and Unabashed Arrogance* (Berkeley, CA: Ten Speed Press, 2011).

4. Acitelli, *Audacity of Hops,* 4, 10–11.

5. Maureen Ogle, *Ambitious Brew: The Story of American Beer* (Orlando, FL: Harcourt, 2006), 265.

6. David Kamp, *The United States of Arugula: The Sun-Dried, Cold-Pressed, Dark-Roasted, Extra Virgin Story of the American Food Revolution* (New York: Broadway Books, 2006) 122–65. Kamp's book offers a twentieth-century perspective on this change, as does Harvey Levenstein's *Revolution at the Table: The Transformation of the American Diet* (New York: Oxford University Press, 1988). For the specifics of the counterculture movement of the 1960s–1970s and the specific

changes in California, see Warren Belasco, *Appetite for Change: How the Counter-culture Took on the Food Industry,* 2nd ed. (Ithaca, NY: Cornell University Press, 2007); and Joyce Goldstein, *Inside the California Food Revolution: Thirty Years That Changed Our Culinary Consciousness* (Berkeley: University of California Press, 2013).

7. For the broader post–World War II affluence and the expansion of the con-sumer society, see Lizbeth Cohen, *A Consumers' Republic: The Politics of Mass Con-sumption in Postwar America* (New York: Vintage, 2003); Andrew Hurley, *Diners, Bowling Alleys, and Trailer Parks: Chasing the American Dream in the Postwar Con-sumer Culture* (New York: Basic Books, 2001); and Susan Strasser, Charles McGovern, and Matthias Judt, eds., *Getting and Spending: European and American Consumer Societies in the Twentieth Century* (Cambridge: Cambridge University Press, 1998).

8. Thomas Pinney, *A History of Wine in America: From Prohibition to the* Present (Berkeley: University of California Press, 2005), 224–52; James T. Lapsley, *Bottled Poetry: Napa Winemaking from Prohibition to the Modern Era* (Berkeley: University of California Press, 1996), 181–209. For the general expansion of viticulture and winemaking across the United States, see Pinney, *History of Wine in America,* 253–338; and Paul Lukacs, *American Vintage: The Rise of American Wine* (Boston: Houghton Mifflin, 2000), 252–90. For the origins and growth of the Oregon story, see Paul Pintarich, *The Boys Up North: Dick Erath and the Early Oregon Winemakers* (Portland, OR: Graphic Arts Center, 1997).

9. Acitelli, *Audacity of Hops,* 24.

10. For the definitions of *craft beer* and *microbrewery,* see "Craft Beer Industry Market Segments," Brewers Association, accessed August 1, 2015, https://www.brewersassociation.org/statistics/market-segments/.

11. Fred Eckhardt, "Eighty Beers Remembers: Part Two," *All About Beer Maga-zine* 27, no. 6 (January 2007).

12. Acitelli, *Audacity of Hops,* 37.

13. Acitelli, *Audacity of Hops,* 41.

14. Hindy, *Craft Beer Revolution,* 32–33. Fred Eckhardt, "Craft Beer—State of the Union, 2010," *All About Beer Magazine* 31, no. 6 (January 2011). Jackson's first, most influential work for early American craft brewers was *The World Guide to Beer: The Brewing Styles, the Brands, the Countries* (1977). But he would later publish many more books and countless articles on the topic of beer and whiskey. For exam-ples of his works on beer, see Michael Jackson, *The World Guide to Beer: The Brewing Styles, the Brands, the Countries* (Englewood Cliffs, NJ: Prentice-Hall, 1977); Michael Jackson, *The Pocket Guide to Beer* (New York: Putnam, 1982); Michael Jackson, *The New World Guide to Beer* (Philadelphia: Running Press, 1988); Michael Jackson, *Michael Jackson's Beer Companion: The World's Great Beer Styles, Gas-tronomy, and Traditions* (Philadelphia: Running Press, 1997); and Michael Jackson, *Ultimate Beer* (New York: DK, 1998).

15. Ogle, *Ambitious Brew,* 300.

16. Hindy, *Craft Beer Revolution,* 28–29.

17. Eckhardt, "Craft Beer."

18. Fred Eckhardt, "A Selected Chronology of Early Craft/Micro Brewers in the United States and Canada," *All About Beer Magazine* 31, no. 2 (May 2010).

19. Grossman, *Beyond the Pale*, 29–38. For a different perspective on the Sierra Nevada story, see Robert Stacey Burton, *Hops and Dreams: The Story of the Sierra Nevada Brewing Co.* (Chico, CA: Stansbury, 2010).

20. Informal interview with Fred Bowman, Art Larrance, John Foyston, Tim Hills, and Eliza Canty-Jones by the author, Bagdad Theater & Pub, Portland, OR, August 30, 2010.

21. Grossman, *Beyond the Pale*, 77–79.

22. Grossman, *Beyond the Pale*, 77.

23. Hop dealers often use a "hop aroma wheel" to sell their wares. The wheels contain the following categories: grassy, fruity, citrus, stone fruit, floral, spicy, tobacco/earthy, cedar, herbal, tropical fruit, and pine.

24. Acitelli, *Audacity of Hops*, 90; Frank J. Prial, "America's New Regional Beers: A Growing Number of Small Breweries Are Bringing Back the Flavors of Handcrafted Beer," *New York Times*, May 15, 1988. As another testament to the celebrated status of the hops, *Zymurgy* magazine published a special issue in 1990s in which Charlie Papazian wrote of his love affair with hops in the introduction, and which includes articles by Michael Jackson, Alfred Haunold, Bert Grant, and many others. See *Zymurgy* 13, no. 4 (1990). *Zymurgy* came out with another special issue on hops in 1997.

25. Tom Brennan, "Blitz Gets Brewhouse Mate as Pabst Beer Comes to Town," *Oregonian,* May 27, 1979, 54.

26. Hindy, *Craft Beer Revolution,* 57; Eckhardt, "Craft Beer."

27. Acitelli, *Audacity of Hops,* 85.

28. Angelo De Ieso, foreword to *Portland Beer: Crafting the Road to Beervana,* by Pete Dunlop (Charleston, SC: History Press, 2013), 7–8; Lisa M. Morrison, *Craft Beers of the Pacific Northwest: A Beer Lover's Guide to Oregon, Washington, and British Columbia* (Portland, OR: Timber Press, 2011), 6, 26–27.

29. For more on the regional arts-and-crafts history, see Lawrence Kreisman and Glenn Mason, *The Arts and Craft Movement in the Pacific Northwest* (Portland, OR: Timber Press, 2007).

30. Dunlop, *Portland Beer,* 78–79.

31. Fred Eckhardt, "Winemaker Ponzi Takes Up Beer-Making," *Oregonian,* October 23, 1984.

32. Karl Ockert, interview by the author, BridgePort Brewery, Portland, OR, August 16, 2008; John Foyston, "Tales of the Pioneers," *Oregonian,* June 29, 2010.

33. Ockert, interview.

34. Ockert, interview.

35. John Foyston, "When It Changed," *Oregonian,* April 29, 2004, Living, E01. John Foyston, "A Bridgeport Timeline," *The Beer Here* (blog), oregonlive.com, published August 11, 2009.

36. Foyston, "Tales of the Pioneers"; Fred Eckhardt, "'Long Friendship with Beer Leads to Brewery in Portland," *Oregonian,* March 11, 1986.

37. Fred Eckhardt, "Best, Worst Beers Available Here Listed," *Oregonian,* January 14, 1986.

38. Foyston, "Tales of the Pioneers."

39. Fred Eckhardt, "English Expert Cites Portland as U.S. Microbrewing Leader," *Oregonian,* September 30, 1986.

40. Art Larrance, interview by the author, Raccoon Lodge and Brewpub, Portland, OR, September 18, 2008; Ockert, interview.

41. Dunlop, *Portland Beer,* 76.

42. Dunlop, *Portland Beer,* 87.

43. Dunlop, *Portland Beer,* 100–101.

44. For more on the Oregon Brewers Festival, see http://www.oregonbrewfest .com.

45. Dunlop, *Portland Beer,* 102.

46. Dunlop, *Portland Beer,* 105.

47. Eckhardt, "Winemaker Ponzi Takes Up Beer-Making."

48. Fred Eckhardt, "The Hophead's Guide to Hops," *All About Beer Magazine* 26, no. 4 (September 1, 2005).

49. Quoted in Brian Yaeger, "Beer Guide 2014: An Oral History of BridgePort India Pale Ale," *Willamette Week,* February 5, 2014.

50. Richard Sassaman and Michael Dolan, "Strange Brew," *Air and Space Magazine* (August–September 1992), 13; Lucy Burningham, "A Hop and a Sip to Fresh Ales," *New York Times,* October 21, 2009. Also see Acitelli, *Audacity of Hops;* and Hindy, *Craft Beer Revolution.*

51. Joh. Barth & Sohn, *Hops . . . / . . .* (multiple years) (Nuremberg: Joh. Barth & Sohn, 1980–99); Al Haunold, interview by the author, Oregon State University, Corvallis, June 29, 2009.

52. James C. Flanigan, "NW Hop Growers Upset at End of Federal Marketing Controls," *Oregonian,* January 3, 1986; Herman Goschie and Vernice Goschie, interview with the author, Silverton, OR, August 18, 2008; Haunold, interview.

53. John Annen, interview with the author, Mt. Angel, OR, March 11, 2008.

54. Barth & Sohn, *Hops;* Haunold, interview.

55. "From Corporate to Craft" (interviews with Karl Ockert, Art Larrance, Jamie Floyd, Christian Ettinger, and Lucy Burningham), *Edible Portland* 24 (Fall 2011): 20–27.

56. Michelle Palacios, interview by the author, Oregon Hop Commission, Hubbard, OR, June 13, 2007.

57. Hop Growers of America, *2009 Statistical Report* (Moxee, WA: Hop Growers of America, 2010), 11.

58. Acitelli, *Audacity of Hops,* xv.

59. Tom Sharkey, interview with the author, The Forgery Brewery, Canterbury, England, May 23, 2013; Peter Darby, interview with author, Wye Hops, China Farm, Canterbury, England, May 21–22, 2013.

1. Brewers Association, "National Beer Sales and Production Data," Brewers Association, accessed, May 19, 2016, https://www.brewersassociation.org/statistics /national-beer-sales-production-data/.

2. Brewers Association, "Number of Breweries: Historical U.S. Brewery Count," Brewers Association, accessed May 19, 2016, https://www.brewersassociation.org /statistics/number-of-breweries.

3. "About Us," Pink Boots Society, accessed October 10, 2015, http:// pinkbootssociety.org/about-us/.

4. For more on the global craft beer movement, see www.craftbeerbrewers .org.

5. Christian Ettinger, interview by the author, Hopworks Urban Brewery, Portland, OR, July 7, 2010.

6. "Facts," Oregon Craft Beer, accessed September 14, 2015, http://oregoncraftbeer .org/facts.

7. "Facts."

8. Herman and Vernice Goschie, interview by the author, Silverton, OR, August 18, 2008; John Foyston, "Fresh Hop Season Is upon Us; Hop Madness This Weekend; Lucky Lab Backyard Hop Harvest Next Thursday," *The Beer Here* (blog), accessed November 4, 2011, last modified August 27, 2010, http://blog.oregonlive.com/the beerhere/2010/08/fresh_hop_season_is_upon_us_ho.html; Roger Worthington, "Indie Hops Organics Update: 2012 Will Be a Big Year for the Big O," indiehops.com, accessed November 4, 2001, last modified April 2, 2011, http://inhoppursuit.blogspot .com/2011/04/indie-hops-organics-update-2012-will-be.html; Roger Worthington, "Indie Hops Goes Green, Commits to 20 Acres Organic," indiehops.com, accessed November 4, 2001, last modified March 26, 2010, http://www.indiehops .com/goschie_organic.asp.

9. For more on the Hillsboro Hops, see http://www.milb.com/index.jsp?sid = t419. For more on the Oregon Hops and Brewing Archives, see http://scarc.library .oregonstate.edu/ohba.html.

10. Hop Growers of America, *2009 Statistical Report* (Moxee, WA: Hop Growers of America, 2010).

11. "Statistics: 2014," Oregon Hop Commission, oregonhops.com, accessed September 25, 2015.

12. For changes in plant patenting, see Noel Kingsbury, *Hybrid: The History and Science of Plant Breeding* (Chicago: University of Chicago Press, 2009); Jane S. Smith, *The Garden of Invention: Luther Burbank and the Business of Breeding Plants* (New York: Penguin, 2009); and Timothy J. Motley, Myree Zerega, and Hugh Cross, eds., *Darwin's Harvest: New Approaches to the Origins, Evolution, and Conservation of Crops* (New York: Columbia University Press, 2006).

13. Chris Colby, "The Bitter End: The Great 2008 Hop Shortage," *Brew Your Own,* Last modified March/April 2008, http://byo.com/grains/item/1476-the-bitter-end-the-great-2008-hop-shortage.

14. John Foyston, "The 2007 Hops Shortage Is Over, the Glut of '09 Is Here," *Oregonian,* October 28, 2009; Dan Richman, "For Drinkers of Craft Beer, Prices May Soon Be Hopping," *Seattle Post-Intelligencer,* October 16, 2007; David Welch, "Hops Shortage Likely to Boost Price of Beer," National Public Radio, last modified November 13, 2007, http://www.npr.org/templates/story/story.php?storyId = 16245024; Pat Muir, "Yakima Warehouse Fire Destroys 4% of Nation's Yield of Hops, *Seattle Times,* October 4, 2006; Matt Brynildson, "Hop Crisis? What Hop Crisis? A Craft Brewer's Thoughts on Surviving the Raw Material Crunch," *Zymurgy* 31, no. 5 (September/October 2008): 22–28; Colby, "The Bitter End."

15. European scientists have taken a lead in questioning how climate change might affect hop agriculture. For example, see Josef Patzak, Vladimír Nesvadba, Alena Henychová, and Karel Krofta, "Assessment of the Genetic Diversity of Wild Hops (*Humulus lupulus* L.) in Europe Using Chemical and Molecular Analyses," *Biochemical Systematics and Ecology* 38, no. 2 (April 2010): 142; Martin Mozny, Radim Tolasz, Jiri Nekovar, Tim Sparks, Mirek Trnka, and Zdenek Zalud, "The Impact of Climate Change on the Yield and Quality of Saaz hops in the Czech Republic," *Agricultural and Forest Meteorology* 149, nos. 6–7 (June 2009): 913–19; and Rex Dalton, "Climate Troubles Brewing for Beer Makers," *Nature,* published online May 2008, http://www.nature.com/news/2008/080502/full/news.2008.799 .html.

16. Writers from across the business and beer spectrum have considered InBev's takeover. For the most complete account, see Julie MacIntosh, *Dethroning the King: The Hostile Takeover of Anheuser-Busch, an American Icon* (Hoboken, NJ: John Wiley & Sons, 2011).

17. Bryant Christie, Inc., "A Report for the Hop Growers of America" (unpublished document acquired by author, July 2008); Chris Guy, "Cows and Hops Could Prove to be a Good Combination," *USDA Blog,* http://blogs.usda.gov/2011/05/17 /cows-and-hops-could-prove-to-be-a-good-combination.

18. Doug Weathers, Don Weathers, and Jamie Floyd, interview by the author, Sodbuster Farms, Salem, OR, June 25, 2013.

19. Weathers, Weathers, and Floyd, interview; Lucy Birmingham, "From Corporate to Craft," *Edible Portland,* Fall 2011, 21–27.

20. See hopunion.com and indiehops.com. Also see Ettinger, interview.

21. David Glassberg, *Sense of History: The Place of the Past in American Life* (Amherst: University of Massachusetts Press, 2001), 3–22.

BIBLIOGRAPHY

Additional materials related to the history of hops and beer can be found on the Oregon Hops & Brewing Archive website: http://scarc.library.oregonstate.edu/ohba.html

UNPUBLISHED SOURCES

Agricultural Experiment Station Records. Record Group 25. Special Collections and Archives Research Center, Oregon State University, Corvallis.

Agriculture Collection, Benton County Historical Society, Philomath, OR.

Agriculture Photographic Collection, Photographic Group PO40. Special Collections and Archives Research Center, Oregon State University, Corvallis.

Annen, John. Interview by the author. Annen Brothers Farm, Mt. Angel, OR, March 11, 2008.

Carmichael, Colin. Papers. Washington State Historical Society, Tacoma.

Coleman, Maureen. Interview by the author. St. Paul, OR, March 12, 2008.

Darby, Peter. Interview by the author, Wye Hops, China Farm, Canterbury, England, May 21–22, 2013.

Ettinger, Christian. Interview by the author. Hopworks Urban Brewery, Portland, OR, July 7, 2008.

Evans, Gale. Oral history by Daniel C. Robertson. Benton County Historical Society, Philomath, OR, April 7, 1982.

Extension and Experiment State Communications. Photograph Group 120. Special Collections and Archives Research Center, Oregon State University, Corvallis.

Floyd, Jamie. Interview by the author. Ninkasi Brewing Company, Eugene, OR, June 25, 2013.

Foyston, John. Oral history by Peter A. Kopp, Tiah Edmunson-Morton, and Tim Hills. McMenamin's Mission Theater, Portland, OR, March 28, 2014.

Goschie, Gayle. Interview by the author. Goschie Farms, Silverton, OR, August 18, 2008.

Goschie, Herman. Oral history by Tim Satler. Benton County Historical Society, Philomath, Oregon, January 23, 1982.

Goschie, Herman, and Vernice Goschie. Interview by the author. Goschie Farms, Silverton, OR, August 18, 2008.

Grim, Amanda. Oral history by Benton County Historical Society. Philomath, OR, April 6, 1982.

Haunold, Alfred. Interview by the author. Oregon State University, Corvallis, March 18, 2008.

———. Interview by the author. Oregon State University, Corvallis, June 29, 2009.

———. Oral history by Keith Peterson. Benton County Historical Society, Philomath, OR, February 1, 1982.

———. Papers. Crop Science Department. Oregon State University, Corvallis.

Havir, Nancy Ann. "An Analysis of Firm Performance in the U.S. Craft Brewing Industry." Ph.D. diss., University of Minnesota, 1999.

Hop History Collection. Independence Heritage Museum, Independence, OR.

Hops Research Council. Records. MSS HRC. Special Collections and Archives Research Center, Oregon State University, Corvallis.

Hudson Cooler, Kathleen E. "Hop Agriculture in Oregon: The First Century." Master's thesis, Portland State University, 1986.

———. Papers. Benton County Historical Society, Philomath, OR.

Hummer, Kim. Interview by the author. National Clonal Germplasm Repository, Corvallis, OR, August 6, 2014.

Hussey, John A. "The Fort Vancouver Farm." National Park Service, n.d. http://www.nps.gov/fova/learn/historyculture/historical-studies.htm.

Jones, Jay Edward. "The Restructuring of the U.S. Brewing Industry and Small Firm Survival." Ph.D. diss., Boston College, 1990.

Joyce, Jack. Interview by the author. Green Dragon Bistro and Pub, Portland, OR, July 2, 2013.

Kasar, Harvey. Oral history by Kathleen Hudson Cooler. Silverton, OR, April 16, 1982.

Kee, Ming. Oral history by Daniel Robertson. Benton County Historical Society, Philomath, OR, April 19, 1982.

Keeler, Elizabeth Louise. "The Landscape of Horticultural Crops in the Northern Willamette Valley from 1850–1920." Ph.D. diss., University of Oregon, 1994.

King, Patrick Joseph. "Labor and Mechanization: The Hop Industry in Yakima Valley, 1866–1950." Master's thesis, Washington State University, 2008.

Kraemer, Thomas. Oral history by Tim Satler. Benton County Historical Society, Philomath, OR, February 2, 1982.

Larrance, Art. Interview by the author. Raccoon Lodge and Brewpub, Portland, OR, September 18, 2008.

Manuscript and Photograph Collection, Aurora Colony Historical Society, Aurora, OR.

Meeker, Ezra. Papers. Washington State Historical Society, Tacoma.

Monroe, Beth. Oral history by Mickey Peterson. Benton County Historical Society, Philomath, OR, January 25, 1982.

Newton, Sidney. Oral history by Kathleen Hudson Cooler. Benton County Historical Society, Philomath, OR, March 24, 1982.

Nickerson, Gail. Oral history by Peter A. Kopp and Tiah Edmunson-Morton. Corvallis, OR, August 4, 2014.

Ockert, Karl. Interview by the author. BridgePort Brewery, Portland, OR, August 16, 2008.

Oliphant, Jim. Interview by the author. National Clonal Germplasm Repository, Corvallis, OR, August 6, 2014.

Oregon Hop Commission. Records. Oregon Hop Commission, Hubbard.

Oregon Hop Growers' Association. Records. MSS OHGA. Special Collections and Archives Research Center, Oregon State University, Corvallis.

Pacific Northwest Promotional Pamphlets Collection, Oregon Historical Society, Portland.

Palacios, Michelle. Interview by the author. Oregon Hop Commission, Hubbard, OR, June 13, 2007.

Photograph Collection, Independence Heritage Museum, Independence, OR.

Photograph Collection, Oregon Historical Society, Portland.

Purvine, Dane. Oral history by Earl Nunnemaker. Benton County Historical Society, Philomath, OR, March 5, 1982.

Salmon, Ernest Stanley. Papers. Imperial College London, London, England.

Sharkey, Tom. Interview by the author. Forgery Brewery, Canterbury, England, May 23, 2013.

Stack, Martin Heidegger. "Liquid Bread: An Examination of the American Brewing Industry, 1865–1940." Ph.D. diss., Department of Economics, University of Notre Dame, 1998.

Staley, Charles. Oral history by Mickey Peterson. Benton County Historical Society, Philomath, OR, April 2, 1982.

Weathers, Doug, Don Weathers, and Jamie Floyd. Interview by the author. Sodbuster Farms, Salem, OR, June 25, 2013.

Winn, James. Oral history by Earl Nunnemaker. Benton County Historical Society, Philomath, OR, February 5, 1982.

PUBLISHED BOOKS, ARTICLES, AND GOVERNMENT PUBLICATIONS

Abbott, Carl. *The Great Extravaganza: Portland and the Lewis and Clark Exposition.* 3rd ed. Portland: Oregon Historical Society, 2004.

———. *Greater Portland: Urban Life and Landscape in the Pacific Northwest.* Philadelphia: University of Pennsylvania Press, 2001.

———. *How Cities Won the West: Four Centuries of Urban Change in Western North America.* Albuquerque: University of New Mexico Press, 2008.

Acitelli, Tom. *The Audacity of Hops: The History of America's Craft Beer Revolution.* Chicago: Chicago Review Press, 2013.

Akins, C. Melvin. *Archeology of Oregon*. Portland, OR: US Department of the Interior, Bureau of Land Management, 1993.

Alamillo, Jose M. *Making Lemonade out of Lemons: Mexican American Labor and Leisure in a California Town, 1880–1960*. Urbana: University of Illinois Press, 2006.

Anderson, Edgar. *Plants, Man and Life*. Boston: Little, Brown, 1952.

Aron, Cindy S. *Working at Play: A History of Vacations in the United States*. New York: Oxford University Press, 1999.

Association against the Prohibition Amendment. *A Criticism of National Prohibition*. Washington, DC: Association against the Prohibition Amendment, 1926.

Baggarley, Maud Ellen. "Joy: A Hop-Yard Story." *Young Woman's Journal* 17, no. 11 (November 1906): 496–99.

Bailey, L. H. *The Country-Life Movement in the United States*. New York: Macmillan, 1911.

Barker, Graeme. *The Agricultural Revolution in Prehistory: Why Did Foragers Become Farmers?* Oxford: Oxford University Press, 2009.

Baron, Stanley. *Brewed in America: The History of Beer and Ale in the United States*. Boston: Little, Brown, 1962.

Barth, Heinrich Johann, and Christiane Klinke. *The History of the Family Enterprise: Joh. Barth & Sohn, Nuremberg*. Translated by Ernest Sinauer, Chris Krason, and Alan Ross. Nuremberg: Joh. Barth & Sohn, 1994.

Barth, Heinrich Johann, Christiane Klinke, and Claus Schmidt. *The Hop Atlas: The History and Geography of the Cultivated Plant*. Nuremberg: Joh. Barth & Sohn, 1994.

Barth, Joh. & Sohn. *Hops . . . / . . .* (multiple years). Nuremberg: Joh. Barth & Sohn, 1909–2015.

Bassil, N. V., B. S. Gilmore, J. M. Oliphant, K. E. Hummer, and J. A. Henning. "Genic SSRs for European and North American Hop (*Humulus lupulus L.*)." *Genetic Resources and Crop Evolution* 55, no. 7 November 2008: 959–69.

Bauters, Merja. *Changes in Beer Labels and Their Meaning: A Holistic Approach to the Semiosic Process*. Helsinki: International Semiotics Institute at Imatra, 2007.

Beckham, Stephen Dow. *Oregon Indians: Voices from Two Centuries*. Corvallis: Oregon State University Press, 2006.

Belasco, Warren. *Appetite for Change: How the Counterculture Took on the Food Industry*. 2nd ed. Ithaca, NY: Cornell University Press, 2007.

Berg, Laura, ed. *The First Oregonians*. 2nd ed. Portland: Oregon Council for the Humanities, 2007.

Bexell, J. A., and E. B. Lemon. *The Oregon Farmer: What He Has Accomplished in Every Part of the State*. Portland: Oregon State Immigration Commission, 1913.

Birmingham, Lucy. "From Corporate to Craft." *Edible Portland,* Fall 2011, 21–27.

Black, John D. *Agricultural Reform in the United States*. New York: McGraw-Hill, 1929.

Bowen, William A. *The Willamette Valley: Migration and Settlement on the Oregon Frontier*. Seattle: University of Washington Press, 1978.

Bowers, William L. *The Country Life Movement in America, 1900–1920.* Port Washington, NY: Kennikat Press, 1974.

Bowler, Peter J. *The Mendelian Revolution: The Emergence of Hereditarian Concepts in Modern Science and Society.* Baltimore: Johns Hopkins University Press, 1989.

Boyd, Robert, ed. *Indians, Fire, and the Land in the Pacific Northwest.* Corvallis: Oregon State University Press, 1999.

Brands, H. W. *The Reckless Decade: American in the 1890s.* Chicago: University of Chicago, Press, 2002.

Bressman, E. N. "Developing New Varieties of Hops." *Science,* n.s., vol. 74, no. 1912 (August 21, 1931): 202–3.

Brooks, S. N, C. E. Horner, and S. T. Likens. *Hop Production.* Washington, DC: US Department of Agriculture, Agricultural Research Service, 1961.

Brooks, S. N., C. E. Horner, S. T. Likens, and C. E. Zimmermann. "Registration of Cascade Hop (Reg. No. 1)." *Crop Science* 12, no. 3 (1972): 394.

Brown, Elspeth H., Catherine Gudis, and Marina Moskowitz, eds. *Cultures of Commerce: Representation and American Business Culture, 1877–1960.* New York: Palgrave Macmillan, 2006.

Brown, J. F. *Guinness and Hops.* London: Arthur Guinness Son, 1980.

Brown, Richard Maxwell. *Strain of Violence: Historical Studies of American Violence and Vigilantism.* New York: Oxford University Press, 1975.

Bsumek, Erika Marie. *Indian-Made: Navajo Culture in the Marketplace, 1868–1940.* Lawrence: University Press of Kansas, 2008.

Bugos, Glenn E., and Daniel J. Kevles. "Plants as Intellectual Property: American Practice, Law, and Policy in a World Context." *Osiris,* 2nd ser., 7 (1992): 74–104.

Bulmer, Martin. *The Chicago School of Sociology: Institutionalization, Diversity, and the Rise of Sociological Research.* Chicago: University of Chicago Press, 1984.

Bunting, Robert. "The Environment and Settler Society in Western Oregon." *Pacific Historical Review* 64, no. 3 (August 1995): 413–42.

———. *The Pacific Raincoast: Environment and Culture in an American Eden, 1778–1900.* Lawrence: University of Kansas Press, 1997.

Burgess, Abraham Hale. *Hops: Botany, Cultivation, and Utilization.* New York: Interscience, 1964.

Burkhardt, D. C. Jesse. *Backwoods Railroads: Branchlines and Shortlines of Western Oregon.* Pullman: Washington State University Press, 1994.

Burton, Rob. *Hops and Dreams: The Story of the Sierra Nevada Brewing Co.* Chico, CA: Stansbury, 2010.

Calagione, Sam. *Brewing Up a Business: Adventures in Beer from the Founder of Dogfish Head Craft Brewery.* Hoboken, NJ: Wiley, 2011.

Callenbach, Ernest. *Ecotopia: The Notebooks and Reports of William Weston.* Berkeley, CA: Banyan Tree Books, 1975.

Campbell, Robert B. "Newlands, Old Lands: Native American Labor, Agrarian Ideology, and the Progressive-Era State in the Making of the Newlands Reclamation Project, 1902–1926." *Pacific Historical Review* 71, no. 2 (May 2002): 203–38.

Carney, Judith Ann. *Black Rice: The African Origins of Rice Cultivation in the Americas*. Cambridge, MA: Harvard University Press, 2001.

Carney, Judith Ann, and Richard Nicholas Rosomoff. *In the Shadow of Slavery: Africa's Botanical Legacy in the Atlantic World*. Berkeley: University of California Press, 2010.

Carson, Rachel. *Silent Spring*. Boston: Houghton Mifflin, 1962.

Chambers, Clarke A. *California Farm Organizations: A Historical Study of the Grange, the Farm Bureau, and the Associated Farmers, 1929–1941*. Berkeley: University of California Press, 1952.

Chan, Sucheng. *This Bittersweet Soil: The Chinese in California Agriculture: 1860–1910*. Berkeley: University of California Press, 1986.

Channing, Alice, Ellen Nathalie Matthews, and Mary E. Skinner. *Child Labor in Fruit and Hop Growing Districts of the Northern Pacific Coast*. Washington, DC: US Department of Labor, Children's Bureau, 1926.

Chapman, Alfred C., ed. *The Hop and Its Constituents: A Monograph on the Hop Plant*. London: Brewing Trade Review, 1905.

Christy, Ralph D. and Lionel Williamson, eds. *A Century of Service: Land-Grant Colleges and Universities, 1890–1990*. New Brunswick, NJ: Transaction, 1992.

Clark, Norman H. *The Dry Years: Prohibition and Social Change in Washington*. Rev. ed. Seattle: University of Washington Press, 1988.

Clinch, George. *English Hops: A History of Cultivation and Preparation for the Market from the Earliest Times*. London: McCorquodale, 1919.

Cochrane, Willard W. *The Curse of American Agricultural Abundance: A Sustainable Solution*. Lincoln: University of Nebraska Press, 2003.

Cohen, Lizabeth. *A Consumers' Republic: The Politics of Mass Consumption in Postwar America*. New York: Vintage, 2003.

Cole, Katherine. *Voodoo Vintners: Oregon's Astonishing Biodynamic Winegrowers*. Corvallis: Oregon State University Press, 2011.

Collins, Robert M. *More: The Politics of Economic Growth in Postwar America*. Oxford: Oxford University Press, 2000.

Commission on Industrial Relations. *The Seasonal Labor Problem in Agriculture, Industrial Relations: Final Report and Testimony Submitted to Congress by the Commission on Industrial Relations*. Vol. 5. Washington, DC: Government Printing Office, 1916.

Conkin, Paul K. *A Revolution down on the Farm: The Transformation of American Agriculture since 1929*. Lexington: University Press of Kentucky, 2008.

Cookson, J. S., and Ann Lawton. "Hop Dermatitis in Herefordshire." *British Medical Journal* 2 (August 15, 1953): 376–79.

Cooper, Alan, Chris Turney, Konrad A. Hughen, Barry W. Brook, H. Gregory McDonald, and Corey J. A. Bradshaw. "Abrupt Warming Events Drove Late Pleistocene Megafaunal Turnover." *Science* 349, no. 6248 (2015): 602–6. Published online July 23, 2015; http://science.sciencemag.org/content/349/6248/602.long.

Cordle, Celia. *Out of the Hay and into the Hops: Hop Cultivation in Wealdon Kent and Hop Marketing in Southwark, 1744–2000*. Hatfield, UK: University of Hertfordshire Press, 2011.

Corzine, Nathan Michael. "Right at Home: Freedom and Domesticity in the Language and Imagery of Beer Advertising, 1933–1960." *Journal of Social History* 43, no. 3 (Summer 2010): 843–46.

Cox, LaWanda F. "The American Agricultural Wage Earner, 1865–1900: The Emergence of a Modern Labor Problem." *Agricultural History* 22, no. 2 (April 1948): 95–114.

Cronon, William. *Changes in the Land: Indians, Colonists, and the Ecology of New England*. New York: Hill and Wang, 1983.

———. *Nature's Metropolis: Chicago and the Great West*. New York: Norton, 1991.

———. "A Place for Stories: Nature, History, and Narrative." *Journal of American History* 78, no. 4 (March 1992): 1347–76.

———, ed. *Uncommon Ground: Rethinking the Human Place in Nature*. New York: Norton, 1995.

Crosby, Alfred. *The Columbian Exchange: Biological and Cultural Consequences of 1492*. Westport, CT: Greenwood, 1972.

———. *Ecological Imperialism: The Biological Expansion of Europe, 900–1900*. Cambridge: Cambridge University Press, 1986.

Cross, Coy F. *Justin Smith Morrill: Father of the Land-Grant Colleges*. East Lansing: Michigan State University Press, 1999.

Currier, Susan Lord. "Some Aspects of Washington Hop Fields." *Overland Monthly* 32 (December 1898): 541–44.

Danbom, David B. *The Resisted Revolution: Urban America and the Industrialization of Agriculture, 1900–1930*. Ames: Iowa State University Press, 1979.

Daniel, Cletus. *Bitter Harvest: A History of California Farmworkers, 1870–1941*. Ithaca, NY: Cornell University Press, 1981.

Daniels, Roger. *Coming to America: A History of Immigration and Ethnicity in American Life*. New York: HarperCollins, 1990.

Darby, Peter. "The History of Hop Breeding and Development." *Brewery History* 121 (Winter 2005): 94–112.

Davis, Edward L. "Morphological Complexes in Hops (*Humulus lupulus L.*) with Special Reference to the American Race." *Annals of the Missouri Botanical Garden* 44, no. 4 (November 1957): 271–94.

Davis, H. L. *Honey in the Horn*. 1935. Reprint, Moscow: University of Idaho Press, 1992.

Delcourt, Hazel R., and Paul A. Delcourt. *Quaternary Ecology: A Paeloecological Perspective*. London: Chapman and Hall, 1991.

Delcourt, Paul A., and Hazel R. Delcourt. "Paleoclimates, Paleovegetation, and Paleofloras of North American North of Mexico During the Late Quaternary." Chap. 4 of *Flora of North America*, vol. 1, ed. Flora of North America Editorial Committee. New York and Oxford: Flora of North America Editorial Committee, 1993.

DeLoach, D. B. *Outlook for Hops from the Pacific Coast.* Washington, DC: US Department of Agriculture, Bureau of Agricultural Economics, 1948.

Deloria, Philip J. *Playing Indian.* New Haven, CT: Yale University Press, 1998.

DeLyser, D. Y., and W. J. Kasper. "Hopped Beer: The Case for Cultivation." *Economic Botany* 48, no. 2 (April–June 1994): 166–70.

Dennison, S. R., and Oliver MacDonagh. *Guinness, 1886–1939: From Incorporation to the Second World War.* Cork, Ireland: Cork University Press, 1998.

Denny, Mark. *Froth!: The Science of Beer.* Baltimore: Johns Hopkins University Press, 2009.

Department of Commerce and Labor, Bureau of the Census. *Thirteenth Census of the United States, 1910.* Washington, DC: Government Printing Office, 1912.

DiGirolamo, Vincent. "The Women of Wheatland: Female Consciousness and the 1913 Hop Strike." *Labor History* 34, no. 2 (1993): 236–55.

Duffin, Andrew P. *Plowed Under: Agriculture and Environment in the Palouse.* Seattle: University of Washington Press, 2007.

Dunlap, Thomas R. *DDT: Scientists, Citizens, and Public Policy.* Princeton, NJ: Princeton University Press, 1981.

———. *Nature and the English Diaspora: Environment and History in the United States, Canada, Australia, and New Zealand.* Cambridge: Cambridge University Press, 1999.

———. *Saving America's Wildlife: Ecology and the American Mind, 1850–1990.* Princeton, NJ: Princeton University Press, 1988.

Dunlop, Pete. *Portland Beer: Crafting the Road to Beervana.* Charleston, SC: History Press, 2013.

Edwards, G. Thomas, and Carlos A. Schwantes, eds. *Experiences in a Promised Land: Essays in Pacific Northwest History.* Seattle: University of Washington Press, 1986.

Edwardson, John R. "Hops: Their Botany, History, Production and Utilization." *Economic Botany* 6, no. 2 (April–June 1952): 160–75.

Egan, Timothy. *The Good Rain: Across Time and Terrain in the Pacific Northwest.* New York: Knopf: 1990.

Ehret, George. *Twenty-Five Years of Brewing with an Illustrated History of American Beer.* New York: George Ehret, 1891.

Ekbladh, David. *The Great American Mission: Modernization and the Construction of an American World Order.* Princeton, NJ: Princeton University Press, 2010.

Ellis, Clyde. "Five Dollars a Week to Be a 'Regular Indian': Shows, Exhibitions, and the Economics of Indian Dancing, 1880–1930." In *American Indian Culture and Economic Development in the Twentieth Century,* ed. Brian Hosmer and Colleen O'Neil. Boulder: University Press of Colorado, 2004.

Ellis, Richard J. *Democratic Delusions: The Initiative Process in America.* Lawrence: University of Kansas Press, 2002.

Evans, Sterling. *Bound in Twine: The History and Ecology of the Henequen-Wheat Complex for Mexico and the American and Canadian Plains, 1880–1950.* College Station: Texas A&M Press, 2007.

Fadeley, Nancie. "Hopping into History: Oregon's Hop Industry." *Oregon Business Journal* 13, no 12 (December 1990): 65.

Federal Cooperative Extension Service, Oregon State College. *Oregon's First Century of Farming: A Statistical Record of Achievements and Adjustments in Oregon Agriculture, 1859–1958.* Corvallis: Federal Cooperative Extension Service, Oregon State College, 1959.

Federico, Giovanni. "Not Guilty? Agriculture in the 1920s and the Great Depression." *Journal of Economic History* 65, no. 4 (December 2005): 949–76.

Feldman, Herman. *Prohibition: Its Economic and Industrial Aspects.* New York: Appleton, 1927.

Fessenden, Thomas G. *The New England Farmer: Containing Essays, Original and Selected, Relating to Agriculture and Domestic Economy. With Engravings, and the Prices of Country Produce.* Vol. 2. Boston: William Nichols, 1824.

Fiege, Mark. *Irrigated Eden: The Making of an Agricultural Landscape in the American West.* Seattle: University of Washington Press, 1999.

Filmer, Richard. *Hops and Hop Picking.* Oxford: Shire Publications, 2011.

Findling, John E., and Kimberley D. Pelle, eds. *Historical Dictionary of World's Fairs and Expositions, 1851–1988.* New York: Greenwood Press, 1990.

Fitzgerald, Deborah. *Every Farm a Factory: The Industrial Ideal in American Agriculture.* New Haven, CT: Yale University Press, 2003.

Freeman, Otis W. "Hop Industry of the Pacific Coast States." *Economic Geography* 12, no. 2 (April 1936): 155–63.

French, H. T., and C. D. Thompson. *Agriculture: Potatoes, Roots.* Corvallis: Oregon Agricultural Experiment Station, 1893.

———. *The Hop Louse.* Oregon Agricultural Experiment Station Bulletin no. 24. Corvallis, OR: Agricultural College Printing Office, 1893.

Fujita-Rony, Dorothy B. *American Workers, Colonial Power: Philippine Seattle and the Transpacific West, 1919–1941.* Berkeley: University of California Press, 2003.

Gamboa, Erasmo. *Mexican Labor and World War II: Braceros in the Pacific Northwest, 1942–1947.* Seattle: University of Washington Press, 2000.

Gardner, Bruce L. *American Agriculture in the Twentieth Century: How It Flourished and What It Cost.* Cambridge, MA: Harvard University Press, 2002.

Gaston, Joseph. *The Centennial History of Oregon, 1811–1912.* 4 vols. Chicago: S. J. Clarke, 1912.

Geer, Theodore Thurston. *Fifty Years in Oregon: Experiences, Observations, and Commentaries upon Men, Measures, and Customs in Pioneer Days and Later Times.* New York: Neale, 1912.

Gibson, James R. *Farming the Frontier: The Agricultural Opening of the Oregon Country, 1786–1846.* Seattle: University of Washington Press, 1985.

Glassberg, David. *Sense of History: The Place of the Past in American Life.* Amherst: University of Massachusetts Press, 2001.

Glickman, Lawrence B. *A Living Wage: American Workers and the Making of Consumer Society.* Ithaca, NY: Cornell University Press, 1997.

Goble, Dale D., and Paul W. Hirt, eds. *Northwest Lands, Northwest Peoples: Readings in Environmental History.* Seattle: University of Washington Press, 1999.

Goldstein, Joyce. *Inside the California Food Revolution: Thirty Years That Changed Our Culinary Consciousness.* Berkeley: University of California Press, 2013.

Goodwyn, Lawrence. *The Populist Moment: A Short History of the Agrarian Revolt in America.* Oxford: Oxford University Press, 1978.

Gosling, F. G. *Before Freud: Neurasthenia and the American Medical Community.* Urbana: University of Illinois Press, 1987.

Gottlieb, Robert. *Forcing the Spring: The Transformation of the American Environmental Movement.* Washington, DC: Island Press, 1993.

Great Britain. Parliament. House of Commons. *Report from the Select Committee on Hop Duties; Together with the Proceedings of the Committee, Minutes of Evidence and Appendix.* London, 1857.

———. *Reports from the Select Committee on the Hop Industry. Together with the Proceedings of the Committee, Minutes of Evidence, and Appendix Home Work to Hop Industry.* London, 1908.

Great Britain. Tariff Commission. *The Tariff Commission.* Vol. 3, *Report of the Agricultural Committee.* London: Tariff Commission, 1906.

Green, Frank L. *Ezra Meeker—Pioneer: A Guide to the Ezra Meeker Papers in the Library of the Washington State Historical Society.* Tacoma: Washington State Historical Society, 1969.

Gross, Emanuel. *Hops: In Their Botanical, Agricultural, and Technical Aspect and as an Article of Commerce.* Translated by Charles Salter. London: Scott, Greenwood, 1900.

Grossman, Ken. *Beyond the Pale: The Story of Sierra Nevada Brewing Co.* Hoboken, NJ: Wiley, 2013.

Habeck, James R. "The Original Vegetation of the Mid-Willamette Valley, Oregon." *Northwest Science* 35, no. 2 (May 1961): 65–77.

Hall, Greg. *Harvest Wobblies: The Industrial Workers of the World and Agricultural Laborers in the American West, 1900–1930.* Corvallis: Oregon State University Press, 2001.

Hallberg, Milton C. *Economic Trends in U.S. Agriculture and Food Systems since World War II.* Ames: Iowa State University Press, 2001.

Hampton, Richard, Ernest Small, and Alfred Haunold. "Habitat and Variability of *Humulus lupulus* var. *lupuloides* in Upper Midwestern North America: A Critical Source of American Hop Germplasm." *Journal of the Torrey Botanical Society* 128, no. 1 (January–March 2001): 35–46.

Hannigan, Robert E. *The New World Power: American Foreign Policy, 1898–1917.* Philadelphia: University of Pennsylvania Press, 2002.

Hantke, Ernst. "Pacific Hops and Fertilizers." *Transactions of the American Brewing Institute* 2 (September 1902–September 1904): 226–33.

Harmon, Alexandra. *Indians in the Making: Ethnic Relations and Indian Identities around the Puget Sound.* Berkeley: University of California Press, 1998.

Harrison, Leonard V., and Elizabeth Laine. *After Repeal: A Study of Liquor Control Administration.* New York: Harper and Brothers, 1936.

Hathway, Marion. *The Migratory Worker and Family Life: The Mode of Living and Public Provision for the Needs of the Family of the Migratory Worker in Selected Industries of the State of Washington.* Chicago: University of Chicago Press, 1934.

Haunold, Alfred. "Polyploid Breeding with Hop *Humulus lupulus L."* *Technical Quarterly* (Master Brewers Association of America) 9, no. 1 (1972): 36–40.

———. "A Trisomic Hop, *Humulus lupulus* L." *Crop Science,* 8 (July–August, 1968): 503–6.

———. "Use of Triploid Males for Increasing Hop Yields." *Crop Science,* 15 (November–December 1975): 833–40.

Haunold, Alfred, C. E. Horner, S. T. Likens, S. N. Brooks, and C. E. Zimmerman. "One-Half Century of Hop Research by the U.S. Department of Agriculture." *Journal of the American Society of Brewing Chemists* 43, no. 2 (Summer 1985): 123–26.

Haunold, Alfred, C. E. Horner, S. T. Likens, D. D. Roberts, and C. E. Zimmerman. "Registration of Willamette Hop (Reg. No. 6)." *Crop Science* 16 (September–October 1976): 739.

Haunold, Alfred, C. E. Horner, and Gail. B. Nickerson. "Registration of Early Flowering Downy Mildew Resistant Triploid Hop Pollinators (Reg. Nos. 9, 10, 11, 12)." *Crop Science* 22 (November–December 1982): 1262.

Haunold, Alfred, S. T. Likens, C. E. Horner, Gail B. Nickerson, and C. E. Zimmerman. "Columbia and Willamette, Two New Aroma-Type Hop Varieties." *Brewer's Digest,* 52, no. 11 (November 1977): 36–39.

Haunold, Alfred, S. T. Likens, C. E. Horner, C. E. Zimmerman, and D. D. Roberts. "Registration of Columbia Hop (Reg. No. 5)." *Crop Science* 16 (September–October 1976): 738–39.

Haunold, Alfred, S. T. Likens, and G. B. Nickerson. "Development of Zero-Alpha Hop Genotypes." *Crop Science* 17 (March–April 1977): 315–19.

Haunold, Alfred, S. T. Likens, G. B. Nickerson, and R. O. Hampton. "Registration of Nugget Hop." *Crop Science* 24 (May–June 1984): 618.

Haunold, Alfred, and G. B. Nickerson. "Registration of Mt. Hood Hop." *Crop Science,* 30 (1990): 423.

Haunold, Alfred, G. B. Nickerson, U. Gampert, and P. A. Whitney. "Registration of 'Liberty' Hop." *Crop Science* 32 (1992): 1071.

Haunold, Alfred, and Charles E. Zimmerman. "Pollen Collection, Crossing, and Seed Germination of Hop. *Crop Science* 14 (October 1974): 774–76.

Hayden, Brian, Neil Canuel, and Jennifer Shanse. "What Was Brewing in the Natufian?: An Archaelogical Assessment of Brewing Technology in the Epipaleolithic." *Journal of Archaeological Method and Theory* 20, no. 1 (March 2013): 102–50.

Hays, Samuel P. *Beauty, Health, and Permanence: Environmental Politics in the United States, 1955–1985.* Cambridge: Cambridge University Press, 1987.

———. *Conservation and the Gospel of Efficiency: The Progressive Conservation Movement, 1890–1920.* Cambridge, MA: Harvard University Press, 1959.

Hedges, James B. *Henry Villard and the Railways of the Northwest.* New Haven, CT: Yale University Press, 1930.

Hedrick, U. P. *A History of Horticulture in America, to 1860*. New York: Oxford University Press, 1950.

Heffernan, Hilary. *The Annual Hop: London to Kent*. Stroud, Gloucestershire, England: Chalford, 1996.

———. *Voices of Kent's Hop Gardens*. Stroud, Gloucestershire, England: Tempus, 2000.

Heise, Ursula K. *Sense of Place and Sense of Planet: The Environmental Imagination of the Global*. Oxford: Oxford University Press, 2008.

Henderson, Silas Milton. *Some Hop-Drying Studies*. Berkeley: Division of Agricultural Sciences, University of California, 1958.

Hicks, John D. *The Populist Revolt: A History of the Farmer's Alliance and the People's Party*. Minneapolis: University of Minnesota Press, 1931.

Hieronymus, Stan. *For the Love of Hops: The Practical Guide to Aroma, Bitterness and the Culture of Hops*. Boulder, CO: Brewers Publications, 2013.

Highsmith, Richard M., Jr., and Elbert E. Miller. "Open Field Farming in the Yakima Valley, Washington." *Economic Geography* 28, no. 1 (January 1952): 74–87.

Hill, D. D., and D. E. Bullis. *Second Annual Report of Investigations to Develop Hop Grades*. Corvallis, OR: Brewer's Hop Research Institute, 1941–42.

Hindy, Steve. *The Craft Beer Revolution: How a Band of Microbrewers Is Transforming the World's Favorite Drink*. New York: Palgrave Macmillan, 2014.

Hinman, Herbert R. *1992 Estimated Costs of Producing Hops in the Yakima Valley*. Pullman: Cooperative Extension, Washington State University, 1992.

Hoerner, G. R. *Brewers Hop Research Institute: Circular no. 1*. Chicago: Brewers' Hop Research Institute, 1940.

———. *Downy Mildew of Hops*. Corvallis: Oregon State Agricultural College Extension Service, 1933.

Hoerner, G. R., and Frank Rabak. *Production of Hops*. U.S. Department of Agriculture, Farmer's Bulletin no. 1842. Washington, DC: Government Printing Office, 1940.

Hoff, O. P. *Sixth Biennial Report of the Bureau of Labor Statistics and Inspector of Factories and Workshops of the State of Oregon, 1915*. Salem, OR: State Printing Department, 1914.

Hofstadter, Richard. *The Age of Reform: From Bryan to F.D.R.* New York: Vintage Books, 1955.

Holmes, George K. *Hop Crop of the United States, 1790–1911*. Washington, DC: US Department of Agriculture, Bureau of Statistics, 1912.

Hop Growers of America. *2009 Statistical Report*. Moxee, WA: Hop Growers of America, 2010.

Hop Industry Productivity Team. *The Hop Industry: Report of a Visit to the U.S.A. and Canada in 1950 of a Productivity Team Representing the Hop Industry*. London and New York: Anglo-American Council on Productivity, 1951.

Hornsey, Ian. *A History of Beer and Brewing*. Cambridge: Royal Society of Chemistry, 2003.

Horst, E. Clemons. "Hop Growing in California." *California's Magazine* 1, no. 1 (July 1915): 565–68.

———. "The New Dried Vegetable Industry." In California State Board of Agriculture, *Statistical Report of the for the Year 1918*, 154. Sacramento: California State Printing Office, 1919.

Hosmer, Brian, and Colleen O'Neil, eds. *Native Pathways: American Indian Culture and Economic Development in the Twentieth Century*. Foreword by Donald L. Fixico. Boulder: University Press of Colorado, 2004.

Hough, J.S., D.E. Briggs, R. Stevens, and T.W. Young. *Malting and Brewing Science*. Vol. 2, *Hopped Wort and Beer*. London: Chapman & Hall, 1982.

Hulse, David, Stan Gregory, and Joan Baker, eds. *Willamette River Basin Planning Atlas: Trajectories of Environmental and Ecological Change*. Corvallis: Pacific Northwest Ecosystem Research Consortium and the Oregon State University Press, 2002.

Hummer, K.E., and J.A. Henning, eds. *Proceedings of the First International Humulus Symposium*. Leuven, Belgium: ACTA Horticulturae, 2004.

Hunn, Eugene S., with James Selam and family. *Nch'i-wána, "The Big River": Mid-Columbia Indians and Their Land*. Seattle: University of Washington Press, 1990.

Hurley, Andrew. *Diners, Bowling Alleys, and Trailer Parks: Chasing the American Dream in the Postwar Consumer Culture*. New York: Basic Books, 2001.

Hurt, R. Douglas. *Problems of Plenty: The American Farmer in the Twentieth Century*. Chicago: Ivan R. Dee, 2002.

Hyde, Anne Farrar. *An American Vision: Far Western Landscape and National Culture, 1820–1920*. New York: New York University Press, 1990.

Ingraham, Aukjen T. "Henry Weinhard and Portland's City Brewery." *Oregon Historical Quarterly* 102 (2001): 180–96.

Isenberg, Andrew C. *The Destruction of the Bison: An Environmental History, 1750–1920*. Cambridge: Cambridge University Press, 2000.

Jackson, Michael. *Michael Jackson's Beer Companion: Revised and Updated*. Philadelphia: Courage Books, 2000.

———. *Michael Jackson's Beer Companion: The World's Great Beer Styles, Gastronomy, and Traditions*. Philadelphia: Running Press, 1997.

———. *The New World Guide to Beer*. Philadelphia: Running Press, 1988.

———. *The Pocket Guide to Beer*. New York: Putnam, 1982.

———. *Ultimate Beer*. New York: DK, 1998.

———. *The World Guide to Beer: The Brewing Styles, the Brands, the Countries*. Englewood Cliffs, NJ: Prentice-Hall, 1977.

Jacoby, Karl. *Crimes against Nature: Squatters, Poachers, Thieves, and the Hidden History of American Conservation*. Berkeley: University of California Press, 2014.

Jamieson, Stuart. *Labor Unionism in American Agriculture*. 1946. Reprint, New York: Arno Press, 1976.

Jenkins, Freda W. *You Picked What?* Salem, OR: Freda W. Jenkins, 1993.

Jensen, Joan M. *Promise to the Land: Essays on Rural Women*. Albuquerque: University of New Mexico Press, 1997.

Jessett, Thomas E., ed. *Reports and Letters, 1836–1838, of Herbert Beaver, Chaplain to the Hudson's Bay Company and Missionary to the Indians at Fort Vancouver.* Portland, OR: Champoeg Press, 1959.

Johannsen, Kristin. *Ginseng Dreams: The Secret World of America's Most Valuable Plant.* Lexington: University Press of Kentucky, 2006.

Johansen, Dorothy O., and Charles M. Gates, *Empire of the Columbia: A History of the Pacific Northwest.* New York: Harper & Row, 1967.

John I. Haas, Inc. *Illustrated Presentation of Yakima Golding Hop Farms and Other John I. Haas, Inc. Hop Growing Enterprises.* Washington, DC: John I. Haas, 1960.

Johnson, Henry H. *Ballads of the Farm and Home.* Elkhart, IN: Mennonite Publishing Company, 1902.

Johnston, Robert D. *The Radical Middle Class: Populist Democracy and the Question of Capitalism in Progressive Era Portland, Oregon.* Princeton, NJ: Princeton University Press, 2003.

Judd, Richard. *Common Lands, Common People: The Origins of Conservation in Northern New England.* Cambridge, MA: Harvard University Press, 1997.

Kamp, David. *The United States of Arugula: The Sun-Dried, Cold-Pressed, Dark-Roasted, Extra Virgin Story of the American Food Revolution.* New York: Broadway Books, 2006.

Kay, Bob. *U.S. Beer Labels (1950 and Earlier).* Vol. 1, *The Western States.* Batavia, IL: Bob Kay Beer Labels, 2007.

Kazin, Michael. *The Populist Persuasion: An American History.* Ithaca, NY: Cornell University Press, 1998.

Kellett, John R. *The Impact of Railways on Victorian Cities.* Toronto: University of Toronto Press, 1969.

Kerr, Norwood Allen. *The Legacy: A Centennial History of the State Agricultural Experiment Stations, 1887–1987.* Columbia: Missouri Agricultural Experiment Station, University of Missouri, 1987.

Kingsbury, Noel. *Hybrid: The History and Science of Plant Breeding.* Chicago: University of Chicago Press, 2009.

Kirk, Ruth, and Carmela Alexander, *Exploring Washington's Past: A Road Guide to History.* Rev. ed. Seattle: University of Washington Press, 1995.

Klingle, Matthew W. *Emerald City: An Environmental History of Seattle.* New Haven, CT: Yale University Press, 2008.

Knoedelseder, William. *Bitter Brew: The Rise and Fall of Anheuser-Busch and American's Kings of Beer.* New York: Harper Collins, 2012.

Koch, Greg, Steve Wagner, and Randy Clemens. *The Craft of Stone Brewing Co.: Liquid Lore, Epic Recipes, and Unabashed Arrogance.* Berkeley, CA: Ten Speed Press, 2011.

Koetter, Uwe, and Martin Biendl. "Hops (*Humulus lupulus*): A Review of Its Historical and Medicinal Uses." *HerbalGram* 87 (2010): 44–57.

Kollin, Susan, ed. *Postwestern Cultures: Literature, Theory, Space.* Lincoln: University of Nebraska Press, 2007.

Kopp, James J. *Eden within Eden: Oregon's Utopian Heritage.* Corvallis: Oregon State University Press, 2009.

Krakowski, Adam. "A Bitter Past: Hop Farming in Nineteenth-Century Vermont." *Vermont History* 82, no. 2 (Summer/Fall 2014): 91–105.

Kreisman, Lawrence, and Glenn Mason. *The Arts and Craft Movement in the Pacific Northwest.* Portland, OR: Timber Press, 2007.

Kulikoff, Allan. *From British Peasants to Colonial American Farmers.* Chapel Hill: University of North Carolina Press, 2000.

Kurlansky, Mark. *Cod: A Biography of the Fish That Changed the World.* New York: Walker, 1998.

———. *Salt: A World History.* New York: Walker, 2002.

Laird, Pamela Walker. *Advertising Progress: American Business and the Rise of Consumer Marketing.* Baltimore: Johns Hopkins University Press, 1998.

Landis, Paul. "The Hop Industry: A Social and Economic Problem." *Economic Geography* 22, no. 2 (January 1939): 85–94.

Lang, H. O., ed. *History of the Willamette Valley, Being a Description of the Valley and Its Resources, with an Account of Its Discovery and Settlement by White Men, and Its Subsequent History; Together with Personal Reminiscences of Its Early Pioneers.* Portland, OR: Himes and Lang, 1885.

Lansing, Jewel. *Portland: People, Politics, and Power, 1851–2001.* Corvallis: Oregon State University Press, 2003.

Lapsley, James T. *Bottled Poetry: Napa Winemaking from Prohibition to the Modern Era.* Berkeley: University of California Press, 1996.

Larsen, Dennis M. *Slick as a Mitten: Ezra Meeker's Klondike Enterprise.* Pullman: Washington State University Press, 2009.

Lawrence, Margaret. *The Encircling Hop: A History of Hops and Brewing.* Sittingbourne, Kent, UK: SAWD, 1990.

League of Nations. Economic Committee. "Hops: Report of the Delegation of the Economic Committee on the Meeting of Experts on Hops" (February 22–24, 1932). Geneva: League of Nations, Economic Committee, 1932.

Lears, T. J. Jackson. *Fables of Abundance: A Cultural History of Advertising in America.* New York: Basic Books, 1994.

———. *No Place of Grace: Antimodernism and the Transformation of American Culture.* New York: Pantheon, 1981.

Lebeda, A., P. T. N. Spencer-Phillips, and B. M. Cooke. *The Downy Mildews: Genetics, Molecular Biology and Control.* Dordrecht, Germany: Springer, 2008.

Lee, Erika. *At America's Gates: Chinese Immigration During the Exclusion Era, 1882–1943.* Chapel Hill: University of North Carolina Press, 2003.

Levenstein, Harvey. *Revolution at the Table: The Transformation of the American Diet.* New York: Oxford University Press, 1988.

Lewis and Clark Centennial Exposition Commission. *Report of the Lewis and Clark Centennial Exposition Commission for the State.* Portland, OR: Lewis and Clark Centennial Exposition Commission, 1906.

Limerick, Patricia Nelson. *The Legacy of Conquest: The Unbroken Past of the American West*. New York: Norton, 1987.

Limerick, Patricia Nelson, Clyde A. Milner II, and Charles E. Rankin, eds. *Trails: Toward a New Western History*. Lawrence: University Press of Kansas, 1991.

Linkon, Sherry Lee, and John Russo. *Steeltown U.S.A.: Work and Memory in Youngstown*. Lawrence: University Press of Kansas, 2002.

Lukacs, Paul. *American Vintage: The Rise of American Wine*. Boston: Houghton Mifflin, 2000.

Lytle, Mark H. *The Gentle Subversive: Rachel Carson, Silent Spring, and the Rise of the Environmental Movement*. New York: Oxford University Press, 2007.

"Machine-Picked vs. Hand-Picked Hops: A Revolution in the Hop Industry." *Scientific American*, supplement, no. 1770 (December 4, 1909): 361–62.

MacIntosh, Julie. *Dethroning the King: The Hostile Takeover of Anheuser-Busch, an American Icon*. Hoboken, NJ: John Wiley & Sons, 2011.

MacLean, Annie Marion. *Wage Earning Women*. Introduction by Grace H. Dodge. New York: Macmillan, 1910.

———. "With Oregon Hop Pickers." *American Journal of Sociology* 15, no. 1 (July 1909): 83–95.

———. *Women Workers and Society*. Chicago: A. C. McClurg, 1916.

Maezials, Frank T., John Foster, Mamie Dickens, and Adolphus W. Ward. *The Life of Charles Dickens*. New York: University Society: 1908.

Malone, Michael P. *James J. Hill: Empire Builder of the Northwest*. Norman: University of Oklahoma Press, 1996.

Marantz, Robin. *The Monk in the Garden: The Lost and Found Genius of Gregor Mendel, the Father of Genetics*. Boston: Houghton Mifflin, 2000.

Marshall, Emma Seckle. "Hop Culture in Oregon." *Sunset*, November 1903–April 1904, 243.

Marx, Leo. *The Machine in the Garden: Technology and the Pastoral Ideal in America*. New York: Oxford University Press, 1964.

Masterson, Jane. "Stomatal Size in Fossil Plants: Evidence for Polyploidy in Majority of Angiosperms." *Science* 264, no. 5157 (April 1994): 421–24.

Matt, Susan J. *Keeping Up with the Joneses: Envy in American Consumer Society, 1890–1930*. Philadelphia: University of Pennsylvania Press, 2003.

May, Dean L. *Three Frontiers: Family, Land, and Society in the American West, 1850–1900*. Cambridge: Cambridge University Press, 1994.

McCartney, Paul T. *Power and Progress: American National Identity, the War of 1898, and the Rise of American Imperialism*. Baton Rouge: Louisiana State University Press, 2006.

McClelland, Gordon T., and Jay T. Last. *California Orange Box Labels: An Illustrated History*. Santa Ana, CA: Hillcrest Press, 1985.

McDowell, R. H. *Hops*. Nevada Agricultural Bulletin 35. Reno: Nevada Agricultural Experiment Station, 1896.

McGahan, A. M. "The Emergence of the National Brewing Oligopoly: Competition in the American Market, 1933–1958." *Business History Review* 65, no. 2 (Summer 1991): 229–84.

McGovern, Patrick E., Stuart J. Fleming, and Solomon H. Katz. *The Origins and Ancient History of* Wine. Amsterdam: Gordon and Breach, 1996.

McKay, Floyd J. "Green Beans, Green Cash: Alderman's Post World War II Teenage Workforce." *Oregon Historical Quarterly* 11, no. 3 (Fall 2010): 372–86.

McKibbon, Bill. *The End of Nature.* New York: Random House, 1989.

McWilliams, Carey. *California: The Great Exception.* Foreword by Lewis H. Lapham. Berkeley: University of California Press, 1998.

———. *Factories in the Field: The Story of Migratory Farm Labor in California.* Boston: Little, Brown, 1939.

Meacham, Sarah Hand. *Every Home a Distillery: Alcohol, Gender and Technology in the Colonial Chesapeake.* Baltimore: Johns Hopkins University Press, 2009.

Mead, Rebecca J. *How the Vote Was Won: Woman Suffrage in the Western United States, 1868–1914.* New York: New York University Press, 2004.

Meeker, Ezra. *The Busy Life of Eighty-Five Years of Ezra Meeker: Ventures and Adventures.* Seattle: Ezra Meeker, 1916.

———. *Hop Culture in the United States: Being a Practical Treatise on Hop Growing in Washington Territory from the Cutting to Bale.* Puyallup, WA: Ezra Meeker, 1883.

———. *Ox-Team Days on the Oregon Trail.* Edited by Howard R. Driggs. Yonkers-on-Hudson, NY: World Book Co., 1922.

Meeks, Eric V. "The Tohono O'odham, Wage Labor, and the resistant Adaptation, 1900–1930." *Western Historical Quarterly* 34, no. 4 (Winter 2003): 468–89.

Meinig, D.W. *The Great Columbia Plain: A Historical Geography, 1805–1910.* Seattle: University of Washington Press, 1968.

Meier, Gary, and Gloria Meier. *Brewed in the Pacific Northwest: A History of Beer Making in Oregon and Washington.* Seattle: Fjord Press, 1991.

Miller, Char, ed. *Cities and Nature in the American West.* Reno: University of Nevada Press, 2010.

Miller, Elbert E., and Richard M. Highsmith, Jr. "The Hop Industry of the Pacific Coast." *Journal of Geography* 49, no. 2 (1950), 63–77.

Mittelman, Amy. *Brewing Battles: A History of American Beer.* New York: Algora, 2008.

Moore, Kathleen Dean. *Rachel Carson: Legacy and Challenge.* Albany: State of New York University Press, 2008.

Morrison, Dorothy. *Outpost: John McLoughlin and the Far Northwest.* Portland: Oregon Historical Society Press, 1999.

Morrison, Lisa M. *Craft Beers of the Pacific Northwest: A Beer Lover's Guide to Oregon, Washington, and British Columbia.* Portland, OR: Timber Press, 2011.

Morse, Katherine. *The Nature of Gold: An Environmental History of the Klondike Gold Rush.* Seattle: University of Washington Press, 2010.

Mosher, Randy. *Tasting Beer: An Insider's Guide to the World's Greatest Drink.* North Adams, MA: Storey, 2009.

Motley, Timothy J., Myree Zerega, and Hugh Cross, eds. *Darwin's Harvest: New Approaches to the Origins, Evolution, and Conservation of Crops.* New York: Columbia University Press, 2006.

Mott, C. W. *All about Fruit and Hop Raising, Dairying and General Farming, Lumbering, Fishing and Mining in Western Washington.* St. Paul, MN: C. W. Mott, 1907.

Mozny, Martin, Radim Tolasz, Jiri Nekovar, Tim Sparks, Mirek Trnka, and Zdenek Zalud. "The Impact of Climate Change on the Yield and Quality of Saaz hops in the Czech Republic." *Agricultural and Forest Meteorology* 149, nos. 6–7 (June 2009): 913–19.

Murdock, Catherine Gilbert. *Domesticating Drink: Women, Men, and Alcohol in America, 1870–1940.* Baltimore: Johns Hopkins University Press, 1998.

Musicant, Ivan. *Empire by Default: The Spanish-American War and the Dawn of the American Century.* New York: H. Holt, 1998.

Myrick, Herbert. *The Hop: Its Culture and Cure, Marketing and Manufacture.* 1899. Reprint, New York: O. Judd, 1904.

Nash, Gerald D. *The American West Transformed: The Impact of the Second World War.* Bloomington: Indiana University Press, 1985.

Nash, Linda. *Inescapable Ecologies: A History of Environment, Disease, and Knowledge.* Berkeley: University of California Press, 2006.

Nash, Wallace. *The Settler's Handbook to Oregon.* Portland: J. K. Gill, 1904.

National Wholesale Liquor Dealers' Association of America. *The Anti-Prohibition Manual: A Summary of Facts and Figures Dealing with Prohibition.* Cincinnati: National Wholesale Liquor Dealers' Association of America, 1911.

Needham, F. E. *Hop Statistics.* Salem, OR: Needham-Taylor, 1940.

Neve, R. A. *Hops.* London: Chapman and Hall, 1991.

"New Help for Hop Crops: Sterility," *Oregon's Agricultural Progress* 21, no. 20 (Fall 1974): 14.

Newton, Sidney W. *Early History of Independence, Oregon.* Salem, OR: Sidney W. Newton, 1971.

Ngai, Mae M. *Impossible Subjects: Illegal Aliens and the Making of Modern America.* Princeton, NJ: Princeton University Press, 2004.

Niete, Warren. "Enforcing Oregon's State Alcohol Monopoly: Recollections from the 1950s." *Oregon Historical Quarterly* 115, no. 1 (Spring 2014): 90–105.

Northern Pacific Railroad. *The Northern Pacific Railroad: Sketch of Its History; Delineations of Its Transcontinental Line; Its Features as a Great Through Route from the Great Lakes to the Pacific Ocean; Its Relations to the Chief Water Ways of the Continent; and, a Description of the Soils and Climates of the Region's Traversed by It as to Their Adaptability to Agricultural Production; with Descriptive and Statistical Exhibits of the Counties on and Near Its Line in Minnesota and Dakota (for the Information of Those Seeking New Homes and Profitable Investments).* Chicago: Rand, McNally, 1882.

Nugent, Walter T. K. *The Tolerant Populists: Kansas Populism and Nativism.* Chicago: University of Chicago Press, 1963.

Ogle, Maureen. *Ambitious Brew: The Story of American Beer.* Orlando, FL: Harcourt, 2006.

Ohmann, Richard. *Selling Culture: Magazines, Markets, and Class at the Turn of the Century.* London: Verso, 1996.

Oliver, Garrett. *The Oxford Companion to Beer.* Oxford: Oxford University Press, 2012.

Oregon Agricultural Experiment Station. *100 Years of Progress: The Oregon Agricultural Experiment Station Oregon State University, 1888–1988.* Corvallis: Oregon State University, Agricultural Experiment Station, 1990.

Oregon State Agricultural College. *Annual Report of the President of the Board of Regents of the State Agricultural College to the Governor of Oregon, Legislative Assembly, Eighteenth Regular Session, 1895.* Salem: Oregon State Agricultural College, 1894.

———. *Annual Report of the Oregon Agricultural College and Experiment Station for the Year Ending June 30, 1989.* Salem: Oregon State Agricultural College, 1899.

Oregon State Board of Agriculture. *The Resources of the State of Oregon: A Book of Statistical Information Treating upon Oregon as a Whole, and by Counties.* 3rd rev. ed. Salem: Oregon State Board of Agriculture, 1898.

Oregon State Board of Horticulture. *Fourth Biennial Report of the Oregon State Board of Horticulture to the Legislative Assembly, Nineteenth Regular Session.* Salem: Oregon State Board of Horticulture, 1897.

———. *Third Biennial Report of the Oregon State Board of Horticulture to the Legislative Assembly, Eighteenth Regular Session.* Salem: Oregon State Board of Horticulture, 1894.

Oregon State College. *Oregon's First Century of Farming: A Statistical Record of Achievements and Adjustments in Oregon Agriculture, 1859–1958.* Corvallis: Federal Cooperative Extension Service, Oregon State College, 1959.

Oreskes, Naomi, and Eric M. Conway. *Merchants of Doubt: How a Handful of Scientists Obscured the Truth on Issues from Tobacco Smoke to Global Warming.* New York: Bloomsbury Press, 2010.

Orsi, Richard J. *Sunset Limited: The Southern Pacific Railroad and the Development of the American West, 1850–1930.* Berkeley: University of California Press, 2005.

Parker, Hubert H. *The Hop Industry.* London: P. S. King and Son, 1934.

Parker, William B. *The Red Spider on Hops in the Sacramento Valley of California.* Washington, DC: US Department of Agriculture, Bureau of Entomology, 1913.

Parsons, James T. "Hops in Early California Agriculture." *Agricultural History* 14, no. 3 (July 1940): 110–16.

Patterson, Mark, and Nancy Hoalst-Pullen, eds. *The Geography of Beer: Regions, Environment, and Societies.* New York: Springer, 2014.

Patzak, Josef, Vladimír Nesvadba, Alena Henychová, and Karel Krofta. "Assessment of the Genetic Diversity of Wild Hops (*Humulus lupulus* L.) in Europe Using Chemical and Molecular Analyses." *Biochemical Systematics and Ecology* 38, no. 2 (April 2010): 136–45.

Paul, Rodman. "The Great California Grain War: The Grangers Challenge the Wheat King." *Pacific Historical Review* 27, no. 4 (November 1958): 331–48.

Pauly, Philip J. *Fruits and Plains: The Horticultural Transformation of America.* Cambridge, MA: Harvard University Press, 2007.

Pearce, Helen Ruth. *The Hop Industry in Australia.* Carlton, Australia: Melbourne University Press, 1976.

Percival, John. "The Hop Plant." In *The Hop and Its Constituents: A Monograph on the Hop Plant,* ed. Alfred C. Chapman. London: Brewing Trade Review, 1905.

Peterson del Mar, David. *Oregon's Promise: An Interpretive History.* Corvallis: Oregon State University Press, 2003.

Pfaff, Christine E. *Harvests of Plenty: A History of the Yakima Irrigation Project, Washington.* Denver: US Department of the Interior, Bureau of Reclamation, 2002.

Phillips, Sara T. *This Land, This Nation: Conservation, Rural America and the New Deal.* Cambridge: Cambridge University Press, 2007.

Pielou, E. C. *After the Ice Age: The Return of Life to Glaciated North America.* Chicago: University of Chicago Press, 1991.

Pinney, Thomas. *A History of Wine in America: From Prohibition to the Present.* Berkeley: University of California Press, 2005.

Pintarich, Paul. *The Boys Up North: Dick Erath and the Early Oregon Winemakers.* Portland, OR: Graphic Arts Center, 1997.

Piott, Steven L. *Giving Voters a Voice: The Origins of the Initiative and Referendum in America.* Columbia: University of Missouri Press, 2003.

Pisani, Donald J. *From the Family Farm to Agribusiness: The Irrigation Crusade in California and the West, 1850–1931.* Berkeley: University of California Press, 1984.

———. *Water and American Government: The Reclamation Bureau, National Water Policy, and the West, 1902–1935.* Berkeley: University of California Press, 2002.

Pollan, Michael. *The Botany of Desire: A Plant's-Eye View of the World.* New York: Random House, 2001.

———. *Second Nature: A Gardener's Education.* New York: Grove Press, 1991.

Pomeroy, Earl. *The Pacific Slope: A History of California, Oregon, Washington, Idaho, Utah, and Nevada.* Foreword by Elliot West. Rev. ed. Reno: University of Nevada Press, 1991.

Postman, J., K. Hummer, E. Stover, R. Krueger, P. Forsline, L. J. Grauke, F. Zee, B. Irish, and T. Ayala-Silva. "Fruit and Nut Genebanks in the USDA National Plant Germplasm System." *HortScience* 41, no. 5 (2006): 1188–94.

Powell, Douglas Reichert. *Critical Regionalism: Connecting Politics and Culture in the American Landscape.* Chapel Hill: University of North Carolina Press, 2007.

Prescott, Cynthia Culver. *Gender and Generation on the Far Western Frontier.* Tucson: University of Arizona Press, 2007.

Putney, Clifford. *Muscular Christianity: Manhood and Sports in Protestant America, 1880–1920.* Cambridge, MA: Harvard University Press, 2001.

Raibmon, Paige. *Authentic Indians: Episodes of Encounter from the Late Nineteenth-Century Northwest Coast*. Durham, NC: Duke University Press, 2005.

———. "The Practice of Everyday Colonialism: Indigenous Women at Work in the Hop Fields and Tourist Industry of the Puget Sound." *Labor: Studies in Working Class History of the Americas*, 3, no. 3 (2006): 23–56.

Rainger, Ronald, Keith Benson, and Jane Maienschein, eds. *The American Development of Biology* (Philadelphia: University of Pennsylvania Press, 1988).

Rasmussen, Wayne D. *Taking the University to the People: Seventy-Five Years of Cooperative Extension*. Ames: Iowa State University Press, 1989.

"R. D. Ginther, Workingman Artist and Historian of Skid Row." *California Historical Quarterly* 54, no. 3 (Fall 1975): 263–71.

"Recreation in the Oregon Hop Fields." *Recreation* 7 (1923): 646–47.

Reisner, Marc. *Cadillac Desert: The American West and Its Disappearing Water*. New York: Viking, 1986.

Reuss, Carl F., Paul H. Landis, and Richard Wakefield. *Migratory Farm Labor and the Hop Industry on the Pacific Coast with Special Application to Problems of the Yakima Valley, Washington*. Rural Sociology Series in Farm Labor, no. 3. Pullman: State College of Washington Agricultural Experiment Station, 1938.

Rich, E. E., ed. *The Letters of John McLoughlin, from Fort Vancouver to the Governor and Committee*. 2nd ser., *1839–44*. Toronto: Champlain Society, 1943.

Richie, Donald A. *Electing FDR: The New Deal Campaign of 1932*. Lawrence: University of Kansas Press, 2007.

Richter, Amy G. *Home on the Rails: Women, the Railroad and the Rise of Public Domesticity*. Chapel Hill: University of North Carolina Press, 2005.

Riesman, David. *The Lonely Crowd: A Study of the Changing American Character*. New Haven, CT: Yale University Press, 1961.

Robbins, William G. *Colony and Empire: The Capitalist Transformation of the American West*. Lawrence: University Press of Kansas, 1994.

———, ed. *The Great Northwest: The Search for Regional Identity*. Corvallis: Oregon State University Press, 2001.

———. *Hard Times in Paradise: Coos Bay, Oregon, 1850–1986*. Seattle: University of Washington Press, 1988.

———. *Landscapes of Conflict: The Oregon Story, 1940–2000*. Seattle: University of Washington Press, 2004.

———. *Landscapes of Promise: The Oregon Story, 1800–1940*. Seattle: University of Washington Press, 1997.

Robbins, William G., and Katrine Barber. *Nature's Northwest: The North Pacific Slope in the Twentieth Century*. Tucson: University of Arizona Press, 2011.

Ronnernberg, Herman. *Beer and Brewing in the Inland Northwest: 1850–1950*. Moscow: University of Idaho Press, 1993.

Rorabaugh, W. J. *The Alcoholic Republic: An American Tradition*. New York: Oxford University Press, 1979.

Ross, Lester A. *Fort Vancouver, 1829–1860: A Historical Archeological Investigation of the Goods Imported and Manufactured by the Hudson's Bay Company*. Part 1.

Vancouver, WA: Office of Archeology and Historic Preservation/Fort Vancouver National Historic Site, 1976.

Rowley, William D. *The Bureau of Reclamation: Origins and Growth to 1945*. Vol. 1. Denver: US Department of the Interior, Bureau of Reclamation, 2006.

———. *M. L. Wilson and the Campaign for the Domestic Allotment*. Lincoln: University of Nebraska Press, 1970.

Rumbarger, John J. *Profits, Power, and Prohibition: Alcohol Reform and the Industrializing of American, 1800–1930*. Albany: State University of New York Press, 1989.

Rybáček, Václav. *Hop Production*. Amsterdam and New York: Elsevier, 1991.

Sackman, Douglas Cazaux. *Orange Empire: California and the Fruits of Eden*. Berkeley: University of California Press, 2005.

Salmon, E. S. *Four Seedlings of the Canterbury Golding*. Canterbury, England: Wye College, 1944.

Saloutos, Theodore. *The American Farmer and the New Deal*. Ames: Iowa State University Press, 1982.

Sassaman, Richard, and Michael Dolan. "Strange Brew." *Air and Space Magazine*, August–September 1992, 13.

Schlebecker, John T. *Whereby We Thrive: A History of American Farming, 1607–1972*. Ames: Iowa State University Press, 1975.

Schlosser, Eric. *Fast Food Nation: The Dark Side of the All-American Meal*. Boston: Houghton Mifflin, 2001.

Schoenfeld, William A., John Marshall, Jr., and Paul C. Newman. *A Compendium of Hop Statistics (of interest to the Pacific Coast States)*. Portland, OR, 1930.

Schwantes, Carlos Arnaldo. *The Pacific Northwest: An Interpretive History*. Rev. ed. Lincoln: University of Nebraska Press, 1996.

———. "Protest in a Promised Land: Unemployment, Disinheritance, and the Origin of Labor Militancy in the Pacific Northwest, 1885–1886." *Western Historical Quarterly* 13, no. 4 (October 1982): 373–90.

Self, Edwin B. *Limbo City*. New York: Herald, 1946.

Simmons, Jack. *The Railways of Britain: An Historical Introduction*. London: Macmillan, 1968.

Singer, Michael. *The Legacy of Positivism*. New York: Palgrave Macmillan, 2005.

Skaggs, Jimmy M. *The Great Guano Rush: Entrepreneurs and American Overseas Expansion*. New York: St. Martin's Press, 1995.

Small, Ernest. "A Numerical and Nomenclatural Analysis of Morpho-Geographic Taxa of Humulus." *Systematic Botany* 3, no. 1 (Spring 1978): 37–76.

Smith, J., J. Oliphant, and K. E. Hummer. "Plant Exploration for Native Hop in the Southwestern United States." *Plant Genetic Resources Newsletter* 147 (September 2006): 29–37.

Smith, Jane S. *The Garden of Invention: Luther Burbank and the Business of Breeding Plants*. New York: Penguin, 2009.

Smith, Jean Edward. *FDR*. New York: Random House, 2007.

Sneider, Allison L. *Suffragists in an Imperial Age: U.S. Expansion and the Woman Question, 1870–1929*. New York: Oxford University Press, 2008.

Soluri, John. *Banana Cultures: Agriculture, Consumption, and Environmental Change in Honduras and the United States.* Austin: University of Texas Press, 2005.

Sosnowski, Vivienne. *When the Rivers Ran Red: An Amazing Story of Courage and Triumph in America's Wine Country.* New York: Palgrave Macmillan, 2009.

Spirn, Anne Whiston. *Daring to Look: Dorothea Lange's Photographs and Reports from the Field.* Chicago: University of Chicago Press, 2008.

S. S. Steiner, Inc. *Steiner.* Rev. ed. Atlantic Highlands, NJ: S. S. Steiner, 2004.

———. *Steiner's Guide to American Hops.* New York: S. S. Steiner, 1973.

Starr, Kevin. *Americans and the California Dream, 1850–1915.* New York: Oxford University Press, 1973.

———. *Inventing the Dream: California through the Progressive Era.* New York: Oxford University Press, 1985.

Starrs, Paul F. "The Navel of California and Other Oranges: Images of California and the Orange Crate." *California Geographer* 28 (1988): 1–41.

St. Benedict's Abbey. *Benedictine Hop Growers of the Willamette Valley.* Mount Angel, OR: Benedictine Press, 1935.

Steingraber, Sandra. *Having Faith: An Ecologist's Journey to Motherhood,* New York: Berkley Books, 2003.

———. *Living Downstream: An Ecologist's Look at Cancer and the Environment.* Reading, MA: Addison-Wesley, 1997.

Stelzle, Charles. *Why Prohibition!* New York: George H. Doran Company, 1918.

Stockberger, W. W. *Growing and Curing Hops.* U.S. Department of Agriculture, Farmer's Bulletin 304. Washington, DC: Government Printing Office, 1907.

Strasser, Susan, Charles McGovern, and Matthias Judt. *Getting and Spending: European and American Consumer Societies in the Twentieth Century.* Cambridge: Cambridge University Press, 1998.

Street, Richard Steven. *Beasts of the Field: A Narrative History of California Farmworkers, 1769–1913.* Stanford, CA: Stanford University Press, 2004.

Sulerud, George L. *An Economic Study of the Hop Industry in Oregon.* Corvallis: Agricultural Experiment Station, Oregon State Agricultural College, 1931.

Tamura, Linda. *The Hood River Issei: An Oral History of Japanese Settlers in Oregon's Hood River Valley.* Foreword by Roger Daniels. Urbana: University of Illinois Press, 1993.

Tartar, H. V., and B. Pilkington. *Hop Investigations.* Bulletin no. 114. Corvallis: Oregon Agricultural College Press, 1913.

Taylor, George H., and Alexi Bartlett. *The Climate of Oregon: Climate Zone 2, Willamette Valley.* Corvallis: Oregon Climate Service, Oregon State University, 1993.

Taylor, Paul S. "Migratory Agricultural Workers on the Pacific Coast." *American Sociological Review* 3, no. 2 (April 1938): 225–32.

"Three Weeks with the Hop-Pickers." *Littell's Living Age,* 5th ser., vol. 20, no. 1750 (December 29, 1877): 789–98.

Thrush, Coll. *Native Seattle: Histories from the Crossing-Over Place.* Seattle: University of Washington Press, 2007.

Tobias, Henry J., and Charles E. Woodhouse. *Santa Fe: A Modern History, 1880–1990*. Albuquerque: University of New Mexico Press, 2001.

Tomlan, Michael A. *Tinged with Gold: Hop Culture in the United States*. Athens: University of Georgia Press, 1992.

Tuan, Yi-Fu. *Space and Place: The Perspective of Experience*. Minneapolis: University of Minnesota Press, 1977.

———. *Topophilia: A Study of Environmental Perception, Attitudes, and Values*. Englewood Cliffs, NJ: Prentice-Hall, 1974.

Tucker, Abigail. "Dig, Drink, and Be Merry." *Smithsonian*, July/August 2011, 38–48.

Udall, Joshua A., and Jonathan F. Wendel. "Polyploidy and Crop Improvement," *The Plant Genome (A Supplement to Crop Science)* 1 (Nov. 2006): S3-S4, S6.

Uekötter, Frank. *The Age of Smoke: Environmental Policy in Germany and the United States, 1880–1970*. Pittsburgh: University of Pittsburgh Press, 2009.

Unger, Richard W. *Beer in the Middle Ages and the Renaissance*. Philadelphia: University of Pennsylvania Press, 2004.

United States Brewers' Association. *The 1918 Yearbook of the United States Brewers' Association*. New York: United States Brewers' Association, 1919.

U.S. Bureau of the Census. *Agriculture: Hops: With Data on Acreage and Production in Selected States and Counties; and Trade Census 11th, 1890 Bull. no. 143*. Washington, DC: US Bureau of the Census, 1891.

U.S. Bureau of Foreign Commerce. *Report upon the Commercial Relations of the United States with Foreign Countries for the Year 1877*. Washington, DC: Government Printing Office, 1878.

U.S. Congress. *Congressional Record: Proceedings and Debates of the . . . Congress* (multiple years). Washington, DC: Government Printing Office, 1874–1986.

U.S. Congress. House. Committee on Agriculture. "Permitting Producers of Hops to Enter into Marketing Agreements under Agricultural Adjustment Act" 75th Congress, 1st Session, House of Representatives, Report no. 1298. United States, Congress, House, Committee on Agriculture, 1937.

U.S. Congress. Senate. Committee on Agriculture and Forestry. "Amend the Agricultural Adjustment Act so as to Include Hops as a Basic Agricultural Commodity: A Report (to accompany S. 626)." Washington, DC: Government Printing Office, 1935.

U.S. Department of Agriculture. *Outlook for Hops from the Pacific Coast*. Washington, DC: Government Printing Office, 1948.

———. *Yearbook of the United States Department of Agriculture* (multiple years). Washington, DC: Government Printing Office, 1895–1920.

———. *Hops: By States, 1915–69*. Statistical Bulletin no. 469. Washington, DC: US Department of Agriculture, Statistical Reporting Service, 1971.

U.S. Department of Commerce. Bureau of the Census. *Sixteenth Census of the United States: 1940: Agriculture, Volume 1, First and Second Series, State Reports, Part 6: Statistics for Counties, Farms and Farm Property, with Related Information for Farms and Farm Operators, Livestock and Livestock Products, and Crops*. Washington, DC: Government Printing Office, 1942.

U.S. Department of Commerce and Labor. Bureau of the Census. *Thirteenth Census of the United States, Taken in the Year 1910, Abstract of the Census.* Washington, DC: Government Printing Office, 1913.

U.S. Department of the Interior. Census Office. *Report on the Productions of Agriculture at the Tenth Census (June 1, 1880), Embracing General Statistics and Monographs on Cereal Production, Flour-Milling, Tobacco Culture, Manufacture and Movement of Tobacco, Meat Production.* Washington, DC: Government Printing Office, 1883.

———. *Report upon the Statistics of Agriculture; Compiled from Returns at the Tenth Census.* Washington, DC: Government Printing Office, 1883.

U.S. Foreign Agricultural Service. *Developing Foreign Markets for U.S. Farm Products.* Washington DC: Government Printing Office, 1957.

U.S. Government Printing Office. *Agriculture: Hops.* Washington, DC: Government Printing Office, 1891.

———. *Growing and Curing Hops.* Washington, DC: Government Printing Office, 1907.

———. *Hop Culture in California.* Washington, DC: Government Printing Office, 1900.

———. *Hops in Principal Countries: Their Supply, Foreign Trade, and Consumption, with Statistics of Beer Brewing.* Washington, DC: Government Printing Office, 1907.

———. *Necessity for New Standards of Hop Valuation.* Washington, DC: Government Printing Office, 1909.

U.S. Patent and Trademark Office. *Index of Patents Issued from the United States Patent and Trademark Office.* Washington, DC: United States Patent and Trademark Office, 1870–2000.

U.S. Senate. Committee on the Judiciary. *Proposing an Amendment to the Constitution Prohibiting the Sale, Manufacture, and Importation of Intoxicating Liquors: Hearings before a Subcommittee of the Committee on the Judiciary of the United States on S.J. Res. 88 and S.J. Res. 50, 63rd Cong., 2nd sess., 1914.* Washington, DC: Government Printing Office, 1914.

U.S. Sixtieth Congress. "Tariff Hearings before the Committee on Ways and Means of the House of Representatives, Sixtieth Congress." Washington, DC: Government Printing Office, 1908.

Vale, Thomas R. *Fire, Native Peoples, and the Natural Landscape.* Washington, DC: Island Press, 2002.

Vaught, David. *Cultivating California: Growers, Specialty Crops, and Labor, 1875–1920.* Baltimore: Johns Hopkins University Press, 1999.

Vitek, William, and Wes Jackson, eds. *Rooted in the Land: Essays on Community and Place.* New Haven, CT: Yale University Press, 1996.

Wallace, Henry A., and E. N. Bressman. *Corn and Corn Growing.* Des Moines, IA: Wallace, 1923.

Walth, Brent. *Fire at Eden's Gate: Tom McCall and the Oregon Story.* Portland: Oregon Historical Society Press, 1998.

Warren, Louis S. *Buffalo Bill's America: William Cody and the Wild West Show.* New York: Alfred A. Knopf, 2005.

———. *The Hunter's Game: Poachers and Conservationists in Twentieth-Century America.* New Haven, CT: Yale University Press, 1997.

Warman, Arturo. *Corn and Capitalism: How a Botanical Bastard Grew to Global Dominance.* Translated by Nancy Westrate. Chapel Hill: University of North Carolina Press, 2003.

Washington Hop Commission. *Kiss of the Hops.* Yakima, WA: Franklin Press, 1980.

Wheatland Historical Society. *Wheatland.* Chicago: Arcadia Publishing, 2009.

White, Doctor E. *Ten Years in Oregon: Travels and Adventures of Doctor E. White and Lady West of the Rocky Mountains.* Compiled by A. J. Allen. Ithaca, NY: Mack, Andrus, 1848.

White, Richard. *"It's Your Misfortune and None of My Own": A History of the American West.* Norman: University of Oklahoma Press, 1991.

———. *The Middle Ground: Indians, Empires, and Republics in the Great Lakes Region, 1650–1815.* Cambridge: Cambridge University Press, 1991.

———. *The Organic Machine.* New York: Hill and Wang, 1995.

———. *Railroaded: The Transcontinentals and the Making of Modern America.* New York: Norton, 2011.

———. *The Roots of Dependency: Subsistence, Environment, and Social Change among the Choctaws, Pawnees, and Navajos.* Lincoln: University of Nebraska Press, 1983.

Withycombe, James. *Annual Report of the Office of Experiment Stations for the Year Ended June 30, 1911.* Washington, DC: US Department of Agriculture, 1911.

Wolmar, Christian. *Fire and Steam: A New History of the Railways in Britain.* London: Atlantic, 2007.

Woodson, C. E. *Benton County, Oregon: Illustrated.* Corvallis, OR: Benton County Citizens' League, 1905.

Woodward, C. Vann. *The Burden of Southern History.* Baton Rouge: Louisiana State University Press, 1960.

Worster, Donald. *Nature's Economy: A History of Ecological Ideas.* Cambridge: Cambridge University Press, 1994.

———. *Rivers of Empire: Water, Aridity, and Growth of the American West.* New York: Oxford University Press, 1985.

———. *Under Western Skies: Nature and History in the American West.* New York: Oxford University Press, 1992.

Wunderlich, Gene. *American Country Life: A Legacy.* Lanham, MD: University Press of America, 2003.

Wye College Department of Hop Research. *Annual Report, 1953.* Wye, England: Wye College Department of Hop Research, 1953.

———. *Annual Report, 1963.* Wye, England: Wye College Department of Hop Research, 1963.

Wyman, Mark. *Hoboes: Bindlestiffs, Fruit Tramps, and the Harvesting of the American West.* New York: Hill and Wang, 2010.

Wythe, George. *The United States and Inter-American Relations: A Contemporary Appraisal.* Gainesville: University of Florida Press, 1964.

Yakima Golding Hop Farms. *Illustrated Presentation of Yakima Golding Hop Farms, Dedicated to Consumers of Hops Everywhere.* Yakima, WA: Republic, 1940.

Young, Allen M. *The Chocolate Tree: A Natural History of Cacao.* Washington, DC: Smithsonian Institution Press, 1994.

Zeigler-McPherson, Christina A. *Americanization in the States: Immigrant Social Welfare Policy, Citizenship, and National Identity in the United States, 1908–1929.* Gainesville: University Press of Florida, 2010.

Zimmerman, C.E., S.T. Likens, A. Haunold, C.E. Horner, and D.D. Roberts. "Registration of Comet Hop." *Crop Science* 15 (January–February 1975): 98.

NEWSPAPERS, TRADE JOURNALS, AND MAGAZINES

All About Beer Magazine
The American Magazine or Frank Leslie's Popular Monthly
Benton Country Herald
Brew Your Own
The Capital Press
Country Life in America
The Enterprise File
Eugene City Herald
Gazette Times
The Hop Press: A Memorandum of What's Brewin'
Hop Report
The Hopper
Hop Stocks
The Independence Enterprise
The London Times
Mark Lane Express Agricultural Journal
The Oregonian
Oregon Business Journal
Oregon Hop Grower
Oregon Native Son and Historical Magazine
The Oregon State Journal
The Overland Monthly
The New York Times
The Pacific Rural Press
Pacific Hop Grower

Pioneer and Democrat
Statesman Journal
The Tacoma Ledger
Transactions of the American Brewing Institute
The West Shore
The Willamette Week
Zymurgy

INDEX

DDT, 147–148, 243n57

Deschutes Brewery (Bend OR), 183–184; Mirror Pond Pale Ale, 184

Dickens, Charles, 56, 57, 131

distilled liquor: and colonial era, 17; homemade, and Prohibition, 103; temperance movement and move to lager from, 18; underground distilleries, and Prohibition, 103, 105–106

distribution: competition for, and corporate mergers, 138; and laws allowing craft beer revolution, 177; post-Prohibition regulatory system and, 116, 177; preexisting networks, and brewery survival of Prohibition, 114

Dominic, Fr. (Mt. Angel Abbey), 163

Donation Land Act (1850), 29, 42, 85

Douglas, David, 25–26

downy mildew disease (*Pseudoperonospora humuli*), *122–123*; Cascade hop variety and, 163; in Europe, 100–101, 123–124; grower associations and attempts to address, 122–124; hybrid hop varieties sought for resistance to, 89, 90, 123–124, 134–135, 146–147; overview, 132–133; in PNW, 111; prevention of, 123; quality of hops affected by, 134; treatments for, 111, 146; twenty-first-century and, 192; in WV, 111; Yakima Valley arid climate as protection against, 136; yields affected by, 122–123, 124, 136

Drake, Bill, *The Cultivator's Handbook of Marijuana*, 164

drying and curing hops: buildings/kilns for, 14, 41, *42*; conversion of dryers to vegetables, World War I and, 98, 228n23; disappearance of, from landscape, 194; improvements to systems of, 132, 146, 221n26; overview of process, 41; scientific research on, 89; shared facilities for, 41; sulfur added during, 41, 45, 89, 208–209n29; wild gathering and, 11

Durst Brothers Hop Ranch, 106–107

Durst family, 84, 107

Early Cluster hop variety, 44; downy mildew disease and, 123; harvest season and, 50

Eastern Oregon Brewing Company, 96

Eckhardt, Fred, 176, 177, 182–183, 184-185; *Treatise on Lager Beers,* 176

E. Clemens Horst Company, 62, 82, 85; labor strike, 130; mechanical hop harvester patented by, 77, 144; and research, 87

Eighteenth Amendment. *See* Prohibition

E. Meeker Company, 44, 45, 47

England. *See* Great Britain

Ettinger, Christian, 189

Eugene (OR), 35, 190

Europe: as beer-making center of the ancient world, 9; big beer funding networking between hop growers, 141; colonial era and, 15–18; and downy mildew disease, 100–101, 123–124; early cultivation of hops, 11–15; and labor/harvest culture, 14, 54–57, 213nn10–13; and Marshall Plan, 236n5; mechanical harvester, adoption of, 146, 242n39; weather and loss of hop crops, 192; World War I and disruption of hop/beer production in, 99–100; World War II and disruption of hop supply, 134. *See also* European disdain for American hops; Europeanization; European noble aroma hops

European disdain for American hops: aroma and flavor issues, 17, 120, 165–166; and big beer, 18, 39, 240n22; colonial era and, 16, 17; craft beer revolution and reversal of, 178–179, 187–188; cultivation quality and, 17; and Haunold, ridicule of, 167; and Horst's donation of hop plants for research, 87; late–nineteenth century and overcoming of, 83; post-Prohibition era and, 120

Europeanization: brewing, 15; crops, 15; of the globe, 15, 200–201n27; and PNW, 25; of WV, 28–33

European noble aroma hops: American-grown substitutes for, 170–171, 179, 184; big beer and use of, 18, 139, 169, 192, 240n22; Cascade hop variety with characteristics of, 163–164, 165–166, 167, 179; craft beer revolution and, 178,

European noble aroma hops *(continued)*
179; definition of, 1; early cultivation
and, 12; and federal research support,
lack of, 90
Evans, Gale, 66
export of hops: colonial era and, 17; early–
twentieth century and, 44, 91–92;
late–nineteenth century and, 37–38,
43–44, *44*, 82–83, 223–224n49; of new
American hybrids, 188; post-
Prohibition era and, 118–119, 232n26;
price competition and, 135; tariffs on
import to Britain, 223–224n49. *See
also* import of hops
extract or pelletized form of hops, 1, 170,
191–192

Farnham Pale hop variety, 13, 16
federal government: containment policy,
productivity and, 236n5b; industrial-
scale agriculture favored by policy of,
143; irrigation projects, 136, 142; World
War II and mandate for beer donations
to military, 118, 138. *See also* federally
supported research; New Deal; U.S.
Congress
federally supported research: agenda of, at
variance with brewer agenda, 90;
Government Printing Office circulars,
88, 225n66; Hop Research Council,
168; origins of, 87–88. *See also*
agricultural experiment stations
Feldman, Herman, 101
fermentation, Pasteur and, 86
fertilizers: chemical/synthetic, 146, 147;
computerized application of, 187;
organic methods, 89, 146, 148, 242n41.
See also cultivation of hops
Firmat, Irene, 184
Fish, Gary, 184
Fobert, Frank, 163
food culture: mid-twentieth-century trend
to homogeneity in, 139, 175; turn to
local/whole/seasonal, 175
Food Distribution Administration, 117–118
Fore, R. E., 158
Fort Vancouver: agricultural focus of,
24–26; alcohol consumption limits at,

26–27; and brewery, lack of, 27–28;
expansion into PNW via, 27–29, 34;
and hop cultivation/sale, 26–28, 36
Foyston, John, 185
France: big beer–sponsored visits by
farmers and representatives to PNW,
141; climate of WV resembling, 22;
colonists from, and brewing, 15; and
fur trade, 24; hops cultivation in,
11–12, 99, 136; World War I and, 99
French-Canadians, expansion from Fort
Vancouver into WV, 28–29
French Prairie (OR), 28
fruits: California branding of, 78, 79; and
Fort Vancouver agriculture, 25;
harvest/labor needs of, 49–50, 215n44;
as healthful and moral, temperance
movement and, 79; physical exertion of
picking, 54
Fuggle hop variety: Anheuser-Busch as
customer of, 169; as English aroma hop,
48, 153
Fuggle hop variety grown in PNW: and
downy mildew disease, *122*, 123, 146–
147, 153; export of, 83; harvest season
and, 50; as less prolific than Cluster, 45;
polyploidy experiments on, 169,
248nn51–52; replacing Cluster due to
downy mildew disease, 159; as second
choice of brewers, 48–49; yields of, 153
Fuller, Andrew, 45–46
Full Sail Brewing Company, 184
Fulton, Charles William, 95
fur trade, 24–25, 26

Gambrinus Brewery, 35, 96, 102
Gaston, Joseph, 85; *Centennial History of
Oregon,* 92
gender: and appeal of mass produced bland
beer, 139; and transformation/
expansion of beer culture, 13
German-Americans: as target of
prohibition campaigns, 94; as targets
of nativist xenophobia, 94, 95
German hop varieties: Hallertauer-
Mittelfrüh, 12, 163, 165, 167, 170–171;
Spalter, 12, 170; Tettnanger, 48, 170–
171, 178

Germany: acreage devoted to hops, 118;
Anheuser-Busch hop contracts and,
192; big beer–sponsored visits by
farmers and representatives to PNW,
141; colonists from, and brewing, 15;
competition with PNW as world's
leading hops producer, 3; and downy
mildew disease, 101; hop-breeding
collaboration, 158; immigration to USA
and brewing culture, 18; and post–
World War II hop cultivation, 136; and
scientific research, 87; spread of hop
cultivation, 12–13; twenty-first-century
hop production in, 188, 192, 197–
198n4; World War I and decline in hop
production, 99, 228n26. *See also*
Bavaria; German-Americans; German
hop varieties

Ginther, Ronald Debs, paintings by, 130,
131, 235n75

Golding hop variety, 13, 185

Goschie, Carl, 145

Goschie Farms, 145–146, 190, 193

Goschie, Gayle, 190, *191*

Goschie, Herman, 145, 150, 151–152, 163

Goschie, Vernice, 145, 150

grains: and agricultural crisis of 1920s, 101;
as commodity, 38, 39; early adoption of
mechanization and, 221n27; price
supports for, 121; producers of, and
anti-Prohibition movement, 95–96;
regionally grown in PNW, 36. *See also*
malts

Grand Ronde Reservation, 66, *67*

Grant, Bert, 180–181, 183, 184, 186

Great Britain: ales, 16–17, 18, 24, 27; Bass
Brewery, 84; big beer–sponsored visits
by farmers and representatives to PNW,
141; climate of WV resembling, 22;
colonists from, and brewing, 15–16;
craft beer revolution and, 188; decline of
hop cultivation in, 82–83, 99–100,
135–136, 223–224nn49,51; early
cultivation of hops in, 13; Farnham Pale
(Canterbury Whitebine) hop variety,
13, 16; Fuggle hop variety, 48, 153, 169;
and fur trade, 24; Golding hop variety,
13, 185; Guinness Brewery, 83, 132, 136,

223–224n49,51; and harvest time/labor
supply, 14, 54–57, 213nn10–3; hop-
breeding collaboration, 158, 159, 167;
hop-breeding program (Brewer's Gold
and Bullion), Wye College, 89–90, 91,
123–124, 146–147, 153–154, 158, 159, 169;
Hop Industry Productivity Team (1950
tour), 135–136, 137, 146, 236n5;
importing WV hops, 43, 83, 99–100;
and Marshall Plan, 136, 236n5;
prohibition movement in, 93–94; and
scientific research, 87; seeded hops as
unavoidable in, 237n8; twenty-first-
century hops production in, 197–
198n4; weather, and 2006 hop shortage,
192; World War I and, 99–100

Great Depression, 101, 111, 112, 125, 127, 130,
186. *See also* New Deal

Great Falls Brewing (Montana), 138

Great Lakes Brewing Company
(Cleveland), 179

Grim, Amanda, 61

grocery stores, beer retail in: competition
for limited space in, 138; craft beer
revolution, 173; legalization of, 116

Gross, Emanuel, 48

Grossman, Ken, 178–179, 184, 186

grower associations: community created by,
150; convention of 1953, 142; downy
mildew disease, attempts to address,
122–124; early-twentieth-century
establishment of, 119–120; hopping-ratio
decreases, as issue, 138–139, 142, 150;
industrialization and modernization of
operations, 142–143; marketing
agreements (price supports), 121–122,
142, 186; marketing and grower
education, 149; mechanization of
harvest, need for, 142; mid–twentieth
century establishment of, 138–139,
149–152; promotion of domestic hops,
120; quality control improvement
efforts, 120–121; research, collaboration
with, 149, 151–152, 170. *See also* grower-
dealers; scientific research

grower-dealers: European, in PNW, 84, 85,
119, 143; hops contracts, competition
for, 43, 83–84; Emil Clemens Horst as,

hop aphids, 45, 136, 146, 147, 148

Hop Bowl, 127

hop-breeding programs: alpha acid content, increase in (see alpha acids); collaboration in breeding programs, 147, 155, 158, 158–160, 163, 167–168; early cultivation and, 12; import of rootstocks and, 44–45; lagging of USA in, 90–91, 151; as organic practice, 147, 152; and plant breeding traditions, 154–155, 169, 242n43, 244n4, 248n52; plant patents and privatization of, 191–192; for resistance to downy mildew disease, 89, 90, 123–124, 134–135, 146–147; E. S. Salmon/Wye College (Brewer's Gold and Bullion), 89–90, 91, 123–124, 146–147, 153–154, 158, 159, 169. See also Agricultural Experiment Station (Corvallis): hop-breeding program

Hop Culture in the United States: A Practical Treatise on Hop Growing in Washing Territory from the Cutting to Bale (E. Meeker), 46–47

hop dryers. See drying and curing hops

"hop fever," 38, 49, 153

Hop Fiesta, 124–127, *126*, 132, 144

hop gardens, as British term, 41

Hop Growers of America, 149, 192

hop houses. See drying and curing hops

Hop Investigations, 89

Hopper, The: The Hop Grower's Magazine, *126*, 142, 143, 144; chemicals covered in, 147–148; list of Oregon hop farmer members of growers association, 237–239n10; "A Substitute for Nicotine," 148

hoppiness, craft beer revolution and: amount of hops used, 179; Anchor's Liberty Ale and, 176; and aroma hop American hybrids, 176, 178–179, 184–185; and blandness, rejection of, 176, 178; Cascade hop variety and, 176, 178–179, 184, 185, 186, 188, 191; European noble aroma hop varieties and, 178, 179; fresh-hopping recipes, 185; global interest in, 187–188; high-alpha hop varieties and, 179; and hop as signature ingredient, 2, 184–185, 186–187; hop as symbol, 2, 185; "hop wars," 185, 187, 190; India pale ale and, 181, 184–185; organic hops and, 190; print culture and, 251n24; and reversal of European disdain for American hops, 178–179, 187–188; Sierra Nevada Pale Ale and, 178; as West Coast style, 185

Hop Queen, 125, *126*, 127

Hop Research Council, 150, 168

hops (*Humulus lupulus L.*), 2, *11*; and angiosperm revolution, 6–7; antibacterial properties of, 1, 10, 193; clarification properties of, 1; as dioecious, 6; extract or pelletized form of, 1, 170, 191–192; fresh/raw, 8, 185; landraces (local/regional varieties of), 6, 7–8, 12, 163; moralist rhetoric on, 78–79, 100, 222n34; nonbeer uses of, 8, 10, 11, 158, 192–193, 197n1; as perennial, 6, 200n17; pickability/pluckability, 167; preferred growing conditions of, generally, 6, 23; rhizomes, 12–13; ripening times, differences in, 51; spread of, from Asia to Americas, 23–24, 202n8; stabilization properties of, 1; as "wolf of the willow" (*lupus salictarius*), 10, 199–200n12; yellow resin (lupulin) of, 1, 10, *11*, 54, 140, 240n22. See also alpha acids; cultivation of hops; European disdain for American hops; European noble aroma hops; hop-breeding programs; wild hops; *specific hop varieties*

hop shortages: of 1882, weather and, 38; of 2006, weather and other factors, 192; and release of Cascade hop variety, 165

Hops Marketing Board (Great Britain), 136

Hop Special (train), 52, 62, *63*

Hop Union (Hopunion), 187, 193

Hopworks Urban Brewery (Portland), 189; Gayle's Pale Ale, 190, *190*

hopyards, as term, 41. See also cultivation of hops

Horner, Chester E. (Jack), 159, 164–165, 166–167

Horsebrass pub (Portland), 183

Horseshoe Bay Brewery (British Columbia), 178

Horst, Emil Clemens, *81*; and branding of WV, 121; and contract with Guinness Brewery, 82–83, 132, 223–224nn49,51; death (and life) of, 132–133; as grower-dealer, 43, 44, 72, 81–84, 85, 100, 119, 132, 141, 223nn44–45; and harvest-labor conditions, 108–109; as "hop king," 81; and knowledge production, 82, 87, 223n47; on prosperity during prohibition, 101; and regulation of labor conditions, 107; and research, 87, 89, 124, 132; and technology, 77, 132, 144; and World War I, 98. *See also* E. Clemens Horst Company

Horst Ranch (Independence OR), 155

Horstville (CA), 82, 223n45

Horticultural Society of London, 26

Hubbard (OR), 190

Hudson's Bay Company: and Fort Vancouver, 24–28; of presence into WV, 27–29

Humulus lupulus L.. See hops (*Humulus lupulus L.*)

Humulus scandens (prev. *Humulus japonicas*), 6, 156

Humulus yunnanensis, 6, 198–199n2

Hurt, R. Douglas, 143

hybridization. *See* hop-breeding programs

Idaho: adoption of Galena hop variety, 172; agricultural station of, and hop-breeding (Parma), 159, 163, 167–168; Cluster hop variety and, 172; state prohibition law, 93; Talisman hop variety and, 163; twenty-first-century hops production in, 3. *See also* Pacific Northwest region

immigrants and immigration: and lager-beer consumption, increase in, 17–18; retrenchment of white xenophobia and racism toward, 106. *See also* labor, racial and ethnic diversity of

import of hops: Cold War and avoidance of, 139; colonial/frontier America and, 16, 26; European market volatility, 169; from Germany, World War I and reduction of, 99; Midwest macro-breweries and, vs. locally grown, 18; nineteenth-century PNW and, 36; post-Prohibition era and, 120

imports: colonial/frontier era and, 24, 27; of raw materials for hop industry, 44, 47; of raw materials for rebuilding of Europe after World War I, 99; of rootstocks, 44–45

InBev, 192, 193

Independence (OR), *63*, 75, 190; Hop Fiesta, 124–127, *126*, 132, 144

"Indian watching," 66

India pale ale (IPA), 180–181, 184–185; bottles found in archeological digs, 27; origins of, 15–16

Indiehops, 193

indigenous population. *See* American Indians

industrial agriculture: efficiency and productivity as focus of, 142–143, 236n5; as "factories in the field," 142–143, 241n31; federal government policy as favoring, 143; food culture and rejection of, 175

industrial agriculture, hop grower adoption of: big beer and demand for, 141, 143; chemical/synthetic fertilizers, 146, 147; chemical/synthetic pest and disease control, 137, 147–149, 243n57; financial risks of, 135, 137, 141, 145–146; and hopping-ratio decreases, worries about, 139; irrigation projects and, 136, 142; mechanical harvesters, adoption of, 137, 143–146, *144–145*; mechanization, general requirements for, 146; mid–twentieth century and necessity of, 135, 137, 142–143, 149; monocropping, 135, 137, 150; Oregon Hop Commission establishment and, 149–152; population of farmers, reduction of, 134, 143; size increases in farms, 137, 143, 146. *See also* big beer

industrialization: and immigration, 74; and Prohibition passage, 93–94. *See also* commercialization and professionalization

National Clonal Germplasm Repository–
Corvallis, 172
National Wholesale Liquor Dealers
Association of America, 96
Nation, Carrie, 97
nativism: diversity of wv farmers as
protection against, 95; Prohibition and,
94; and World War I, 94, 106
near beer, 97, 99
Needham, E., 119
Needham, Frank, 124
Needham-Taylor and Company, 119
Nevada: hops research in, 88; mining
towns of, 32
Neve, Ray A., 159, 167, 248n51
New Albion Brewery (Sonoma CA),
175–176, 178, 180; New Albion Ale, 176
New Belgium Brewing (Fort Collins CO),
179
New Deal: Agricultural Adjustment Act
(price supports), 121; dam-building
projects, 136; Farm Security
Administration, 235n70; hop-
marketing agreement (price supports),
121–122, 142, 186; migratory workers,
plight of, 128, 129–131, 234–
235nn67,68,70,74,75; National
Recovery Act, 128–129, 234–235n68;
repeal of Prohibition, 113
New England region: commercial hop
cultivation and, 17; craft beer
revolution and, 174; rum production
and, 17
New York State: alcohol production,
generally, 17; colonial era cultivation
in, 17; European trade firms expanding
to, 84; harvest time, 57–58, 214n24,
215n44; nineteenth-century
cultivation of hops, 37; and repeal of
Prohibition, 113; twenty-first-century
hops production in, 197n4
New York Times: on harvest season, 51; on
repeal of Prohibition, 112–113; on
research, 87; on state of hop industry,
211n57
New Zealand: hop-breeding collaboration,
158; import of rootstocks from, 45
Nickerson, Gail, 167

nicotine (tobacco), pest control with, 45,
47, 146
Ninkasi Brewing Company, Tricerahops
Double IPA, 190
Ninkasi, Hymn for (Sumerian), 9, 173
Noakes, C. F., 120–121
noble hops. *See* European noble aroma
hops
Northern Brewer hop variety, 90
North Pacific Brewery (Astoria), 35
North West Company (Canada), 24
Nugget hop variety, 170, 171, 172, 179, 186
nursery and horticultural supply,
Lewelling and Meek (nineteenth
century), 30–31, 38
"nutritive tonics," 99, 102–103

oast house, 14, 41. *See also* drying and
curing hops
Ockert, Karl, 181, 182
Oliphant, Jim, 23–24
Olympia (WA), and nineteenth-century
cultivation of hops, 36
Olympia Brewing Company (Tumwater
WA), 180; Applju (soda), 103; and
corporate mergers, 115; and
Prohibition, 102, 103
Oregon: anti-Prohibition movement in,
95; Brewpub Bill (1985, HR 1044), 183;
Chinese immigrant population of,
217n72; craft beer revolution and
number of breweries in, 190; early
temperance laws, 26–27, 29;
environmental advantages of, 32;
homesteading of, 29; Japanese
immigrant population of, 217n72;
mid-twentieth-century and ongoing
importance of hop production to, 142;
natural resources, importance of,
73–74, 91, 132, 219n8; "Oregon System"
of ballot initiatives, 93, 106, 227n2; and
post-World War I conservatism, 106;
prohibition and, 93, 95, 96–99, 102–
103, 105–106, 112–113, 115, 227n2;
prohibition repeal, 112–113; Seasonal
Employment Commission, 107;
statehood of, 29, 34; women's voting
rights, 93, 227n2. *See also* Agricultural

Pacific Northwest region *(continued)*
  Willamette hop variety, 169, 186;
  anti-Prohibition movement in, 96–98;
  big beer and relationships with hop
  growers, 140–141, 186; breweries in
  nineteenth-century, 34–36, 43; craft
  beer and relationships with hop
  growers, 178, 186, 187–188, 190, 193;
  craft beer revolution and, 174; global
  hop grower visits to, 141; Haunold as
  ambassador for, 170; percentage of
  world's total hops production, 3,
  197n4; Prohibition and, 102–106; rain
  as defining climate of, 22; and
  transnational conglomerates, 192–193.
  See also Oregon; Washington;
  Willamette Valley (OR); Yakima
  Valley (WA)
Pacific Rural Press, 39, 47
Painter, William, 86
Papazian, Charlie, 176, 177, 179, 251n24;
  Complete Joy of Home Brewing, 176;
  Zymurgy, 176, 251n24
Pasteur, Louis, *Studies on Beer,* 86
patents: enhancement to existing systems,
  77, 221n26; mechanical hop harvester,
  77, 132, 144; multi-use driers, 228n23;
  plant patents, 191–192; wire trellis
  dropper, 84
Patterson, Paul L., 142
pelletized or extract form of hops, 1, 170,
  191–192
pests and diseases: bottomland vs. upland
  cultivation and, 75; Comet hop variety
  resistance to, 168; hop aphids, 45, 136,
  146, 147, 148; hybrids offering solution
  to, 89, 90, 123–124, 134–135, 146–147;
  labor conditions and, 131; and
  nineteenth-century hops shortages, 38;
  nineteenth-century lack of, 37; organic
  (traditional) controls for, 45, 111, 146;
  powdery mildew disease, 192; red
  spider mites, *123,* 147, 148; tobacco,
  pest control with, 45, 47, 146;
  verticillium wilt, 147; Willamette hop
  variety resistance to, 169. *See also*
  downy mildew disease
Peterson, Fred, 142

Pink Boots Society, 189
place, sense of, specialty crops and, 3–4,
  193–195
Plant, Marjorie, 125
Pliny the Elder, 10, 199–200n12
Pokorny brothers, 163
pole men/pole pullers, 61, 215n36
polyploidy, 169, 248nn51–52
Ponzi, Dick and Nancy, 181–182
Portland (OR): bar and pub culture of, 181;
  as "Beervana," 3, 4, 184; as "Craft Beer
  Capital of the World," 3, 184, 189; and
  craft beer revolution, 180, 181–185,
  189–190; craft breweries, number of, 3,
  189; do-it-yourself culture of, 181; as
  hop trade center, 91–92; lack of
  independent breweries (early 1980s),
  180; local products valued, 181;
  population of, 91, 107; and Prohibition,
  102, 105–106, *105*; and trade
  infrastructure, 31; world's fair of,
  73–75, 132, 220n10
Portland Brewing Company (craft beer
  revolution): MacTarnahan's Amber
  Ale, 184; Portland Ale, 183
Portland Brewing Company (original),
  102, 115. *See also* Blitz-Weinhard
  Company
powdery mildew disease, 192
prices: and agricultural crisis of 1920s, 101;
  for Cascade hop variety, 166; claims
  for, in print culture, 40; hops as 14th
  most lucrative U.S. crop, 74–75; and
  hop shortage of 1882, 38; of imported
  hops, and decline of British hop
  industry, 135; overproduction and
  lowering of, 49; supports for,
  marketing agreements as, 121–122, 142,
  186; variability of, 48. *See also*
  advertising; grower-dealers; marketing;
  overproduction of hops
print culture: and claims for hop
  cultivation, 40, 208n27; and craft beer
  revolution, 174, 176, 251n24; downy
  mildew addressed in, 124; Government
  Printing Office circulars, 88, 225n66;
  and harvest season, 51, 56–57, 59, 65,
  125; hop cultivation guides, 14; and

186–187; adoption of aroma hops (American hybrids), 171, 187–188, 191; adoption of Cascade hop variety, 165–166, 169, 171, 191; adoption of Nugget hop variety, 170, 171, 186, 191; adoption of Willamette hop variety, 169, 171, 186, 191; agricultural infrastructure, development of, 28–33; as agricultural utopia, 29–30, 39–40, 91, 194–195; big beer contracts for hops, 141, 240n25; big beer cultivating relationships with and among hop growers, 140–141; climate of (windward marine), 22, 23; and craft beer revolution, 186–188, 190–192, 193, 194–195; diversity of hop varieties grown, as new development, 187, 191, 193; as "Eden"/garden, 3, 4, 32, 198n7; Europeanization of, 28–33; flora and fauna of, 22, 30; geological formation of, 19–22, *20–21*; grower-dealers of, 84–85, 224n57; as "Hop Center of the World," 3, 40, 47–48, 72–73, 74–75, 82–83, 116, *117*, 124, 132–133, 140; as "horn of plenty," 29–30; list of hop farmers (1944), 237–239n10; maps of, *19–20*; and marketing agreement (price supports), 186; number of individual hop growers, 134, 237–239n10; percentage of world hop production, 191; postwar diversification of agriculture in, 236n1; Prohibition era and expansion of hops production in, 98–101; rainfall, 22, 30, 192; size of farms, changes in, 42, 137, 143, 190, 209n31; soil of, 19, 22, 23; and transnational conglomerates, 192–193; vanishing traces of hop agriculture, 194–195; wild hops absent from, 23–24; World War I and expansion of hops production in, 99–100; World War II and expansion of markets for, 117–118, 134; Yakima Valley competition with, 136–137, 190–191, 237n9. *See also* Agricultural Experiment Station (Corvallis); craft beer revolution; cultivation of hops; Pacific Northwest region; Portland (OR)

Willamette Week, 184
wire men, 61
wire trellis cultivation system: adoption of, 76, 100, 221n19; advantages of, 76–77; continued visibility of, 194; electrocution risk during harvest, 59; general description of, 76; harvest process with, 61; natural resource commodities needed for, 77, 100; technological improvements to, 100
Withycombe, James, 89
Wolf, Phil, 150
Wolf, Ron, *182*
women: advertising targeted to, 116; American Indian harvest culture and, 66; and craft beer revolution, 189; Donation Land Act (1850) and, 29; grower associations supported by, 150; harvest camp roles of, 57, 63; harvest labor experiences of, 51–53, 61–62, 66; harvest labor, hiring of, 52, 57, 143; as university scientists, 167; voting rights for, 93, 227n2
Women's Christian Temperance Union, 96, *105*
Woodell, Bob, 184
Wood, Isaac ("Uncle Isaac"), 36, 38, 42
Wood, L. D., 78
world fairs: grower-dealers' attendance at, 43–44, *44*; nationalism and empire and, 220n10; of Portland, 73–75, 132, 220n10
World War I, 94, 98–100, 106
World War II, 117–118, 134, 143–144, 146, 147, 148
wort, 1, 9, 179
Wye College (England), 155; collaboration with USDA program, 159, 167; and Hop Industry Productivity Team, 136; hybrid hops species (Brewer's Gold and Bullion), 89–90, 91, 123–124, 146–147, 153–154, 158, 159, 169

Yakima Brewing and Malting Company, 180–181; Grant's Scottish Ale, 180; India pale ale, 180–181
Yakima Chief, 187

Yakima Valley (WA): acreage planted in hops, 136, 237n9; adoption of Cascade hop variety, 165, 172; adoption of Comet hop variety, 168; adoption of high-alpha varieties (e.g. Nugget), 170, 172, 186; climate of, 136; competition with WV, 136–137, 190–191, 237n9; craft beer revolution and, 180–181, 187–188; and downy mildew, 136; irrigation and, 136, 190–191; as largest producers in USA, 137; percentage of world's total hops production, 3, 187–188; and powdery mildew, 192; railroad infrastructure of, 136; seedless hops produced by eliminating male plants, 136–137; and transnational hop-traders, 143; warehouse fire and 2006 hop shortage, 192; yields, 136

Yearbook of the United States Brewers' Association, 87

yeast, 9, 36, 86, 190

yields: of Cascade hop variety, 163, 165; of Cluster hop vs. European noble varieties, 90; of Cluster hop vs. Fuggle, 153; downy mildew disease affecting, 122–123, 124, 136; of hops bred for alpha acid content, 147; in mid-twentieth-century WV, 136, 237n9; in nineteenth-century PNW, 37; in nineteenth-century WV, 37; print culture and claims for, 40, 208n27; in twenty-first-century WV, 190; of Yakima Valley (WA), 136. *See also* prices

Younger, Don, 183

Zeus hop variety, 191–192

Zimmerman, Chuck, 166–167, 168

Zimmerman, Jane, 175